I0032803

Thomas King Chambers

The Indigestions

Or Diseases of the Digestive Organs Functionally Treated

Thomas King Chambers

The Indigestions
Or Diseases of the Digestive Organs Functionally Treated

ISBN/EAN: 9783337035167

Printed in Europe, USA, Canada, Australia, Japan

Cover: Foto ©berggeist007 / pixelio.de

More available books at **www.hansebooks.com**

THE

,INDIGESTIONS;

OR

DISEASES OF THE DIGESTIVE ORGANS
FUNCTIONALLY TREATED.

BY

THOMAS KING CHAMBERS,

HONORARY PHYSICIAN TO H. R. H. THE PRINCE OF WALES, CONSULTING PHYSICIAN AND LECTURER ON
THE PRACTICE OF MEDICINE AT ST. MARY'S HOSPITAL, CONSULTING PHYSICIAN TO
THE LOCK HOSPITAL, AUTHOR OF "LECTURES CHIEFLY CLINICAL," ETC.

SECOND AMERICAN

FROM THE

SECOND AND REVISED LONDON EDITION.

PHILADELPHIA:
HENRY C. LEA.
1868.

PHILADELPHIA.
COLLINS, PRINTER, 705 JAYNE STREET.

PREFACE.

In 1856 I published a small volume on the same subject that I am now again taking up. It has been a good while out of print, for I was not content with it enough to sanction a reproduction in the same shape. But I have always intended to handle it again some time or another. During the last year or so I have been looking over my old notes of cases, and it struck me that it would be interesting to pick out such as bore upon indigestion, to classify them according to the points they illustrate, and see how far they upheld or overthrew my previous views. Then linking them together, after the fashion of a clinical teacher, with a running commentary, I made them tell their own tale, and added such observations as either occurred to me at the time I had the patients under my eye or have flowed from after experience. So grew up, not the new edition I had thought of, but what consistency bids me call a new work. It has therefore a new title, pointing to the different aspect in which the subject is viewed. In the former work it was anatomically, here it is functionally treated.

So pleasant has been the holiday task thus

> "..........to the sessions of sweet silent thought
> To summon up remembrance of things past,"

that I am fain to dwell upon it, and to try to lead others towards the same source of enjoyment by describing the way in which my store has been heaped up. For it is needless to say I did not lean on my memory alone, or the number of trustworthy histories would have been few indeed.

The cases of those who are named as inmates of St. Mary's Hospital in the following pages are copied mainly from the diary kept by the clinical clerks. I have been always used to make this a chief source of teaching. The clerk was instructed to take notes with the sick person before him, and in his own words; and when he read them out at my visit, I added my observations, sometimes in the hospital case books, sometimes in my own. These formed the groundwork on which to build my clinical lectures for the current week. They are irregular in wording, but preserve a fair record of the disease.

The details of private practice have been kept in a shorter and more mechanical way. I make it a rule, to which exceptions need be very few, to write all prescriptions and papers of advice in a copying-book, which makes a duplicate of them by means of transfer paper; and at the back of this transcript I write, usually with the patient before me, his history, at least so far as to explain my reasons for the advice, before I go on to the next page. The periodical indexing of these sheets is an easy job for an hour of weariness; and the whole time consumed is so crumbled up that it is never missed, and neither business nor amusement feels itself robbed.

Some people tell me they can make their notes of the day's work more fully and scientifically when it is over, and they are quiet in their study. I do not like the plan so well. For one thing it interferes with the relaxation needed to keep the mind healthy and broad. That time belongs to rest—*datur hora quieti*—and should not be wasted on labor. An instinctive feeling of the truth of this causes a duty which is put off to such an opportunity to be put off often still further, often altogether. Again, unless an immediate note be made, the new and the strange in the day's experience are stamped in the mind deeper than the common-place, and so they are apt to take up more than their fair share of room in the diary; while personal friendships, the social standing of the patient, and other considerations will sometimes blot out, sometimes unduly brighten our recollections of the case.

To these brethren in art then, as well as to others more emphatically who have hitherto failed to put by any written record at all of their acquired knowledge, I earnestly commend the method. Independently of the advantages to patients of being able to cast back to their antecedents, wherever you see them again and at whatever interval, the not perhaps wholly selfish satisfaction of living over again at will any portion of your professional life is worth tenfold the trouble it gives. Its value to the public is directly proportioned to the value of the individual himself; his experience is a simple fraction of himself.

I cannot expect the reading of my notes to be as agreeable to others as it is to me. Still the gracious way in which even rough clinical sketches are received in the shape of lectures, makes me hope that these studies, being of a quasi-clinical nature, may have some of the same favor shown them. A still higher reward would be that my testimony to the pleasantness of the task should lead others, richer dowered than I am, to unbarn the harvest of their experience in its own living form, instead of merely the distilled essence of it in their opinions.

In this second edition I have added a few cases which seemed more illustrative of the matter in hand than those previously quoted; but I have not altered the form of a running commentary, in which the work was first put before the public, and which readers tell me they find agreeable.

T. K. C.

22B, BROOK STREET, GROSVENOR SQUARE;
May, 1867.

CONTENTS.

THE INDIGESTIONS.

CHAPTER I.

INTRODUCTION.

SECTION I.

THE name " indigestion," or "dyspepsia," speaks to the mind
of the physician of a very large class of morbid phenomena,
various in their nature and appearing under a great variety of
circumstances. There are those who would banish the words
from our nosologies; some because the outward manifestations
are so diverse that it is impossible to bind them together in any
classification of symptoms ; some because the parts of the body
whose morbid states induce indigestion are so doubtful and
numerous that it cannot be brought under any anatomical nomenclature. It is quite true that a definition of it cannot be
given according to the symptoms, and equally true that it cannot be called a lesion of one part or of any set of parts or
tissues. For when symptomatic phenomena are made the principle of classification, those attendant on this morbid state are
so numerous and so discordant that they appear in every class ;
and if the organs that originate diseases are employed to give
them names also, there are very few in the body which are not
sometimes concerned in producing the disorder in question.
Still to the practitioner the name has a distinct meaning, and is
a definite guide to action. No symptomatic, anatomical, or even

2

chemical considerations have prevailed over it, simply because
it indicates a true method of classification, a natural linking of
facts.

The link drawing into one class the morbid phenomena which
are the subject-matter of this volume is *a partial defect in the
necessary supply of that of which the body is built up, before it
arrives at the medium of distribution.*

It is worth while to pause a little over this definition and on
what it implies. The essence of "digestion" consists in absorption from a canal communicating with the external air into a
closed system of tubes where is contained the nutritive fluid.
Preparatory to this absorption is solution, aided by nerves and
muscles; and the end of it is assimilation, or conversion of the
substances received into a like nature with the fluid they float
in. Till this has been done they cannot be used for the nutrition of the body.

The blood is the floating capital lying between assimilation
and nutrition—a treasury liable to continuous drafts from the
latter, and requiring therefore constant supplies from the former
to keep up its efficiency.

"Indigestion," then, or the incompleteness of digestion, is a
defect anterior to constructive assimilation and to the blood,
intervening between life and the new matter by which it seeks
to renew itself.

This explains the fact mentioned before, and familiar to us
all, of indigestion producing such an almost infinite variety of
morbid phenomena as immediate or remote consequences, and
affecting more or less all the functions of the body. It perverts
incipient life at its very source, and therefore perverts all its
future manifestations.

It explains also why morbid anatomy here gives us less help
than in any other classes of disease; why the degeneration of
the viscera found in the dead body have less causal relation to
the phenomena during life than in other cases. The changes
seen by the morbid anatomist are those of solids exhibiting
faulty processes essentially posterior to the blood, dependent
on perverted nutrition, or on mal-directed or arrested destructive assimilation.

Let the use of a negative word as a title to my volume stand as a witness of my persistent belief in what I have made it my chief aim to urge in all teaching of medical theory and practice for some years; namely, that all disease is for the physician essentially a deficiency of life, an absence of some fraction of the individual organization of force, and that all successful medical treatment must aim at a renewal of vital action. I do not feel myself called upon here to repeat the arguments I have used clinically and in print for that faith; but I am more than ever convinced, as years roll on, of the soundness of the principle, and of the safety of applying it to practice.

The knowledge, therefore, of the deficiency can, in my opinion, be gained only through knowledge of that of which it is a deficiency. Physiology is the only guide to the pathologist, and all just views of disease must have a direct reference to health, just as the — must be preceded by an expressed or implied + for it to have an arithmetical power.

In a former work, alluded to in the preface, out of which the present volume has grown, this was timidly shadowed forth in the arrangement of the parts—"Digestion" and "its Derange-ments"—two books with an equal number of corresponding chapters with similar headings, each as it were complementary to the other; the organs being exhibited by the first book in their typical state, by the second in its deviations; as both being in fact equally manifestations of physiological laws. My reasons for now passing over the first part of the subject are mainly a desire to diminish the size of the volume, and partly a feeling that recent special writers on physiology have placed in the hands of our countrymen most of the information which, when I wrote, existed only in foreign periodicals and theses. All that, and more, is now so easy of access in a readable form, that one may be excused from reprinting it.

In the physiological cure of diseases it is almost impossible to exaggerate the importance of the digestive viscera. In every acute case, surgical or medical, the modification of the result produced by our efforts depends almost entirely on how far, how wisely, or how foolishly, these organs are watched over; whether they are well or ill treated, either by the scientific guidance of the skilled physiologist, or by the empirical rules

of the routine practitioner, according to the tradition of the
nurse, or the instinct of the patient. Each of these may be in
their fashion a useful guide; but the first is at least most capable of improvement by labor.

As regards chronic diseases also, science enables us to trace,
by steps more or less distinct, many of them not manifested in
the organs themselves, but affecting the whole body, to an
abnormal state of the digestive viscera. It is needful only to
name gout, tubercle, and anæmia, to engage our closest attention
to their causes.

Whatever value we may attach to the evidence of the dependence of disease on the digestive organs, it is very clear that we
look to them for relief from disease. Out of the six or seven
hundred forms of medicines in habitual use, very few indeed
are not occasionally offered to the stomach for acceptance, and
an overwhelming majority of them are adapted for use only in
this way. If we are still to employ this time-honored agency
in our attempts to cure bodily ailments (and I see no threatening of a change at present), it is surely a matter of great interest
to secure the active working condition of what our forefathers
in anatomy picturesquely called the *portal.* It is waste toil to
try and enter locked doors. A great advantage of paying
special attention to the digestive organs is that, as a rule, they
are more directly curable, and that by their means distant parts,
otherwise out of our control, may be favorably influenced.
The evil of neglecting them is obstinate disobedience of the
body of the patient to any medicine administered.

When a sudden poison or paralysis has fallen on this gate of
entrance, our hands are paralyzed too; the staffs we lean on
fail us. What a cataract of physic used to be poured through
the half-dead bowels in our first epidemics of cholera! I once
found about a drachm of Ipecacuanha safe in the stomach; and
Calomel pills and powders, and Opium, and Cayenne pepper,
were a frequent constituent of the stools; and I verily believe
that ninety-nine hundredths of what was swallowed in the
stage of collapse followed the same fate, or the patients could
not have outlived the poisonous doses which are recorded to
have been given.

Watch a case of typh-fever, and see what immediate improve-

ment follows the shedding of the dead epithelium with which
the mucous membranes have been coated—a change which is
ushered in by what is called the "cleaning of the tongue," but
which foreshadows much more, in fact the cleaning of the whole
intestinal tract. See how immediately on this the poisoned
nervous system begins again to renew its life, and delirium
ceases, as new nervous matter fit for duty is generated. Or
watch another less fortunate case of the same malady, how as
the tongue gets dirtier and drier and browner each morning, the
weakness of the nervous and muscular system increases, and
hope is more and more clouded over. In both instances the
difference between one case and another, between the patients
who are a credit and a joy to us and the patients who continue
to wring our heart with anxiety, mainly lies in the more or less
vitality of the abdominal mucous canal.

We must remember it is of no use to employ the best possible
means of staying the morbid symptoms, unless the absorbents
assimilate sufficient material to replace that which is diseased,
and to remove which we are bestowing our pains. Labor is
wasted in clearing away a worn-out wall, if a new structure
does not take its place. To that end the only path is to insure
the assimilation of food. And to insure the assimilation of food,
the stomach and its colleagues must be in working order. So
that in point of fact the only fair trials of depletory measures
must be connected with feeding, and those who would uphold
the good fame of such expedients must be careful of their
patient's digestion.

But even if diseases of the alimentary organs had not so much
influence as they possess on the duration of life, their extreme
frequency would alone entitle them to attention. Unfortunately,
it has had a contrary effect: medical men are apt to set down
what is so common and inevitable. They neglect indigestion as
unimportant, forgetful that though its removal may not perhaps
in every case lengthen life, yet that it would at all events treble
its value both to the individual and to society.

The digestive tract has not the advantage enjoyed by the respiratory and by the upper part of the urinary apparatus, and
other parts, of being double. An animal has two lungs, two
kidneys, two hemispheres to the brain, two testicles or ovaries,

two sides to his body generally; but only one stomach, and one intestinal canal. A further reason for great caution in preserving the integrity of the latter—there is less to spare for disease to affect. A deposit of tubercle (for instance) the size of a nut in the pulmonary tissue may be neither here nor there, may be never known by its effects. But put it in the peritoneum, or in Peyer's glands, and what a disturbance is produced! A man may lose a leg or an arm, and enjoy life very fairly afterwards; but let him lose the use of his œsophagus or his rectum, and what can he hope for?

This singleness helps to explain the powerful influence which derangement of any one of its parts has not only over the whole tract, but over the whole body and mind. No chain is stronger than its weakest link, and an interruption of the function at one point is an interruption of the whole.

It also has a bearing of considerable importance on the treatment. It is extremely difficult to obtain that rest which is so essential in the management of disease. If you have pneumonia, you may give a holiday to the smitten lung and generally recover with the other; but if you have an equally acute inflammation of the œsophagus or stomach, the danger is great, because necessarily they are in constant use.

Indigestion is a chronic disease. By that I mean that its natural path is straight on from bad to worse, unless from the interposition of some extraneous circumstances of accidental or designed origin foreign to the phenomena of the disease itself. Hence it is not difficult to test upon it the action of a remedy. A cautious observer may from a moderate number of well-considered cases come to a rational conclusion as to its value. On the other hand, the tendency of acute disease, as I understand it, is to progress in a circle towards the recovery of health; each process, however dangerous and abnormal it may be, being a step towards the final arrival at that result, if only the patient's strength hold out.[1] Art, therefore, modifies it much less

[1] Dr. Pierre Petit, in the Preface to his "Commentary on Aretæus," compares acute diseases to race-horses, which run round to the goal, unless they founder on the way.

The Greek primary division of diseases into "acute" and "chronic" is turned to such excellent therapeutical account by Aretæus, that the loss of his intro-

glaringly ; a small experience is sure to lead to fallacies, and it is only on the numerical comparison of a large number of un-prejudged cases that an opinion can be formed.

For this reason the failure of homœopathy in the cure of the digestive function is very conspicuous. I suppose a good third of the private patients who come to me with dyspeptic ailments have tried this system, and confessedly tried it without advantage; though some of the same people still continue to think that in acute disease, in scarlatina, catarrh, measles, rheumatic fever, pneumonia, and the like, their convalescence is brought about by homœopathic drugs.

My object in this chapter has been to point out the import-ance of a skilful management of the digestive organs in disease. I mean in disease generally, and not only that which speci-fically affects those organs alone. Let us not be deceived by the expression "merely symptomatic" sometimes applied to the derangements of digestion where organic changes exist. All parts and functions of the body are so knit together in one to form the great circle of life, that their comparative value to in-dividual existence is more a question of time than of power. The failure of any one shortens the days more or less, and the immediate cause of death is as often "a mere symptom" as an organic change. It is also a serious consideration that in respect of the patient in chronic pathological states this is in reality often the whole duty of the medical adviser. Often, on stating in consultation an opinion that some viscus is chronically de-generated, one is met by the remark, "Well, what is to be done? —we cannot cure that." Very likely not; then let us try and find something else which we can cure. In the great majority of patients this curable something may be found in functional impediments to the entrance of nutriment into the medium of assimilation; and when once nutriment can be got in, a cure is begun. Do not, therefore, let us indulge despair even after it

duction, in which he probably defined acuteness, is much to be regretted. Galen misses the practical point, when he introduces the artificial element of time, pedantically limiting acuteness to a definite number of days. And several of our contemporaries have fallen into the same error; as if there could be any natural difference between an inflammation lasting twenty-one days and an inflammation lasting twenty-two days !

has become certain that the principal viscus which gives a name to the disease is past remedies, and though little can be prescribed for the part mainly affected. It is seldom too late to try and administer to the failing organ the most potent of all remedies, the human blood of the patient himself, made healthy by the means adopted, and flowing in continuously by its natural channels.

I remember, when I was physician to St. Mary's, talking in the wards to the pupils on this subject (as was my frequent habit) when we came upon a new patient, an undergrown and undeveloped girl, the mitral orifice of whose heart was narrowed by rheumatic inflammation in childhood. On her being carried into the hospital, her face was like that of a corpse, and she could not stand without fainting. I presume no sane student expected to see means used for the dilatation of that valve, whose contraction was the source of evil: despair was not an illogical conclusion from the diagnosis I dictated to the clinical clerk, and I was but little surprised to hear behind my back the remark, "This, at all events, is not much of a case for treatment." Yet observation of the functional state of the alimentary canal, indicated by the œdematous tongue and fauces, made me improve the opportunity by expressing an opinion that she would "walk home with color in her cheeks." A month afterwards I was able to say, at a clinical lecture on the case, "This she has been able to do, and the better nourished heart now beats steadily and evenly; though its mitral orifice is as small as ever, if the ear and stethoscope are to be trusted."

And not once, but almost weekly, have I demonstrated to the same class, with respect to many a consumptive, how little it helped us to know that half of the upper lobe of each lung was filled with crude tubercles: pulmonary remedies had been of no benefit; but the reflection that the stomach was secreting an excess of mucus at the same time with the lungs, led to effectual means for the relief of both together.

In the following chapters I intend the sketches I give of morbid phenomena, and the simple classes into which I divide the indigestions, to apply equally, whether they are alone, or whether they are united to obvious and more fatal organic changes.

SECTION II.

There is no pedantry in definitions when their object is to lead the writer and reader better to understand one another. So I shall not shrink from a charge of over-precision when, the first time I use in the following pages a word capable of various interpretation, I often stop to say what I mean by it. And at the present stage of the volume let us not grudge a few pages to dividing the subject, assigning it a nomenclature, and trying to turn that nomenclature to practical profit.

HEALTHY digestion is *quick, complete,* and *easy.* There can be no excess of it, for food cannot be too quickly and completely converted into chyme and taken into the blood, and there is no such thing as too much health and bodily comfort.

In ILL HEALTH digestion is impaired in one or more of these qualities—it becomes *slow, defective,* and *painful.*

We may use Greek words, and call the above-named erring qualities of the digestion *Bradypepsia, Apepsia,* and *Dyspepsia ;* only let it be remembered, that making the old adjectives into new substantives adds no one whit to our knowledge—nay, unless care be taken, runs some risk of being a stumbling-block to its progress. For when we have in this way given a proper name with a capital letter, we are apt to think (like a naturalist with a new butterfly) that we have defined an individual and active motive power, instead of what is really the deficiency of a function. And thus we fall into the errors of our forefathers, whose efforts to destroy their abstract foe the " Disease," instead of restoring the existing patient, led to so much bad practice in the generation now passing away. I shall generally use the English adjectives, but first I will say shortly what I mean by them in this connection.

By digestion being *slow* I mean that the act in some part of the alimentary canal is not completed by the time when the convenience of the individual requires that it should be completed. The stomach may retain so much of a former meal that it is not in a fit state to receive the new one which is needful for the sustenance of the body. Hence arises a want of the natural appetite and (when it is long continued) imperfect nutrition, anæmia, debility, &c. Or, if we attempt to force food too quickly on the unwilling stomach, we have chemical decomposition and defective digestion as consequences.

The average time by which the stomach should have naturally emptied itself varies in different healthy persons from two to four hours. The intestines have extracted all that they are capable of absorbing in eight or nine hours; and the relics of complete digestion are ready for expulsion from a vigorous young person in twenty-four hours.

By *defective* digestion, I would imply that food capable of nourishing the body cannot do so from lack of certain changes which it should naturally undergo in the alimentary canal. It is passed from thence either unaltered or chemically decomposed. There are seen in the feces, either by the naked eye or the microscope, masses of starch, muscular fibre, fat, &c. I have several times had them brought to me, under the idea that they were worms, pieces of intestine, or other foreign bodies. Or else the products of their decay, consisting of various obnoxious gases and acids, are developed in a quantity subversive of social comfort.

Painful digestion may be both defective and slow; but, on the other hand, it not unfrequently also is complete and performed with sufficient quickness. All that it is intended to express by the word is its accompaniment, at some stage of its progress, by feelings varying from slight discomfort to absolute agony.

A few days ago a patient depicted to me very well the first-named sensations by saying, in answer to the question whether he had any pain, no, but that he felt where his stomach was, and knew where his food went to. And on the other hand I have heard the consequences of an ordinary meal described by a theologian (who ought to know) as "the tortures of the damned."

A very practical division of cases of indigestion is derived from the substance which is indigested, namely, which of the chief constituents of the diet, whether (1) *Starchy* and Saccharine, (2) *Albumenoid*, (3) *Fatty*, or (4) *Watery* food most exhibits the failure of the function. This is to be learnt partly from the patient, and partly from observation of the consequences which ensue to the alimentary mass. On it is grounded an important

part of the treatment, namely, the dietetic; and as its indications are simple as well as valuable, it will come first under
consideration.

A very essential step in the cure of indigestion consists in
the removal of its external causes, where these are removable.
The following chapter therefore will be on "Social Habits," by
which I mean causes dependent.on the patient's will.   There is
one obvious practical distinction between these and other causes,
namely, that the discovery of them is the true cure.   You may
bid a man cease from over-eating, over-fasting or smoking, but
you cannot bid him cease from being poor or sorrowful, from
having tubercles in his lungs or an ulcer in his stomach.

The symptoms of these errors of the function may be divided
according to the period of digestion at which they occur; some
being exhibited before the alimentary mass has left the stomach:
some during its way along the intestines ; some after the passage
of the ilio-cœcal nerve.

Now, remark, I avoid calling these phenomena diseases *of* the
salivary or gastric glands, *of* the intestines, or *of* the colon.
They are not so, and must not be treated as such.   The discomforts felt soon after a meal, for example, may be due to organs
far away from the stomach—to the uterus, to the kidneys, to
the lungs—yet they must be called by the same names as when
they are owing to anatomical changes in that part.   So those
occurring later often are traceable not to anything wrong in
the duodenum, ilia, or colon, but to excess of gastric mucus or
deficiency of pepsine.

Let it be understood then, that, when I speak of the *first* stage
of digestion I mean, before the food has passed the pylorus; by
the *second* stage, its transit along the small intestines; by the
third stage, all that takes place beyond the ilio-cœcal valve.

These phenomena almost always require separate treatment
of themselves in addition to the general condition from which
they originate.   Indeed they are such a prominent part of the
ailment that some writers, with a considerable show of reason,
have divided indigestions according as one or other of them is
the distinguishing features, as for instance M. Chomel does into
" *dyspepsie flatulente, dyspepsie gastralgique, dyspepsie boulimique,
dyspepsie acide,*" &c.   There is a certain amount of convenience

in this in respect of very marked cases: the worst of it is that
many invalids exhibit several of the name-giving symptoms all
at once in equal degrees, and it is difficult to assign them their
due place: while in some the dyspepsia (or rather the patient)
is flatulent one day, gastralgic another; at one time has acidity
in excess, at another alkalinity.   I have thought it better therefore in this treatise to discuss each symptom, separately where
it requires separate treatment, or in groups of several together,
where the pathology and therapeusis admit of union.   These
will fill the succeeding chapters.

    Let it be understood that I speak of "Heart-burn," "Acidity,"
"Vomiting," &c., in these chapters as diseases independent of
the cause from whence they arise; referring them indeed to
that cause where it is known, but not placing them in a separate
class where it is unknown.

    A chapter on the influence exercised by the functional de-
rangements of the digestive viscera over the nervous system
will conclude the volume.

CHAPTER II.

INDIGESTION OF VARIOUS FOODS.

Section 1.—Defects of digestion as exhibited in Starchy and Saccharine food. Section 2.—In Albuminoid food. Section 3.—In Fatty food. Section 4.—In Watery food. Section 5.—Treatment based on articles indigested. Section 6.—Treatment based on pathological condition.

SECTION I.

Indigestion of Starch and of Sugar, or Vegetable Food.

IN its original condition, either raw or when broken up by boiling, it does not appear that starch is capable of being absorbed by the alimentary canal. But on its first introduction at the mouth it meets with a fluid capable of converting it into one of the most absorbable of alimentary substances: indeed its change into sugar by the saliva may be fairly regarded as its ceasing to be a part of a plant and becoming a food. The reduction of it makes it into a complementary food duly represented in our type of a perfect diet, milk, by its peculiar form of saccharine matter "sugar of milk."

This metamorphosis begins immediately with the introduction of the morsel into the mouth, and is almost instantaneous in all parts with which it is brought into contact by the action of chewing. A good deal of the secretion is carried down into the stomach and continues its influence, though much retarded by the necessary acidity of the gastric juice. Should any of the starch granules have escaped rupture, or if it be eaten raw, the gastric juice (if strong) dissolves the albuminoid envelope, and sets some more amylaceous matter free. When the mass passes through the pylorus its acidity is neutralized, the action of the remaining saliva recommences on the starch yet uncon-verted or lately set free; and this action is reinforced by the pancreatic juice.

Of the sugar made, a very great part is absorbed in the mouth and gullet, sometimes all of it, for chemists have great difficulty in finding sugar in the stomach. The most essential part of the digestion of vegetable food is doubtless the osmosis of the sugar through the mucous coat of the alimentary canal. A great part of the excess is converted into lactic acid to aid in the solution of flesh food, and the rest is taken up by the intestines.

For the reduction of starch, then, so as to bring it into a condition capable of digestion, it is necessary, first, that the salivary glands should secrete a sufficiency of fluid suited to convert it into sugar; and that not only at the moment of chewing, but that they should go on supplying it as long as any starch remains unconverted. Now, the salivary glands are more exposed to derangement by circumstances external to themselves than any organ not directly subject to the will. Temporary emotion affects them temporarily, and continued emotion affects them chronically. We all know the dry mouth of the coward, the lover, the pitiful, and how the tongue clings to the roof when bad news is brought. We see too how for days, or even for weeks, "bread eaten in sorrow" can hardly be swallowed, so long it takes to moisten the morsel. Even in the healthiest person bodily exertion parches the throat. Again, there is scarce any morbid condition that does not make itself felt in the fauces and seen in the tongue. Numerous analyses of the saliva in inflammatory, and in other affections which seem remote (such as uterine disease, phthisis, chlorosis, ague), show that the changes are marked enough to be detected by chemical examination.[1] These changes are all in the direction pointing to want of vitality: either the animal constituents are deficient, producing a watery saliva; or they are turned sour or putrid by decomposition. It was to have been expected, therefore, that all defects of general vitality should give rise to them.

Besides the saliva, another digestive secretion practically comes into play in the solution of starch, to wit, the gastric juice. Cookery, even when most efficient, rarely ruptures the whole of the amilaceous granules. Many escape in the best, and in

[1] They may be found in Simon's "Chemistry," vol. ii. p. 9 (Sydenham Society's Translation).

bad cookery the majority escape. They cannot, therefore, be affected by the saliva, till their albuminous envelope has been dissolved by the gastric juice. Then they may be converted into grape sugar, either rapidly by the saliva present, or more slowly by the pancreatic and other intestinal secretions. The digestion of amylaceous and saccharine matters without inconvenience requires also the œsophagus to be throughout in a normal state; for, as before remarked, in health the greater part of the sugar made should be absorbed before it arrives at the stomach. In the lighter cases of dyspepsia, and in the temporary dyspepsia of excess, it is probably this power which principally fails; and the consequent symptoms probably arise from the presence of an excess of sugar thus passed on and apt to ferment in the intestinal canal.

Requiring so much, and so much that is soon affected by outward circumstances, the digestion of starchy food may be easily understood to suffer the first, the most completely, and the most commonly. Let us take as an example the simplest of all forms of deficient vitality, the consequences of an imperfect supply of that which vitality most requires for its manifestation, albumenoid food and pleasurable emotion. How familiar is such a case as the following, upon which I fall on turning over only a very few leaves of my hospital note-book!

CASE I.—Caroline P—, aged 49, a musician's widow, who formerly with her husband earned an easy subsistence by her profession, had been for a year living (as she called it) on needlework and looking on life with despair. On her admission to St. Mary's in April, 1860, she is described as a long-faced cachectic woman with sharp features, thin lips and hair still brown. Her complaint was of pain and an intolerable feeling of weight at the epigastrium coming on an hour after eating, flatulence[1] either breaking off from the mouth or afterwards distending the abdomen. So that she almost dreaded to take food. But what was this food? Bread, potatoes, tea, sometimes a bit of bacon, hardly ever a bit of meat. However, she had got not to care, eating at all was so distasteful to her. When ordered two meals of fresh meat daily, she could not indeed at first relish it without half a glass of port wine to wash it down. But that disrelish soon passed away. In a fortnight she had re-

[1] By "flatulence" I mean the superabundant collection of gas in the alimentary canal, whether it pass away upwards or downwards, or remain in the abdomen.

covered her appetite, and moreover was able to take even vegetable food. In
another week she was well enough to leave the hospital.

Gastric juice is required to rid the digestion of starch by dis-
solving the albuminous envelopes of the granules: gastric juice
is a highly animalized fluid; to make it animalized fresh blood
is a *sine quâ non:* and the material of which that is quickest
made is meat food. Such is the rationale of these satisfactory
cases.

Where the starvation and consequent anæmia fall by accident
on a patient less acclimatized by gradual grinding inevitable
poverty to them, they produce symptoms not probably more
fatal, but more marked and more painful to the sufferer. For
the nervous system retains its sensitiveness—its instinctive
courage, as it were—and complains loudly of the partial death
. forced upon it. Here is an example, illustrative of a more
sensitive indigestion also following inanition.

CASE II.—Mr. V—, aged 50, a periodical writer, usually well paid and
usually healthy, from loss of employment during the spring and summer of
1866 was reduced to great straits, and for long periods at a time had barely
enough food to keep body and soul together. After a time he ceased to feel
hunger, and the principal warning he had of starvation was the feeling of an
extraordinary lack of muscular power. He did not perceive his intellectual
energies flag at all. After he got into luck again he found it impossible to
take advantage of the plenty poured around him. Eating anything was
followed in ten minutes or a quarter of an hour by intense pain in the epigastrium extending downwards over the hypogastric region. The sensation was
" as if the navel were gripped with an iron hand." The mildest liquids, such
as tea, brought on this agony quite as much as the most insoluble viands or
strongest stimulants. He was much troubled with flatulence after vegetables,
both upwards and downwards. Meat did not indeed induce so much flatulence,
but its ingestion was followed almost immediately by the intense pain at the
epigastrium. If he could bear this, he digested it pretty completely. Then
he noticed how emaciated he grew, so that the balance testified to a loss of
two stone of flesh. He was not conscious of this during the starvation. Now,
too, he became sensible of a failure of mental power, especially in the hours
when the flatulence oppressed him most. When I saw him in October, he
presented a pitiable aspect of distress and emaciation; his abdominal pain
was growing worse and worse, his urine was of extremely low specific gravity
(I have lost the note of the exact figure).

I prescribed for him a draught of Hydrocyanic Acid with Battley's Sedative Liquor (four minims of the first and five of the second), to be taken a
few minutes before each of the daily meals. In three weeks' time he came

again to report favorable progress. He had gained flesh, lost his pain in a great measure, and was able to take a meal often even without the aid of the draught. But he still had some degree of gaseous distension of the abdomen, cognizable by the hand and eye; still his brain was not so clear for work as could be wished. I gave him my usual Quinine and Strychnine mixture, with three minims of Hydrocyanic Acid and a grain and a half of Iodide of Potassium, to take twice a day, under the use of which he got quite strong again.

Observe here how the digestion of vegetables was *slow* and *defective*, while that of meat was only *painful*. For the cure meat was a needful medicine, and therefore the first thing to be done towards the cure, was artificially and by drugs to stay this inconvenient pain. To have stayed the pain without supplying the meat would have been merely palliative or allopathic treatment; to make the anæsthesia a stepping-stone to the formation of gastric juice out of albuminous food was strictly restorative.

The Iodide of Potassium in the last prescription was designed to promote the activity of the salivary glands. I have found that it does so in my own person, and it has seemed to me that the saliva is not merely increased by the addition of water, as the urine is by nitre and other neutral salts, but that it converts more starch into sugar by the augmentation of its special ferment. But it has also seemed to me that this action of the Iodide of Potassium was a very temporary one, that it lasted but a few days, and that a continuance of the drug could not be trusted to for producing a continued effect. We must therefore use it just at the right moment, and not waste it by beginning too soon or pushing it too long.

The use of larger doses does not obviate the difficulty, and I therefore do not give them. The larger dose in the accompanying case was given for another purpose.

CASE III.—Mr. Wm. S—, an artist of middle age, was my patient in the autumn of 1866 for *alopecia* of doubtful origin. Among other experiments I tried the effect of Iodide of Potassium in doses of thirty grains a day. It did no good to the disease of the skin or nails, but the patient at first took a great fancy to it because it enabled him, he said, to take his meals of mixed food without consequent discomfort. He could digest even potatoes and beer. After a fortnight or so, however, he was forced to confess that his stomach was not what it was in youth, and that he was not to be exempted from the "care for his meat and diet" which the Son of Sirach considers characteristic of a good man of intellect.

3

The appetite is a very important part of the animal functions. The loss of it is familiarly known to be a consequence, but it is also often a cause of bad health. And the mode in which it affects the health seems to be by diminishing the supply of albumenoid food, and so indirectly the quantity of gastric juice.

CASE IV.—In November, 1861, I admitted to St. Mary's a pleasing, well-grown girl of twenty, named Margaret C——. She was extremely pale, and so weak that she could hardly raise herself in bed; yet no organic disease or pathological state of any of the solid parts of the body could be discovered to account for her condition. She attributed her illness to the close smell of a workshop in which she had been employed, which spoiled her appetite, making her at first crave for unwholesome things, and reject what was set before her, afterwards destroying the desire for victuals altogether. She observed it did not seem to make the other apprentices ill, who were not so squeamish. Perhaps the accident of her having been more tenderly nurtured, and so more sensitive than her companions—perhaps the responsibility of being placed over them as forewoman—contributed to this result. The end was, that the loss of appetite induced anæmia, swelled ankles, amenorrhœa, and a certain amount of hysteria.[1] Then, though removed from the original cause of the disease, she took a disgust to all articles of food, complained that vegetables especially produced flatulence, and had to come into the hospital, she was so weakened. When put upon the diet of milk and Liquor Calcis every two hours, with a pint of beef-tea in divided doses daily, she digested it without inconvenience, but at first without relish. After two days she was able to take an egg, and after twelve days of gradual additions of this

---

[1] Dr. Reynolds remarks, in the Introduction to his *System of Medicine*, that "by 'hysteria' is conveyed some meaning or none at all, and, when the former, a meaning as various in character as are the individuals who use the word." I take this opportunity of saying that by it I intend no reference to the womb (ὑστέρα), from which the Greeks derived the word in deference to a false theory of its origin ; which unfortunate derivation often leads to a bad practice. By it I mean a peculiar state of the individual in which the emotions exercise a sway over the functions of mind and body usually subject to voluntary control, and produce spasmodic contractions and paralysis of involuntary muscles.

Its pathology must be sought for in that puzzling part of the circle of life which lies between spirit and matter. We know so little about the chain which connects the two, that its links are reckoned by us as few and short, we have no names for any of them ; and in default of names for even the healthy functions, we must not expect an accurate nomenclature for their aberrations. So that the most we can do in trying to classify forms of hysteria is to trace how near their origin lies to one or other extremity of the series of vital actions which are interfered with ; what relation their phenomena bear on the one hand to mind ; and what on the other to body. We shall thus have set in a natural series the varieties of the disease, with pure insanity at the one end, and epilepsy traceable to organic lesion at the other.

sort she arrived at the full allowance of "ordinary diet"—mutton, porter, beef-tea, and milk. She lost her flatulence, and soon picked up her good looks.

For medicine she had Iron three times a day, and pills of Aloës and Myrrh, at first every night, afterwards not so often.

In a clinical lecture given at the time I laid the condition of the patient at the door of the digestive canal. The circumstances to which I have traced the illness act, directly or indirectly, both on this part. In the first place the mental exertion involved in an untoward responsibility thrown on a conscientious person would lessen the life of the involuntary muscles which carry along the mass of food through the alimentary canal. We know well how heavy our food lies in our stomach if called to a serious case just after a hearty meal; how long it is in leaving it, and how enervated the parietal muscles feel. Then at the same time that mental causes thus slackened the mechanics of nutrition, the unwholesomeness of the air in the close shop where the patient was employed poisoned the mucous membranes, diminishing the vitality of the epithelium, and causing them to be abnormally covered with a thick layer of mucus. By this tenacious coating the entrance of alimentary substances into the veins and absorbents was impeded, and the owner pined in the midst of plenty. So all the usual signs of starvation followed. First irregular hunger—by no means a constant companion of chronic deprivation of food, yet sometimes present as here; then loss of appetite—a much more frequent phenomenon; then paleness, languor, weariness; then anasarcous œdema, and all the other more marked symptoms of anæmia in their order.

The deficiency of digestive power was exhibited principally in respect of potatoes, pastry, sugar, porter, all of which produced flatulence and pain at the epigastrium, and were in consequence at first omitted from the dietary. She digested easily from the beginning small quantities of milk and beef-tea. It is obvious that if I had written down ever so many "full diets," one to whom the very sight of food was an abomination would have gained nothing by it; she would simply have gone without each meal. So I directed no meals at all to be taken, and no solid food; but a cup of milk with a third part of lime-water in it to be given as medicine every two hours, and a pint of beef-tea in divided doses during the day. I remarked to the students that our object must be to make the patient a carnivorous animal in order to make her omnivorous.

After a short time, shorter perhaps than one can always hope for, the selected nitrogenous food restored the digestive juices to their full powers, and she was able to take without pain, even the most difficult to digest of all aliments, a hospital potato.

Sometimes the condition which originates the indigestion is an organic and incurable degeneration. Yet in spite of the persistent cause growing of course worse and worse, the indigestion may not proceed beyond that of starch. For instance, Bright's disease of the kidneys may have this effect.

Case V.—Maria R—, aged 52, had suffered for four winters from cough and shortness of breath, which went away when the weather became mild. Her appetite was generally good, and she did not suffer from indigestion. In the spring of 1861, as she got better from her winter attack, she found she could not get about from swelling of the legs. This anasarca grew worse, she got ascites as well, and came into St. Mary's. The urine was found to be albuminous. It was not till she had been in the ward above a week that she complained of bad nights from flatulence, and a sensation of fulness in the epigastrium, which shortly after dinner became pain. Hydrocyanic Acid and Soda did not relieve this, but the change from ordinary diet to fish without potatoes did so immediately. The patient remained in the hospital more than three months, for abscesses and sloughs formed in the cellular tissue of her legs from distension, and she was a long time on a water-bed. She finally got well enough of the dropsy to be discharged, but never was able to digest well, though no more severe symptoms were produced than those detailed.

Of a more common form of the dyspepsia accompanying albuminuria I find an instance a few pages further on in the case-book.

Case VI.—Frances S.—, aged 63, a superannuated governess, was admitted June 7, 1861, with general dropsy of three weeks' duration, dependent on albuminuria. She had for some time suffered a good deal from flatus, occurring soon after meals, and followed by vomiting of the ingesta. When, for a few days from time to time, the victuals were retained, griping and diarrhœa were often the consequence.

In these sort of cases it is to be observed that the symptoms are much more severe, not confined to amylaceous food, and producing the phenomena of vomiting, which will be a matter for future consideration. If vomiting fails to be produced, diarrhœa often takes its place.

Disease of the heart produces sometimes a very marked indigestion. This, I think, mostly occurs where it is aggravated by an atonic condition of the arteries, so that the abdominal circulation is especially obstructed.

CASE VII.—Mr. James L—, aged 63, came under my care March 11th, 1867. He had for a year past been growing very wheezy and short-winded. On examination of the heart, it exhibited a tumbling action; its stroke was uneven and irregularly intermittent; the sounds were distant and not accompanied by morbid murmur. The pulse at the wrist was more irregular and intermittent than the heart. The urine was albuminous, of the spec. grav. 1.008 when warm from the bladder, and was free from tube-casts after standing.

For the last few weeks he had been unable to sleep at nights, suffering much from flatulence. This did not arise from overloading the stomach, for his appetite was gone. The skin had become yellower than usual, indeed the yellowness of his complexion was visible through the fixed russet-apple tint of hearty old age which appeared in his cheeks. I prescribed two grains of Aloes and Myrrh pill every night, and two tablespoonfuls twice a day of the following mixture:—

> R.—Ferri et Quiniæ Citratis, gr. 40.
> Tincturæ Digitalis, ½ fl. oz.
> Ætheris, ½ fl. oz.
> Aquæ, fl. oz. 15.

On the 19th, he said he was twice the man he was in both breath and bowels. And indeed he looked so, for the skin was clearer, his appetite and sleep were better, and the pulse was nearly regular. The urine was 1.015 in sp. grav., but still albuminous. So that all improvement was traceable to the heart, not to the kidneys.

The amelioration in this case seemed due to the tonic action of the Digitalis on the arteries, combined with the arterialization of the blood by Iron.

Sometimes the organic disease producing amylaceous dyspepsia is not one of these tissue-changes which are capable of anatomical demonstration, and with which we are familiar from museums and drawings, but those more recondite alterations in the nervous system which are provisionally termed functional. For example—

CASE VIII.—James L—, a bricklayer, aged 32, was admitted to St. Mary's under my care in August, 1857, principally for a loss of power and of sensibility in the legs, due to some extraordinary exertion six weeks previously. Besides this he complained of pain at the epigastrium, not increased by pressure, but only by taking food. His tongue was dryish and white, his pulse feeble. When this was observed, potatoes were omitted from his dietary, which had previously contained them, and the discomfort ceased. He afterwards had Quinine and full meat diet, with an extra allowance of bread to take the place of potatoes, and remained in hospital altogether three weeks, without any return of the epigastric pain.

Here subtraction of diet was possible and effectual, and its effect confirmed the diagnosis. Potatoes, as usually cooked, are probably the most objectionable article of food which can be presented to a weak digestion. The starch granules are but half ruptured, and are held together by cellular tissue, so that they are reduced by mastication only into small pellets, which require long soaking in gastric juice before they can be broken up. Mashing them expedites this process very much, but even then a good deal escapes digestion. Mixing them when mashed with meat gravy is a further expedient which is also useful.

The relation of muscular exhaustion to indigestion was incidentally alluded to in the last case. There are several forms of its exhibition. The most common is that which we see more as members of general society than as physicians, namely, where in a healthy person some extraordinary exertion brings on a temporary inability to digest all food more or less, but especially vegetable.

Case IX.—A healthy young woman, after being on horseback eight hours in a hot sun, first took a short nap, and then sat down to dinner, of which she partook moderately, having but little appetite. In the evening she suffered much from tightness across the chest ; in the night was unable to sleep for flatulence coming up in eructations, and rolling about the bowels ; she vomited, went to sleep, and the next day was well.

Sometimes the disturbance proceeds further—

Case X.—In 1863, during my autumn holiday in Switzerland, I had sauntered up from Zermatt, and spent the afternoon on the Riffelberg. I remember it well, for it is the last Alpine climb I shall ever have. A little before dinner I was joined by a gentleman whom I had seen some days before stepping out well on the Theodule glacier, and who was evidently used to the high Alps. He had now just come from the ascent of Monte Rosa in a shorter time than usual, and both he and the guides confessed to being tired. He dined moderately well, and after dinner proposed to accompany me back to Zermatt. We set off accordingly, but long before we arrived at our destination he was taken with flatulence and eructations, which shortly led to gripings and diarrhœa, with extraordinary explosions of wind. I felt very glad that darkness had cleared the road of all wayfaring spectators, and that we had a guide with us. However, we got down all right, and the next day he was none the worse.

Sometimes the diarrhœa is later in its supervention—

CASE XI.—S. G. S—, after a tedious ride across the Sierra in Andalusia, vomited a luncheon of bread and cheese which he had eaten. Feeling a want of appetite and a disgust to meat, he afterwards dined lightly on soup, bread, and vegetables. In the night and next morning he had frequent eructations, and in the afternoon of the day following the exertion, griping and diarrhœa, with much flatulence. Throughout the attack the urine was very high-colored, being as dark as porter, though not scanty. A day's rest set all to rights.

In these temporary attacks there is a general feverishness of the system; not, indeed, marked by rigors, but by anorexia, high-colored urine, thirst, and dryness of throat. They are not evidences of bad health, for they arise only in consequence of such exertions as no one need take without they like. But as a rule I doubt if such violent exercise contributes to good health, especially in middle-aged persons, and it is not a wise employment of those short holidays which we Londoners can afford ourselves.

Another form, more suited for medical treatment, is where the indigestion of starchy food recurs habitually after even moderate exertion, an exertion such as does not tire the remainder of the body.

CASE XII.—John E—, a sturdy, hard-featured fox-hunter, aged 73, who had lived in the country, farming and riding to hounds all his life, and never been ill before, came to me in November, 1863, complaining that for the two months past, whenever he undertook any of the usual exertion entailed by his active life, he was overwhelmed with flatulence. The abdomen swelled up, and he passed wind first upwards and then downwards; after which he felt relief, if he took rest. He was well so long as he kept quiet, but each fresh occasion for muscular effort brought back the uncomfortable symptoms. Prescribed Quinine and Strychnine, more sedentary habits as suitable to his time of life, and to take luncheon and late dinners, instead of a heavy midday meal and high-tea, as his custom had been.

In this instance it was pretty clear that the defective vital
power of the stomach was the first indication of approaching
old age, and was a warning that habits more suitable to that
inevitable event must be adopted. The older the stomach is,
the less it can bear either long abstinence or the overloading
which is consequent thereon in a person of active pursuits.

Sometimes the agency which brings on chronic indigestion is
of a much more sudden character.

CASE XIII.—William S—, aged 37, came to me August 1, 1861. He said
that ten years before he had exerted himself violently at Epsom on the Derby-
day in hallooing and running. He was suddenly attacked with a severe stitch
in the side and excessive flatulence. He vomited, and that temporarily
carried off the pain. Before that period his digestion had always been perfectly strong, but ever since he has suffered from eructations of tasteless air
from the stomach, within an hour after meals. Recently, after any mental
annoyance, he had had attacks of vomiting. Latterly, too, his general health
had become affected; his muscles were flabby and tremulous, like those of a
spirit-drinker, and his temper had become irritable. There was no pain on
pressing the epigastrium. The appetite was moderate, and the evacuation of
the bowels was daily. After taking Iron for a fortnight, his nervous symptoms were much amended, but he complained of the flatulence, and the bowels
were costive. He was ordered Quinine with Strychnine, and pills of Aloës
and Myrrh, which seemed to suit him well, for I find no note of any future
change of medicine.

It may be presumed from the permanence of a chronic effect
arising out of a temporary cause, that in such cases as this some
structural change takes place in the organ primarily affected.
But what that change is nobody knows. It is not chronic inflammation, ulceration, or thickening; or else pain on pressure,
either sharp or dull, would be found. Perhaps it is a sort of
dilative paralysis, such as occurs in hollow organs like the
bladder from sudden unwonted stretching. But this will be
spoken of hereafter, when we come to speak of the structural
changes of the viscera. At present we have to do with it only
as a cause of the lightest degree of indigestion, the amylaceous.

The use of Aloës and Myrrh (the *Pilula Aloës et Myrrhæ* of
the British Pharmacopœia) has been spoken of as ordered in
this case. It is not merely a purgative, nor would any other
purgative do as well. On the contrary, most purgatives would
probably have been injurious. Gamboge, Senna, Sulphate of

Magnesia, Colocynth, Mercury, and several others which produce elimination of serum and increase secretion generally, do
harm just in proportion to their activity. It seems established,
by the experiment of making them act when injected in a fluid
form into the circulation, that their soluble principles have a
destructive agency upon the blood; whereas the soluble alkaloid
in Aloës (aloine) is a bitter tonic, and the purgative power
resides in its insoluble resin. It is very moderately eliminative
—in small doses it but adds to the solid excreta of the colonic
glands, and elicits matter feculent in smell and of consistent
form—whilst at the same time it strongly restrains by its bracing
bitter the formation of mucus. See its effect on moist piles, how
it dries them up and often makes them smart! And we may
judge from this what its action on the gastro-intestinal mucous
membrane is. At the same time, by the more vigorous peristaltic movement and by the solid mass passed along the gut, the
already existing mucus is cleared away. Aloës is thus employed
strictly as a clearer of the intestinal membrane, and it is joined
with Myrrh, partly to divide it minutely and make a small dose
go further, and partly to give the patient the benefit of the extra
resin.

I am particular in enlarging upon this point from a fear lest
any words of mine should be construed as an encouragement
to an unfortunate tendency, common to both the public and
our profession, towards commencing treatment habitually with
destructive remedies. Some call this "clearing the decks for
action;" in a majority of instances they may be said to throw
overboard much of the best tackling in the ship and loosen her
armor-plates. A so-called "sluggishness of the liver" is a
frequent pretext. In a half-nourished person (and all invalids
are *ex vi termini* but half-nourished) the feces are apt to be
light-colored and scanty, inasmuch as the blood they come
from is light-colored and scanty. Blue Pill gives them immediately a darker hue and increases their quantity, but sadly at
the cost of the patient's strength, while the temporary change
soon passes off. Meat and Iron produce the same result, by
giving them more to be made out of, and this improvement is a
real and permanent one.

I have hitherto said nothing about the influence of the mind in inducing indigestion. Perhaps to that more often than to any other cause the history of it may be traced in the classes of society placed above the chance of physical want. I have no notes of ever having attended any patient suffering only temporarily from this cause; it is scarcely a case for a doctor; but probably every one's personal experience will supply him with an instance of it. More commonly our professional experience supplies us with examples like the following:—

CASE XIV.—My old patient T—, an anxious lawyer aged between 30 and 40, with a young family, complains that whenever he has to see a worrying client (and clients seem to become more worrying than they used to be), his mouth gets dry, his hands and feet get cold, his eyeballs burn, his head gets in a whirl. He goes home to dinner with a pain in his loins, but with a good appetite. His food lies like lead on his stomach, and seems to produce an intense headache. Once in bed, he drops to sleep; but he is woke up, at four in the morning at the latest, by either eructations or wind in the bowels. If this can be passed off he feels somewhat better, and can go to sleep again. I persuaded him to give up his house in London and sleep in the country, which seemed for some time almost to make him a new man, but he still suffers in some degree from his weakness of nervous power, whenever he has any but the most routine business to do.

I suppose such cases will always be common, so long as society demands brain-work, highly rewards it and concentrates in such places as this metropolis. Remark the sequence of events: the mind occupies the whole business of the brain; no nervous energy is left to preside over the secretions; the mouth is dry from lack of saliva, and if we could see them we should probably find the œsophagus and stomach dry also; the amylaceous food is not converted; it lies like a weight at the epigastrium till it undergoes a chemical instead of an organic solution; it ferments, and gives out carbonic acid. In the mean time the tired brain is causing headache, and laying the blame of its pains on the stomach; whereas its own weakness was the cause of all the troubles.

It is a foolish plan for a lawyer to sympathize too deeply with his clients—they do not want his sympathy, they want his help and his reason; and they will get both of a better quality, if he does not make himself ill by overcaring.

Where the nervous system is so irritable, Strychnine does

not seem to avail. I gave it as a tonic in this case, but without
benefit. Charcoal brings temporary relief, but it is a bulky,
troublesome, gritty powder, and in chronic cases a man cannot
go on for ever taking it. Occasional courses of Quinine and
occasional courses of mild Alkalies seem of most use.

Sometimes the determining agency to the stomach is a sudden mental shock, such as this :—

CASE XV.—Edward F—, a man of 40, had had dysentery in 1852, but had
been fairly strong again till a few months before he came to me early in May,
1860, when the failure of a bank completely upset him. He began to suffer
from a feeling of weight at the epigastrium, and of palpitation of the heart
after dinner, which, however, would be relieved by eructation. He had nocturnal flatulence, he lost his marital vigor, and grew thin. His nose too had
got red, which a man of 40 still careful of his appearance does not like. I
saw him again twice in June, by which date some powders of Strychnia and
Pepsine, which I ordered him, and time, seemed to have been effectual in
setting his digestion right, and he was able to gratify a wish to travel on the
Continent.

More commonly the mental distress is of a wearing character
rather than a sudden shock.

CASE XVI.—John C—, aged 55 (Dec. 2, 1862), had been slowly becoming
bankrupt for some years, though in point of fact he was safe from physical
want. He was also unhappy at home through the misconduct of a wife.
During this time he first began to suffer habitually from oppression at the
epigastrium after meals, so as to grow particular in his diet, and from experience to eschew potatoes. After food he felt something " working up and
down" as if flatus was trying to come up. He had grown emaciated, and
lost as much as thirty pounds in weight, and his fecal evacuations were
scanty and irregular.

In the last instances quoted there was no previous state of the
internal viscera which could be suspected of having determined
the weakness to the stomach. But that is rather the exception than the rule. Mental causes are much more powerful
when joined to some previous pathological condition. As in the
following :—

CASE XVII.—Mrs. B—, aged 66, came under my care in July, 1861. She
had had bronchial catarrh with frothy sputa for several winters, but had
suffered little from her stomach. In the early part of the year she had been
nursing a son-in-law, a patient of mine, with pulmonary vomica, and had

naturally experienced much anxiety on his account. This was followed by
pain in the right hypochondrium and a sensation of cramp in the stomach
when it was empty. Vegetable food produced flatulence and was avoided.
Apparently in consequence of that the bowels had become costive. She also
occasionally suffered from water-brash.

Is it not fair to assume that the catarrhal condition which had
affected the pulmonary travelled to the gastric mucous membrane? that the patient, in fact, had a catarrhal diathesis which
by the slowness of digestion resulting from the mental worry
was fixed in the stomach? So that the dyspepsia remained
though the external cause of anxiety was removed.

The chief mode in which chronic catarrh induces dyspepsia
I believe to be by enveloping the food and impeding the gastric
juice from freely mixing with it in the stomach; thus the
starch which had escaped the mixture with saliva by being
unbroken remains still undissolved. The mucus further on in
the ilia and in the colon by its slipperiness and elasticity prevents the muscles of the gut from duly urging forwards the
mass. Hence we have the starch fermenting and generating
gases and morbid acids, relieving pain, indeed, for a certain
period after it is swallowed, but by the above-mentioned
chemical decay producing infinite distress in the later periods
of digestion, failing to afford nutriment in the intestines, and
causing costiveness when it gets lower down, making the feces
lumpy, slimy, and hard.

I shall in a future chapter recur to this connection between
the mucous membrane of the lungs and of the stomach, when I
review the subject of what is called phthisical dyspepsia, where
a permanent pathological condition of the respiratory viscera
produces, or seems to produce, a very obstinate form of derangement of the digestive viscera, as a secondary consequence.

At the latter end of the cholera epidemic of 1854, when the
plague was becoming more general but less fatal, I used to see
a good many such cases as the following :—

CASE XVIII.—Joseph W—, aged 42, a laborer, had an attack of the
prevailing diarrhœa in August. From that time till he came to St. Mary's
on October 27, though his appetite was good, his bowels had never recovered
their healthy action, being always either costive or relaxed. Lumps of undigested vegetable food used to appear in the feces. For the last three weeks

also he had suffered from pyrosis. The tongue was large, flabby, and redder than natural, as if skinned. The epigastrium was tumid and drummy. The kidneys seemed quite to have recovered the choleraic congestion, for the urine was acid, clear, and free from albumen, though (as in most dyspeptics) of low specific gravity, 1.016. He was treated with rest in bed, liquid animal food, Bismuth, warm baths, and Castor Oil. And the treatment seems to have suited, for he was discharged "cured" (which means in hospital language " well enough to go to work") on November 1. But I presume some symptoms remained, for he was ordered to take a store of Bismuth with him.

This is an instance of the partial paralysis of the vital powers of the digestive organs which often succeeded to choleraic diarrhœa—I think more often in moderately mild cases than in the most severe. I suppose because they got about too soon; just as dropsy oftener occurs after slighter scarlet fevers than after dangerous attacks.

It is wrong to class such cases as inflammation of the stomach and bowels, for there are none of the usual accompaniments of inflammation, such as heat of skin, quick pulse, thirst, or even loss of appetite. So that unless the word inflammation is to be made coextensive with disease, it cannot include them.

A return to the district infected by the cholera poison will be apt to cause a relapse in those who have got rid of their diarrhœa by temporary removal, as happened in the following incident of the visitation, which now in trembling hope we call "the *late* epidemic."

CASE XIX.—Mr. V—, aged 35, a chemist and druggist at the east end of London, came to me on the 13th of October, 1866. During the prevalence of the epidemic in the early summer months he had experienced an attack of choleraic diarrhœa, with cramps, &c. On his recovery he was too weak to attend to business, so he left London for a six weeks' holiday, and then felt quite well. On returning to his work, he was careful not to over-exert himself; but still he had no sooner begun to sleep in the house than his rest was disturbed by flatulence and diarrhœa; this morbid looseness of bowels often alternating with costiveness, giddiness, and confusion of ideas. The feces, instead of being formed, homogeneous, and of natural odor, as in the country, contained large lumps of unaltered food, and smelt like rotten flesh. A second absence from home relieved in a few days these symptoms, and again a return brought them back. A third holiday was again effectual, and now in October he was very anxious to get back to work; but still he has a slight threatening of the indigestion of food in the stomach, indicated by loss of appetite and flatulence. My prescription was, a careful abstinence from purgatives, and a draught containing two grains of Quinine and $\frac{1}{16}$th of Strychnia twice a day, which appears to have been at last effectual.

It would seem that the external cause of cholera, whatever its nature or seat may be, remains behind still active, after all those most prominently sensitive to it have been affected, and that it is still poisonous, though in a minor degree. The impression I receive from the facts we possess is, that there is actually a poisonous material very generally diffused during epidemics, and occasionally generated also sporadically at other times; and that most of us very frequently digest it, assimilate it, or otherwise destroy it, as we may do other poisons or drugs, without injury or with very slight injury. Like the eggs of tapeworms, only a small fraction of what is generated, finds a suitable home.

But that is no argument in favor of the evacuant treatment. We have no evidence that the poison, conjectured to exist, and to be of a purgative nature, remains in the body during the continuance of its effects. No one takes for granted that a bullet which has injured a rib is still to be extracted—he turns his attention to curing the wound. To treat cholera by purgatives seems to me very like hoping to counteract an overdose of Elaterium by an overdose of Colocynth. Probably the minor purgative does not increase the risk in severe cases, for it is carried off safely by the major purgative, but I do not see how its use can be justified by rational physiology.

It was on these grounds that I advise an abstinence from purgatives in such cases as last related.

I believe that the indigestion following choleraic diarrhœa is very much kept up by the habitual taking of purgatives.

CASE XX.—T. E. D—, a surgeon, aged 69, who had passed many years in India, consulted me about himself, February 4th, 1867. Though exposed to epidemic and other noxious influences of tropical climates, he had never suffered from them; and this exemption he attributed to a congenital slowness of digestion having kept him from the self-indulgent habits of excess which a few years ago were so common among our countrymen abroad. Last year, however, being at Edinburgh during the invasion of cholera, he was attacked with diarrhœa, which was succeeded first by constipation, and that by a return of diarrhœa, which then assumed a chronic form. Then he caught a severe cold on the chest, on which the abdominal affection ceased. But on the cough being relieved, the looseness of bowels returned.

He described himself as being woke up at four in the morning by a feeling of discomfort about and above the navel, then there was as it were a working or fermentation in the belly generally, and sometimes a burst of wind upwards by eructation, but more commonly a resonant gurgling, and then followed a feeling of relief without any explosion *per anum* necessarily taking place. After breakfast there was again discomfort and flatulence. The stools were either watery or pultaceous, with occasional lumps of ragged matter. The urine was variable, often pale and watery, often thick and highly colored. The pulse had the emptiness and sharpness indicating the inelastic arteries of old age, and was quicker than natural. The palms were hot and dry. The tongue was white and furry. The appetite was good, and the usual diet of the easy classes was taken with satisfaction and without any consequent epigastric distress.

I found that he had a "dinner-pill" of his own, consisting of Colocynth, Ipecacuanha, and Henbane, which he took before dinner, and then he said there was after an action of the bowels before going to bed, and the morning discomfort was alleviated if not prevented. Still it returned again, and he thought it got worse. He therefore had somewhat lost faith in his favorite pill, and having to his surprise experienced much solace from a few drops of Chlorodyne taken at night, came to consult me about it.

I advised him entirely to give up the purgative pill, to take the Chlorodyne for a short period, then to try and do without it, and to have a *bouillon* and a glass of wine for breakfast, instead of the ordinary British fare.

Doubtless the production of indigestion after acute illnesses is most frequent where those illnesses have specifically affected the alimentary canal. Yet we find it follow also other serious diseases, though quite unconnected with that mucous membrane. Thus—

CASE XXI.—Harriet R—, a pale single woman, aged 42, was ill enough to be admitted an in-patient at Saint Mary's, July 12. 1862, a time of year when slight cases are usually kept out. She complained of pain at the epigastrium so severe that she took hardly any solid food, living mostly on tea. The mouth was dry and sticky, her appetite was gone, the urine was of the specific gravity of only 1.010 and scarcely acid. the pulse 108, quick and small, the catamenia had become irregular and almost ceased. This state of innutrition she attributed to a rheumatic fever a year before, since which time she had experienced this pain at the epigastrium caused by the amylaceous food which only she could get. She, however, digested well milk and lime-water, and in a week was eating "ordinary diet."

Still further removed from the stomach is the following origin:—

CASE XXII.—The immediate cause of M. A. S— coming under my care on January 23, 1855, was an attack of cramp-like pain after a meal at which she had eaten both rice and potatoes. But I found that she had suffered after vegetable food for many months, and that this weakness was traced by her to her last child-labor, up to which time she had always been strong. Her stomach was so blown out by the immediate illness, and so painful for several days, that I had to put on some leeches and feed her on milk and lime-water, but I doubt not that a lesser degree of the same condition was habitual to her. She seemed to have been flatulent, and had costive bowels ever since the birth of her last child.

Those in whom dyspepsia is produced by a mucous diathesis of the internal membranes are peculiarly affected by climatic influences. The union of cold and damp found in an ordinary English winter is the most common exciting cause, making familiar such cases as this.

CASE XXIII.—Henry L—, a schoolmaster, first made himself ill at the age of 28 by reading too hard for an University degree. His nervous system was completely prostrated. This was at the end of the summer of 1860. In the succeeding winter he began to feel a weight at the epigastrium after eating, and a strange kind of vertigo, as if the ground were falling away from under his legs. He would have sudden flushes and perspirations. If he could eructate any considerable amount of flatus there was an immediate relief to the symptoms. His nights were disturbed by cramps and by wind rolling about in his bowels, as if they were going to act. But no, they were constipated. He found acid often rising in the throat, and occasionally vomited a mass of stringy mucus. During the summer he was much better, and he was able to play at cricket, but when I first saw him, in December, 1861, his old miseries were returning with double force. He was getting very weak and nervous, his tongue was white and tremulous, his pulse very rapid, and he said that occasionally he was quite hysterical. Iron did not seem to suit him, in spite of the evident anæmia. But Quinine regularly, and *pro-re-natâ* doses of Valerian were of use. He has diligently gone on with occasional courses of this treatment, and suffers very much less. He is quite well every summer, but at the beginning of each Christmas holidays he is threatened with a relapse and comes to be encouraged to ward it off.

In the last case the failure of the nervous system was a cause of the failure of the digestive powers, but the circle of events sometimes turns the other way round in this class of cases which are affected, as now described, by climatic influences.

CASE XXIV.—Mrs. R—, aged 34, the mother of four children, is the wife of a thriving tradesman, and certainly as little exposed to any strain on the mental powers as any one I know. But she is fat, leuco-phlegmatic, subject to leucorrhœa and to mucous discharge with prolapsus of the rectum. She is costive, and was much in the habit, till I told her not, of taking purgatives.

to take toll from each passing delicacy at the table-d'hôte. But you will soon see them returning to tea and toast at the one, and restricting their performances at the other more and more to the vegetables and the maccaroni.

I cannot say I agree with those who attribute to this vegetable diet the comparative freedom from indigestion; such an argument seems to me a confusion of cause and effect. For we all more or less, I suppose, admit the value of meat diet in curing such complaints. The freedom seems to me the rather due to that vigorous condition of the mucous membranes which the climate insures.

Compare, for example, the amount and the intensity of such a disease as Chronic Bronchitis in Italy and in England. At St. Mary's Hospital, London, in the patients admitted between 1853 and 1861 inclusive, 1 in every 32 was a case taken in for Chronic Bronchitis. In the statistics of Milan Hospital which I have compared with them in a little volume published last year, there is only 1 in 8323.[1]

The dryness of the air without excessive heat or cold renders it needless for the mucous membranes to put on their slimy winter coats. They are in a more active condition for the work of absorbing oxygen, digesting, extracting nutriment or water, or whatever else they are required to do. They are filled with blood, and pass it on rapidly with its fresh burden of renewed life to the rest of the body.[2]

[1] "Some Effects of the Climate of Italy," p. 41.

[2] The term by which anatomists have designated this tissue is apt to lead even the most thoughtful of us into a fallacy. Active members of society are named after the work which is their most important occupation. The industry of the lawyer is the administration of the law; the doctor in any faculty ought at any rate to be learned enough to teach; the duty of bishops and others is ἐπισκοπεῖν—to oversee—each their several departments. But the chief work of mucous membrane is *not* to secrete mucus. It is most active when it is not doing so, and its activity is decreased just in proportion to the copiousness of the mucus. Typical health certainly consists in its absence; robust people pass weeks without expectorating; many find their handkerchiefs clean and unfolded after several days in their pockets, spite of all the voluntary and involuntary irritants to which the Schneiderian membrane is subject; and the urinary and intestinal outlets ordinarily contribute only an infinitesimal quantity, which may fairly be attributed to a temporary defect in some fraction of their large area.

The real office of mucous membrane is to offer a passage inwards for oxygen,

The indigestion of starch is in point of fact the indigestion of sugar, for it is into the latter substance that it is converted by either the action of the saliva immediately, or, in lack of that, by the contact with the mucous membranes slowly. It is therefore quite natural to find that any excess of sugar taken ready-made induces discomfort in dyspeptic patients. It is undigested, and the greater part of it undergoes acetic fermentation by the second or third hour after it is eaten. Some perhaps may undergo alcoholic fermentation, and generate the excess of carbonic acid, which fills the alimentary canal with flatus.

During its fermentation it also encourages fermentation in other articles of food, and by its presence oleaginous food is apt to be rendered indigestible also. Great discomfort will sometimes go on for a long time from this cause without being suspected, and cease by the simple expedient of leaving off so little necessary a constituent of diet as sugar.

CASE XXXI.—Edward W—, a gentleman farmer, aged 45, and inclined to corpulence, came to me in March, 1848, complaining of extreme pain running up the back of the sternum in the third hour after almost every meal, but especially after breakfast. This was followed by intense headache and giddiness, so that he feared he was going to have apoplexy. On examination of the stomach it was not painful on pressure, but drummy to percussion close up to the cardiac pulsation.

I gave him a course of Colocynth and Mercurial purgatives, and saw him again in July. The headache and the fear of apoplexy was then relieved, but the dyspeptic pain was as bad as ever, and the tongue was very yellow and thickly coated. I desired him to abstain from sugar and to take his morning tea with a slice of lemon in it.

In March, 1850, he came to me for a gonorrhœa which had become obstinate, and I asked him about his dyspeptic symptoms. He said abstinence from sugar had quite cured them.

In addition to these pains caused by its fermentation, sugar will in some instances cause pain immediately on its ingestion. It has seemed to me most probable that in such cases there is some rawness or local morbid sensitiveness of surface in the primæ viæ, and that the pain is analogous to the peculiar sort of twinge which the presence of sugar will cause in a tooth

water, fat, albumen, and other useful substances, and to defend the less easily renewed tissues beneath it from the deleterious action of external agents. These functions it best fulfils when it is bedewed with a moderate exhalation, and not with mucus.

unnaturally sensitive from caries, or even from neuralgia without solution of continuity. This pain arises too immediately to be due to decomposition. Syrup does not cause it, but only hard sugar.

Indigestion of Albumen and Fibrin.

Grazing animals are obliged to take their food leisurely, so as to mix it up with the secretions of the mouth, and many of them even to bring it up and chew it again, if they would not have it ferment in the bowels and risk a rupture. On the other hand, to beasts of prey the only use of saliva seems to be the keeping their throats moist. They need to chew the morsel only enough to prevent it sticking in the œsophagus. It would appear that while vegetables require for their perfect digestion a perfect condition of the whole alimentary canal as sketched out in the previous section, flesh-meat is at least independent of the salivary glands.

And this observation, drawn from natural history, is quite confirmed by physiological experiment, which finds the peculiar solvent of albumen and fibrin in the gastric glands. These glands the salivary can only aid by affording an aqueous diluent.

There is also this difference between the digestion of starch and of albumen, that whereas normally the former should be rapidly converted into an absorbable substance, and rapidly absorbed in the upper part of the alimentary canal; the latter does not begin to be dissolved till the food has proceeded a considerable distance, and the action is continued for nearly the whole of its course. In the healthy subject, a great portion of the sugar has been taken up before the albumen is affected at all.

As we might have expected, the digestion of animal food is less interfered with by external circumstances, and therefore less frequently interfered with than that of vegetable. Some considerable debilitating action on the nervous system is required to produce even an acute temporary dyspepsia of meat.

And I may observe that it is through the nervous system in almost all instances that proteinous indigestion arises.

We will begin with the least rare shape in which we see the disease. ·

CASE XXXII.—Lucy P—, aged 22, a servant of all work, debilitated previously by a long rheumatic fever and a hard place, was admitted to St. Mary's for scarlatina anginosa on February 22, 1861. The throat was much inflamed and a little ulcerated. On the 2d of March she was ordered ordinary meat diet, but the ingestion of it brought on such severe pain in the epigastrium that it was obliged to be left off.

An example of making the mistake of overhaste in the desire to renew the flesh lost after acute fevers. A good example, because in anginous scarlatina, if in any acute fevers, you would expect the salivary glands to have more particularly suffered. You would expect the indigestion of starch to have been the marked feature. But it was not so, and the deficiency of life manifested was a following of the genus fever, not of the species anginous—the injury to the whole patient took the precedence of the injury to the part.

On that ground I preserved a record of the case, for often as it must have happened in other instances that my patients have been allowed solid meat too soon, I cannot find another note of the fact. And I must trust my memory and my reader's experience to assert that the evil consequences are not confined to scarlatina.

When I say "evil consequences," I do not mean merely the temporary pain, but an attack of feverish indigestion, sometimes of vomiting, which throws the patient back some days.

It is to be remarked that it is not so much the chemical composition as the form of the aliment which renders it improper for incipient convalescents. Through the whole course of a typh-fever a continuous supply of liquid flesh in the shape of beef-tea has been kept up. If the stomach could not digest it the intestines did, and so the patient's strength was sustained. But give him a meal of roast beef, and it rolls about in the stomach till it decays; digestion is impossible, and it causes diarrhœa.

This caution is most requisite in cases where a relapse is

possible. As for example in fever, of either typhus or typhoid type, where the bowels have become inflamed. Here solid meat may bring back the worst features of the disease.

But especially in rheumatic fever there is a painful necessity for restricting the supply of nutriment. If meat be given before the power of fully converting it into living flesh is restored, a semi-conversion into lactic acid takes place. And then a febrile disturbance is produced, which is followed by a return of the rheumatic pains. Or perhaps rheumatic fever really is due to an excess of lactic acid in the blood; and if so the relapse which ensues on the generation of it is readily explicable.

Even when the pains are gone and there is an urgent call for replacing lost flesh, the most suitable diet for supplying it will sometimes bring on their return. The redder and more mus-cular it is, the more it disagrees, and we must very cautiously get back to "ordinary diet," else a risk is run of losing more by a second attack of the disease than is to be gained by haste. Vegetable matter does not expose patients to the same danger, and thus by dint of rice pudding, porridge, gruel, bread, mashed potatoes, and the like, you may try to satisfy the mouths which often loudly complain of starvation. If we cannot by such arguments succeed in staying the appetite, it is our duty to be cruel, or experience will soon convince us of the hurtful effects of solid meat in causing relapses of rheumatic fever.

In acute diseases the condition of the stomach which pre-vents it from digesting meat is merely temporary, and all that is requisite is patience. But where the failure of the organ is chronic the affair is much more serious. A state of anæmia is induced which is a long time in being recovered from.

CASE XXXIII.—Emma Ch—, aged 17, was admitted into St. Mary's, De-cember 21, 1855, in such an extreme state of weakness that she was obliged to be kept in bed. The pulse was 112, there was a systolic bruit with the first sound of the heart; the breathing was very short on slight exertion, the catamenia had ceased, and the complexion had become clear and pale like wax. She complained of extreme lassitude and headache. The bowels were irregular, reported costive on admission, but affected by diarrhœa during the second day of admission after taking broth with meat in it. It appeared that she always avoided meat, that she was disgusted with the sight of it, and that it caused pain in the epigastrium. There was also a painful spot on the dorsum of the tongue, impeding deglutition of solid meat, and this spot and others near it were denuded of epithelium so as to give a marbled aspect to the part. For this reason she had lived on vegetables, gruel, bread, and tea. A gradual return to a meat diet, through beef-tea, eggs beaten up in wine, and cocoa at short intervals, aided by absolute rest, Borax, Iron, and Chalk, restored her so by January 10th that she was able to get up and dress. Her pulse was 80, and firmer, the systolic bruit was not nearly so loud as at first, and she had some color in her cheeks. On the 12th she was able to eat the mixed " ordinary diet" of the hospital. On the 15th she wanted to discharge herself from my care; but when she lay down, I found the systolic bruit was still audible; so I kept her in for a little longer rest, on the excuse of having her vaccinated. She left hospital early in February.

It is impossible for a growing girl to make red blood without meat, and the longer she goes without, the less able is she to digest it; the power of the gastric juice is lowered by the abstinence.

As to the organic cause of the complaint in this instance, I presume one may be allowed to judge of the unseen by the seen, and to conjecture that the state submitted to our sight in the most conveniently visible portion of the alimentary tissue (the tongue) was also present lower down. And we may rationally suppose it to have been worse lower down, for the special gastric functions were defective, while the special oral functions remained unaffected. This superficial aphthous state, where the epithelium is destroyed instead of being raised into blisters, is common in the throat, tonsils, and os uteri, and may by stimulating applications have the edges so raised and the centre so depressed as to look exactly like ulceration. Borax and Quinine soon cure it, and are therefore probably equally good for the stomach under the same circumstances.

In some cases of pulmonary consumption we find the stomach affected with indigestion of albuminous food. It is so easy to trace out a physiological chain of causation from diminished diet, to atrophy, deposit of tubercle and phthisis, that it is commonly too readily assumed that such is the invariable course of events. I do not think so. It seems to me extremely probable that the condition of mucous membrane induced in the lung by the presence of tubercle, may be communicated to the stomach, and that the latter may be in fact an effect, not a cause. Though still it is a most serious effect, which reacts upon and aggravates the pulmonary injury. This is a practical point, for if it is an effect we may have more hopes of curing it, and so arresting the galloping consumption which is so imminent.

CASE XXXIV.—George C—, aged 24, came under my care at St. Mary's, October 19,1855. He was much emaciated, and had occasionally expectorated bloody streaks in the mucus from his bronchi. He had also slight hectic, and profuse nocturnal perspirations. On examination bronchial breathing with sibilant râles was found in the upper part of the thorax on the right side, and that part also was flatter than the corresponding part on the other side. During the last two or three months he had experienced pain during digestion. He had pain in the epigastrium and occasional diarrhœa alternating with constipation. After eating there was a feeling of weight at the pit of the stomach. This was especially noticed after animal food. Milk even caused it if taken without bread.

He got a good deal better on rest. Quinine, and Iron, and was able to take a mutton-chop. But what seemed to do him most good was Cod-liver oil, by the use of which he improved in spirits, in power of taking meat, and in weight. He increased as much as ten pounds avoirdupois between October 29th and November 7th, and was discharged as well on the 9th.

The deficient digestion of animal food in phthisis is a very serious thing. It keeps the patient in such a weak state that fatal effects follow shocks which could otherwise be borne up against.

CASE XXXV.—Thomas H—, aged 25, a tobacco-pipe maker, was admitted to St. Mary's, September 15th, 1852, having suffered from hæmoptysis, cough, and other phthisical symptoms for sixteen months, during which time he had been out-patient to the Brompton and other hospitals. There appeared to be crude tubercles to a moderate amount at the apices of both lungs. He complained of a feeling of coldness at the epigastrium, which increased to pain after meals when they consisted of meat. He had also frequent vomiting; even the broth diet of the hospital brought it on. Hydrocyanic Acid checked it a little, and Bismuth also seemed to deaden the pain in the epigastrium, so that he gained a pound or so in weight, in spite of what we thought was a softening of the tubercles under the left clavicle. He was able to digest fish with less pain than meat.

During this time his father was ill at home with the same complaint as our patient, and on October 13th he had news of his death. He was much affected, but the special symptoms did not seem aggravated, and he left the ward for home at his own desire next morning. Two days afterwards he died quite suddenly; and the immediate cause of decease was reported by his friends to have been grief at the loss of his father.

In this case it is to be observed that vomiting was present,

which is a grave symptom. Its connection with phthisis shall
be discussed in the special chapter to be devoted to that subject.

Other local morbid conditions besides those of the stomach
will sometimes cause indigestion of meat especially of all
victuals.

CASE XXXVI.—In February, 1849, Mr. K— came up from Wiltshire to
be under my care. His complaint was of vomiting, especially of meat. The
morsel seemed to stick at the back of the sternum, to cause a boiling and a
gurgling there, and to be rejected, apparently without arriving at the pit of
the stomach. He was much reduced in strength and flesh by this enforced
abstinence from meat. A fair trial of Prussic Acid was made without success.
But a drachm of Bismuth three times a day deadened the morbid sensibility
of the part affected so far, that he was able to swallow meat, considered himself cured, and returned home.

Here I felt no doubt that the seat of injury was in the
œsophagus, which for some reason or another rebelled against
conveying meat.

I suspect that it is the form, rather than the chemical constitution, of mammalian muscular fibre which causes it to be
objectionable. For in the following case we were enabled to
get animal food, reduced to a liquid shape, conveyed past a sore
spot as easy as vegetable.

CASE XXXVII.—Elizabeth S—, aged 25, died at St. Mary's, March 3d,
1852, of an ulcer of the œsophagus perforating-the pericardium. She had
been in the ward since January 23d, and during that time a great variety of
articles of diet had been tried, to find which easiest would pass into the
stomach. Meat she could never swallow at all, but eggs beaten up with wine
and thick cocoa, she could retain better than even quite fluids. And, indeed,
for some time before the accident which caused death, she got a good deal of
nutriment.

Even the friability of the fibre will make a difference. Thus
in the case of Thomas H., quoted page 57, it is noted that fish
could be borne though red meat was not. I cannot lay my
hand on another similar case, but the idea is familiar.

Yet there are cases when the very softest animal food is
objected to as experimentally proved to cause pain.

CASE XXXVIII.—In October, 1858, Mr. George R—, aged 54, first came under my charge. He was excessively thin and miserable to look at, but I could never discover any organic disease in any part. He said that from a boy he had never been able to eat animal food without great consequent pain. Fluid or liquor made no difference. "Even an egg," he said, caused it, and often brought on eructations of sulphuretted hydrogen, though taken in small quantities. He has often constipation with severe headaches. There was slight pain on pressure of the pyloric region of the epigastrium, a very white (nervous) tongue, and a red nose. What had especially troubled him lately was an impression, whenever he attempted to eat meat, that there was a pin in it. He was always quite aware that it was a delusion, but still could not shake it off. He had never been a spirit-drinker, nor a taker of medical drugs; though, like most dyspeptics, he had given homœopathy a trial. I prescribed Pepsine, by the use of which he had gained a little weight before he left town. I saw him again next August, with respect to an eruption of purpura on the legs, and he said his old failings had got much better, though he could not quite shake off the fancy about the pin.

The habitual indigestion of meat is allied to the indigestion of fat. There are transitional cases between the two, that is to say, cases where there is partial indigestion of both, when taken in the slightest excess, or in certain forms, but no very glaring inconvenience or illness under usual circumstances. We find, for instance, people who can eat mutton easily, but not beef. Now the main difference between mutton and beef lies in the infiltration of the bundles of muscular fibres by fat. Some of these persons are close observers enough even to find that they can eat beefsteak, but not roast beef, which difference is capable of a similar explanation. Perhaps the entire absence of fat in fish may be a part reason why it also is easier digested.

SECTION III.

Indigestion of Fat.

The digestion of fat is quite independent of the salivary and gastric fluids. Hence, even when they are in a morbid condition, and when the digestion is so slow that the meals are detained long enough for the fermentation set up to extend itself to the fatty matters present, and to develop butyric and other oily acids, still sufficient fat is digested to keep up the nutrition of that tissue in the body. Nay, patients with indigestion of starch and albumen will sometimes even get obese, especially if large eaters. Several specimens may be found in this volume.

The most familiar instance of the indigestion of fat is found in that disease which gets its name from the characteristic phenomenon arising out of that indigestion—Phthisis. In tubercular consumption the body wastes away, not because of the destruction of fat being increased, but because of its renewal being arrested.

Its renewal is arrested primarily and directly by any disease which affects the ilia, such as diarrhœa especially, because the ilia are the immediate instruments of its absorption; secondarily by the inefficiency of the secretions which assist in its solution and alkalization, such as the pancreatic juice and bile; in a less degree by the colonic or fecal viscera; and by the other organs of the body just in proportion as they influence indirectly these. As illustrations of the agency of its renewal and non-renewal, compare the following cases.

CASE XXXIX.—In September, 1857, I was called to see Mrs. B—, aged 45, reported as in an incipient stage of consumption. There were old scrofulous scars in her neck, and apparently a moderate deposit of tubercle in both pulmonary apices, indicated by bronchial breathing under each clavicle, and by sibilant, occasionally crepitant râles on the left side. This did not seem to account for the emaciation, and anatomically justified the diagnosis of incipiency. But it appeared that she had tried to take Cod-liver oil and failed; and that the cause of the failure was its induction of diarrhœa; so that the more she took the thinner she got. Also after every meal which included the smallest quantity of any fatty matter diarrhœa followed. The stools were like peasoup, and when she was taking the Cod-liver oil, drops of the oil used to be seen floating on the surface of them. Attempts to change this condition failed, and she never got any better, I know, though I did not attend her to the last.

CASE XL.—Ellen L—, an ill-starred orange-girl in her 19th year, left alone in the world from all her relations having died of consumption, came into St. Mary's Hospital, January 8th, 1856, in the third stage of the same disease. Indeed she had been an invalid long enough for the catamenia never to have appeared, being probably arrested by her ill-health. She was a flabby-faced, strumous-looking girl with grayish-brown eyes, much emaciated, and with the finger-nails curved into claws. She had much cough with blood-tinged expectoration, for which she had been in several hospitals, always with relief. The upper part of the left ribs was flattened, and there was that large crepitation and metallic bubbling to be heard which distinguishes a vomica, while at the right apex there was fine crepitation with tubular breathing. She rested but little, and the lips were livid from the imperfect aeration of the blood in the obstructed pulmonary tissue.

She was anxious to come into the hospital, because she liked Cod-liver oil, took it "with a relish." and got better on it. Her bowels were always costive. She certainly did improve on Cod-liver oil and Quinine. Her hectic abated, she got a good appetite ; the pulse was fuller, and she lost her cough. So at her own request she returned to her occupation on February 16th.

CASE XLI.—In April, 1858, I met in consultation Dr. C. B. Williams on the case of Alice C—, aged 14, who we made out to have a small amount of tubercular disease at the apex of one lung, and in the bronchial glands adjoining. She had a great deal of cough, and was much emaciated. Up to that time she had been most carefully attended to and actively treated. Blisters, Iodine. Iron, tonics, had been assiduously administered. Having made our diagnosis, we wrote out a prescription of numerous materials in accordance with the orthopraxy of the day. During the performance of this duty, I remembered that I had seen our patient (who was the daughter of an intimate friend) at luncheon eating some fat mutton, and it struck me that this capability had never received full play. When, then, the prescription was written I proposed that it should be put away for six weeks, and the patient have no medicine at all. except whatever she fancied to eat. It was a struggle to waste a good prescription, but yet that was agreed to, and in the six weeks so much progress was made by change of air, and scene, and diet, that the father declared that had she taken three globules, to those three globules he could not but have attributed the cure. This was eight years ago, the patient has since been a sea voyage, and is careful during the winter ; the air does not enter her left apex so well as the right, and she is shortish of breath ; but she runs, and rides, and hunts, and dances, and has plenty of flesh on her bones.

CASE XLII.—Just after Lady-day, 1861, Harriet B—, a maiden lady, aged 30, whose "father and mother had both died of decline," was placed under my care by Dr. Buckell of Chichester. She had evidence of a small focus of tubercle in the apex of the left lung, producing pain, dulness, and crepitation (from the partial condensation of the lung round it), but no marked pulmonary ailment. I thought that the quantity of tubercle was slowly increasing from week to week. What she complained of, however, was emaciation and diarrhœa, accompanied by the passage of pus and sometimes streaks of blood in the mucous feces. She was soon relieved of this by appropriate remedies ; and with a store of Hæmatoxylum and Copper was able to go on a long summer visit to some country cousins. I heard of her as going on well, and did not expect to see her again, or to make her case available for science. But as she returned through London in September, proclaiming herself quite stout and hearty, I had an opportunity of examining her chest again. I could then detect by neither percussion nor the ear any disease at all in the lungs. The pulmonary tubercle had become dormant. Two years afterwards I saw her walking briskly through the streets, looking well.

CASE XLIII.—J. B. F—, aged twenty-eight, and actively engaged in retail business when well enough, was put under my care January 29, 1867. The history given me by his medical man (Mr. Skaife of Northampton Square) was as follows. He had been a great consumer of beer, but always in fair health till the beginning of 1865, when he began to pass uric acid gravel in his urine. This continued a year, when by dint of careful dieting it was cured. In the spring of 1866 he had a dysenteric diarrhœa, passing in the stools a great quantity of mucus and occasionally blood for many weeks. He was seen several times by Dr. Brinton in consultation with Mr. Skaife, and cured by Alkalies. This diarrhœa reduced him excessively, so that he lost nearly two stone in weight. He regained his weight, and continued well till December, when he began to cough, and was troubled a good deal with that symptom. In the morning he expectorated transparent mucus with it, but during the day it was dry. Then he began to sweat of nights, and again rapidly to lose weight. So that during the six weeks before I saw him he had lost fifteen pounds, the loss being very often as much as half a pound daily, at which rate it was then going on.

The rapidity of the emaciation drew attention to the urine, which I found, after exertion during the day, clear, full-colored, free from albumen and sugar, and of the specific gravity 1.027.

The chest was narrow and did not expand freely. It was normally resonant. Scattered about, in the upper lobes especially, there were spots in which a full inspiration revealed crackles and occasionally interrupted ("wavy") breath sounds.

There was extreme slowness of digestion, especially of flesh food, and especially at the later periods of the day. Between three and four hours after a meal he would have eructations of nauseous taste and smell like rotten eggs. The quantity taken made no difference; they followed a light snack as often as a heavy dinner. His appetite was fair, but yet he could not eat more than two meals a day on account of these eructations. For they were brought on by anything solid or liquid he took after a two o'clock dinner. Supper especially he could never take, though he had tried all sorts of victuals. The attempt kept him awake all night with flatulence upwards and downwards and feverishness. The bowels acted naturally, but the stools were apt to be fetid.

I prescribed for him $\frac{1}{16}$th of a grain of Strychnia and 2 grains of Quinine twice a day, and a bottle of claret for drink, mixed diet and no tea or supper at present.

On February 6th his weight had become quite stationary, none having been lost since he began the treatment. The specific gravity of the morning urine was 1.022. The night sweats were much diminished and he felt stronger altogether. So much so indeed that, visiting a friend's gymnasium on the 4th, he tried his hand at raising a weight by a pulley. This exertion was followed the next day by severe pain between the sternum and left shoulder, which remained at the time of his coming to me. On examination it appeared due to tenderness at the insertions of the pectoral muscle in the ribs.

A fortnight afterwards he had gained six pounds in weight, and on the 8th of March a pound and a half more. On the 25th of March he was stationary in that respect. The night-sweats had diminished to a mere dampness during the second or morning sleep. He was able to take meat, sometimes fat meat, twice a day, and toasted cheese, without the nidorous eructations which formerly followed.

He was alarmed, it is true, by the appearance in the urine of some fine powdery red deposit, which on examination proved to be uric acid; and he doubted about the wisdom of continuing the high-feeding. But I pointed out how much rather to be chosen it is than the emaciation, which had been really on the verge of carrying him to the grave ; and to overcome it ordered some Hydrochloric acid instead of Citric as a solvent for the Quinine.

Observe the difference between these patients (not picked specimens, but such as are constantly occurring), and observe wherein it lies. In the power of assimilating fat. The first had not the power, and lost her life with enough healthy lung in her chest to have lasted her for many years; the second had the power in extraordinary force, so that she was able to take an excessive quantity of oleaginous nutriment, and so to bear up for a time against a most formidable amount of softening destructive tubercle; the third had not, indeed, any extraordinary power, but she had less amount of disease to bear, and she bore it; the fourth and fifth had lost the power, but regained it, and with it overcame the morbid diathesis.

It is truly by aid of the digestive viscera alone that consumption can be curable. Medicines addressed to other parts may be indirectly useful sometimes, but they more commonly impede the recovery ; whereas aid judiciously given in this quarter is always beneficial and often successful.

The chest is the battle-field of past conflict, the lymphatic duct the drill-ground for new levies of life.

Remark in the orange-girl the costiveness and the amenorrhœa. Both of these are good things in consumption. I do not mean good signs, which they are not, but advantageous in the prolongation of life. For in such a condition, the fat taken in is not exhausted by even the natural drain—abnormally requisite it is abnormally retained.

The effects of Cod-liver oil become less and less a marvel the more we know of physiology. The instinctive desire shown by all nations for an oleaginous diet, and their association of substances of this nature with proverbial ideas of happiness in all ages, show the value of a certain amount of it to man's

comfort. The "butter and honey" of the prophet, used as a
phrase for royal food, and the constant reference in the Bible
to oil as a luxury (though it could have been no rarity in "a
land of oil-olive")—these are sufficient to prove its estimation
among the Hebrews. The Hindoo laborer, when he devours
his gallon of rice for a meal, will spend all the pice he can get
on the clarified butter of the country; and "as good as ghee!"
is his expression of unqualified admiration. It was a mistake
in Baron Liebig to state that oily foods are disgustful to natives
of hot climates. All races of men require them and seek after
them; and the taste of the Esquimaux, so often quoted, depends
mainly on the abundant supply of the article which the sea
places at his disposal, coupled with a scantinesss of other pro-
visions. Throughout mankind there is an instinctive appreciation of the importance of this aliment, independent of accidental
differences of nation or locality. It seems felt to be, as science
shows that it really is, a necessary material for the renewal of
the tissues, and the desire for it becomes synonymous with a
desire for augmented life.

An easily assimilated oil comes, in fact, into the short list of
directly life-giving articles in the pharmacopœia; for it is itself
the material by which life is manifested. Hence, under its use,
beneficial influences are exerted throughout the whole body;
old wounds and sores heal up; the harsh wrinkled skin regains
the beauty of youth; debilitating discharges cease, at the same
time that the normal secretions are more copious; the mucous
membranes become clear and moist, and are no longer loaded
with sticky epithelium; the pulse, too, becomes firmer and
slower—that is to say, more powerful, for abnormal quickness
here is always a proof of deficient vitality. Such are the effects,
perfectly consistent with physiology, of supplying a deficiency
of molecular base for interstitial growth.[1]

But that supply is useless unless the absorbents are fit to
take it up, unless they are prepared (as the fourth case) by
proper tonics for its reception.

To find the easiest assimilated oil, and to prepare the diges-

[1] See "Lectures, chiefly Clinical," by the author, p. 275, &c.

tion for the absorption of oil, are the main problems in the cure of consumption.

Closely allied to this condition is what is called in children "strumous" or "rickety" dyspepsia, because it leads to struma and rickets. Cases are common enough among the ignorant and the poor. The following includes as many of the ordinary typical symptoms, and as few individual peculiarities as any I could select:—

CASE XLIV.—James A—, aged 7, admitted to St. Mary's, August 9th, 1856, had an angular countenance of grave expression, with gray eyes and long fringed eyelashes. The veins of the eyelids and temples were large and conspicuous. The arms and legs were very attenuated, in strange contrast with the swelled and drummy, but flaccid, belly, on which also the parietal veins were enlarged. The skin of the limbs was dry and unrenewed, giving them a dirty look. In bed he was restless, picking his nose, rubbing his anus, fidgeting his head about, kicking off the clothes, and getting into all sorts of odd postures; but when dressed he was preternaturally grave and quiet, and cried when he was roughly touched by any one. He ground his teeth and perspired profusely when asleep.

Though so thin, he was said to have a ravenous appetite. His tongue was pinkish with white spots; he was thirsty. His stools were copious, pale colored, as if entirely deficient in bile; with inky stains in parts, as if Iron had been taken, which, however, was denied. There were no worms in them, though the presence of these parasites had been reported. Their smell was very nauseous, resembling that of the macerating-tub of a dissecting-room.

During a fortnight that he was in hospital, the stools became natural under the use of purgatives, Iron, and meat; and in close ratio to the improvements of the stools, was the patient's increase of flesh.

Examination by the microscope of stools like those shows them to contain lumps of unaffected muscular fibre, undissolved fat, and free oil-globules in great quantity. In such quantity, indeed, that their whitish color might really be due to the emulsioned oil. Fat is here taken down—truly "down" to chemical decay; but not taken up—up to living tissue—by the lymphatics.

The passage of free oil shows the imperfect action of the intestines, the lumps of fat the imperfect action of the stomach. Dr. E. Schröder found that in the healthy stomach of a woman with gastric fistula, adipose tissue which was swallowed became so far disintegrated that the oil was freed from the areolar sacs

5

which contained it, united into drops, and floated free in the
fluids around it.[1]

In the already quoted cases there has been no deficiency of
appetite. Fat is swallowed, but is not absorbed. And this
would seem to depend on the fault of the lower part of the intestinal canal, of the intestines with their lymphatic vessels.
In other patients there is found a disgust to fat and to all that
contains it so great as to induce nausea when the attempt is
made to force the inclination. In these it seems to me probable
that the upper parts of the digestive apparatus, and especially
the pancreas, whose duty is the emulsification of fat, are to
blame. The nausea often takes the aspect of repugnance to
meat, for all flesh is scented by its own peculiar adipose tissue,
and owes to that its distinctive odor, which is unavoidably
associated in the mind of the patient with the meat itself.

CASE XLV.—Miss A. M—, aged 20, was first placed under my care November 3d, 1865. She was excessively emaciated, the checks were hollow,
the abdomen fell in so that the first thing you felt on pressing it was the
spine, and the haunch bones stuck up like the arms of a chair. The space
made dull by the liver on percussion was very small. The skin of the body
was harsh and dry. The bowels were excessively costive, not having acted
for years without strong purgatives or laxative enemata. The mind was unnaturally quiet and retiring; she avoided speaking of her ailments, and would
go and cry in solitude if pestered about them. She expressed an excessive
disgust to animal food, especially if moist or savory. The only chance of
getting her to eat a bit of meat was to have it so dried up that most people
would refuse it. There was no hallucination or morbid fancy about her diet,
but she persisted that she had severe headaches and pain in the abdomen after
it, till such time as it was removed by action of the bowels. Cod-liver oil,
which, naturally, a medical man had ordered for her, had caused such excessive nausea and vomiting, that it was impossible to persist in it. She had
a craving for alcoholic stimulants. With all her atrophy the color had not
left her checks, the lips were full and red, and she retained a peculiar delicate
style of beauty like a hectic consumptive. The urine, too, was of a natural
quantity, color, and smell, and of the mean specific gravity of 1.018. So
there was naught of what could be called anæmia. The muscles, too, were
well nourished and innervated, so that she could walk, and would walk if
permitted, more than was prudent; and it was after these exertions that she
used to urgently ask for wine. The heart and lungs, whose sounds and

[1] Succi Gastrici Humani Vis Digestiva, &c., auctore Ernesto de Schröder.
Dorpati, 1853, p. 30.

motions the skeleton condition of the chest exhibited with anatomical distinctness, were quite healthy.

The only anæmia was amenorrhœa, which had existed for eight months. The history was derived from a most kind and observant step-mother. It appeared that Miss M— and her sisters had, during their father's widowhood, been under the care of a horrible French school-mistress, who with a sort of insane wickedness conceived a hatred to the family, and actually tried to starve them to death to spite the father. One died, but the circumstances were so painful, that I could not cross-examine into particulars. The elder ones are alive and hearty. And this one seemed to get quite strong. She was plump, and the catamenia came on at sixteen. But soon after that she began to fall into her present condition, and the catamenia ceased, as above stated, gradually.

She had tried vegetable tonics, Iron, mineral acids, baths, hydropathic packing, homœopathic remedies, &c. &c., without consequent benefit or injury, so far as I could discover. Also, in passing from one hand to another, usually the purgatives were increased in intensity and variety. Her father, a physician, was growing sceptical of his profession.

I commenced treatment by leaving off all purgatives and giving her unknown to herself, Opium in half-grain doses, with a view of stopping the pain, which she said she always had when the bowels were not open. I took it to be an abnormal sensibility of the intestinal canal, which would not allow the requisite food to lie there a sufficient time. In such a case the orthodox " bowels open once a day" is a diarrhœa. This plan was successful, for with the exception of warm-water enemata twice a week during the winter, she had taken no purgatives since, and now the bowels are open of their own accord, and the Opium pills are left off.

I gave her Pepsine, which seemed quite inert; and then Strychnine, which brought back the pain and curiously prostrated her. It was bad practice.

I have already said Cod-liver-oil was out of the question.

On January 17th I began to educate her gradually to use Dobell's Pancreatic Emulsion, and she has continued to take from one to three doses daily ever since. I hear of her from time to time as slowly improving; any acquaintance who have not seen her for several weeks, invariably remark on her gain of flesh. The appetite is better, and the bowels open naturally three or four times a week. (*July 6th*, 1866.)

Remark how the deficiency in the assimilation of fat (induced probably by the starving alluded to, though not assigned to that by the narrator) made no sign before puberty. Up to that period apparently enough had been taken in for the ordinary purposes of life, but after it the supply was insufficient and the failure in health followed accordingly.

What has puberty to do with fat? Certainly something: in normal health girls before the change naturally dislike fat, but afterwards take to it instinctively. Under ordinary circumstances and with the restraints which society teaches us to lay on our appetites, especially in youth, the instincts are scarcely made apparent; but accidental occurrences will sometimes exhibit their existence in a somewhat unsuspected manner. The following anecdote shows what a strongly marked line can be drawn between the child and the woman in their relish for food, and how full development is not exhibited in this or that organ exclusively, but in the whole person simultaneously. It was narrated to me by the chief actor.

CASE XLVI.—In 1825 or '26 the late Mr. Ridout, a much respected surgeon in the neighborhood of Russell Square, was summoned to St. Albans, to see the apprentices who to the number of sixty were employed in the Abbey silk-mills at that place. A great number of the inmates of the house were suffering from a variety of obscure symptoms of various degrees of intensity. On the examination of the invalids he arrived at the suspicion that their illness depended on the poison of lead, and advised their being treated accordingly. In the mean time, specimens of the water were reserved for analysis, the milk-vessels made of crockery were examined for the metal in question; but nothing deleterious was found, nor had any part of the building been recently painted. Still fresh cases kept occurring, and those who had recovered relapsed, and had colic a second time. The cause of the evil was evidently permanent.

Now the surgeon in ordinary attendance had been loth to agree to the diagnosis which assigned the symptoms to lead-poison, from some connection which seemed to exist between the occurrence of the disorder and the uterine functions. Not only were the catamenia arrested in those attacked, but it was observed, that all the girls under puberty had wholly escaped, while all who had ever menstruated, from the maiden of fourteen to the matron superintendent, were affected in various degrees.

The search was still pursued for the avenue by which the lead had entered the system, and the mystery was at last solved on probing to the bottom of a salting trough in which fat pork was kept. It was found to be lined with the deleterious metal, and to have impregnated the outside of each joint with the poisonous carbonate. Inquiries were then made of the apprentices themselves for some link which would connect this discovery with the anomalous escape of some parties, and injury to others.

It appeared that this fat pork was placed on the table three times a week, but never alone, being always accompanied by some fresh meat, and the girls were at liberty to take which they liked. Now on questioning them it came out, though not previously observed, that the older apprentices and adults always ate the pork, while the little girls—all, that is, under puberty—invariably chose mutton. The disease which had attacked the one and spared the others, was a test of the truth of the statements which they had made.

The newly-acquired desire for fat meat at the age of puberty is a most interesting and curious fact. It is more observable in the female sex, from the deep influence on the vital actions of the whole individual which that change exerts in them, but indications of the same thing may be seen in boys. How shall we associate this fact with what we know of the other corporeal functions of this period? The mere growth of the body in size is much the same before and shortly after puberty. Nor is it easy to conjecture what the evacuation of the catamenia can have to do with oleaginous matters.

There is, however, a change that takes place in the excretion of one organ, which modern chemistry has taught us to allay with the chemical changes of all carboniferous substances in a strict and peculiar manner. It is to the lungs that we would look for assistance in explaining the circumstances before us.

It appears from the researches of MM. Andral and Gavarret,[1] that the excretion of carbonic acid by the lung increases in quantity during childhood very exactly in proportion to growth, the augmentation steadily progressing up to the period of puberty. In boys it would seem but little affected by that new function; but with girls the case is entirely different; there the occurrence of menstruation puts a complete stop to the increase in the amount of carbon thus passing away, and sometimes even causes it to make a retrograde movement. Thus a child of thirteen years of age exhaled 6.3 grammes of carbon hourly; a girl of fifteen years and a half, who had not menstruated, 7.1 grammes; while another, also fifteen years and a half, but in whom the flow was regular, gave out only 6.3, the same quantity as the one two years and a half younger. The same observation was the result of experiments on healthy women of twenty-six, thirty-two, and even forty-five years of age, who still continued to experience their monthly evacuations. After the change of life has occurred, the exhalation of carbonic acid begins to increase again, and in elderly women is much the same as in elderly men. What is still more curious is, that when from either pregnancy or illness the catamenia are stopped, then

<hr>

[1] "Annales de Chimie et de Phys.," vol. viii. p. 129.

temporarily the pulmonary excretion is augmented and occupies a vicarious position in respect to the other functions.

The uterus, then, and vital actions which are expressed by it, play an important part in the decomposition of carbon in the system. When we reflect on this, changes in the digestion which supplies that carbon and changes in the instincts which supply the digestion will not surprise us, when they accompany the radical alteration which the generative organs experience at puberty.[1]

This digression has been so long that it has taken us away from the patient before us. The chief practical point of remark is the importance at that critical period in woman's life of watching over the digestive organs, especially in respect of their appropriation of fat, then so eminently necessary.

A condition leading to the non-assimilation of fat may, like other kinds of indigestion, be brought on by overstrain of the mind, as well as of the body; a fact of the utmost importance in tracing the history and applying it practically to the cure of the patient.

CASE XLVII.—Miss A. D— had an ambitious intellectual governess, who finding her pupil very retentive of learning and persevering, from fourteen to fifteen pressed her forwards in her education with great energy. The nutrition of the mind went on, that of the body was stayed; she was very sharp and learned, but she ceased to grow. The menses appeared for once, and never again. It was also observed that she loathed her food, and anything "rich" (that is, greasy) made her peculiarly uncomfortable afterwards, so that she sometimes threw it up. Her temper became queer and her conscience fanciful, and she distressed herself needlessly about her failing powers of observation and work. It was noticed too that sometimes in reading a kind of cataleptic stiffness would come over her; she stopped for a minute or so as if a parenthesis was snipped out of existence, and then went on with her employment, unconscious for the most part that anything out of the way had happened.

By rest and quiet treatment her wish for food returned, but still she got more and more emaciated, so that at seventeen Dr. Gibbs Blake, under whose care she was, desired my assistance in the case. I found her on March 6th last presenting an appearance very similar in many respects to Case XLV. described a few pages back. The tongue, lips, and cheeks were fully colored; the pulse was firm and rather slow, only 55 when asleep. The temperature of

---

[1] "Gulstonian Lectures," by the Author, in the Lancet, for 1850.

the body at night was 96° Fahr. The heart and lungs were perfectly normal. The urine was clear, of the specific gravity from 1.023 to 1.026. The bowels were open daily. But the skin was harsh and dry, the emaciation was extreme; and the mammæ, which at fifteen and sixteen began to swell, had completely disappeared. The menses had not again shown, though the pudenda were in every other respect developed in proportion to her time of life.

I ordered her the Pancreatic Emulsion, prepared according to Dr. Dobell's plan, in milk. She took it well and profited remarkably, so that when I sent for her to come again to London, in the middle of April, I hesitated at first to shake hands, supposing it to have been a sister of my patient that I saw. The reason for bringing her up to London again was however this—she caught a catarrhal cold, it flew to her stomach, the emulsion nauseated her; and yet she persisted in swallowing it with her innate perseverance. Suddenly while engaged in playing a round game of cards, she went off into an epileptic fit, vomited a quantity of bile, and next morning knew nothing of what had taken place.

The emulsion had done its work, and was beginning to do mischief. It has been left off, and I have heard of no more *contre-temps*.

It is a comfort to find a treatment sometimes during harm, it thereby shows itself capable of doing good.

The dryness and apparently dead dirty aspect of the skin has been noticed in several of the last cases of non-assimilation of fat. The cutaneous imperfection sometimes goes farther and exhibits itself in the shape of eruptions.

CASE XLVIII.—Miss D'O—, aged 24, has never been plump or strong since menstruation was first established. She is excessively thin, her ribs sticking out, and her bust flat, enabling the normal condition of the lungs and heart to be easily proved. She is energetic, pretty, and lively, and her bright eyes and red lips are much admired. The specific gravity of the urine is 1.023, and it deposits lithates on cooling. The bowels are regular. Rich food is disagreeable to her, and cream indigestible, so that she is accused of being fanciful in her diet. Latterly she has suffered great inconvenience from an eruption of an eczematous[1] character on her forehead and skin, and thinks she is thinner and thinner. She gets on well enough when living quite quiet, but animal and spiritual life enjoyed for a season throw her back on each occasion, and she does not pick up her health again. A trip to Italy, with its numerous temptations to bodily and mental exertion, was the last blow. She had taken tonics and homœopathic remedies.

I gave her the Pancreatic Emulsion, which is working apparent benefit.

[1] *Eczema* = "a superficial formation, consisting mainly of serum from the denuded connective tissue of the corium, without external existing causes." I do not take this as a dogmatic definition, for dermatologists are a difficult class to please, but simply to explain what I myself mean by the word.

In a lecture on Pulmonary Consumption in 1862[1] I made
some observations on the connection between that disease and
cutaneous degeneration, apropos of a case of impetigo of the
finger-nails. Perhaps the mal-assimilation of fat has something
to do with it.

By the artificial emulsion of fat with pancreatic juice, we
certainly seem to be put in possession of an easily assimilated
oleaginous material, and a most valuable contribution to the
restorative pharmacopœia. Dr. Dobell has used it more extensively than anybody else, and he is convinced of its superiority
to Cod-liver Oil in consumption. My only fear about it is that
careless or dishonest chemists might manufacture it from tri-
chinous swine, should it chance to be much used; and then,
being unboiled, it would be a possible means of introduction of
that dangerous parasite. Medical men should warn patients to
be careful whom they get it from.

The experiments of Drs. Bidder and Schmidt,[2] and of their
pupil Lenz,[3] have indeed deposed the pancreas from the position
in which it was placed by Bernard[4] as almost the sole actor in
the digestion of fatty substances; but yet it still remains as an
important link in the chain of physiological agencies conducing
to that digestion. And it is fortunately one which we are able
to supply by artificial means. As to the form of preparation it
is much more practically convenient to give (as is done in the
emulsion) the digester and the article to be digested at the same
time, than to divide them, as in the proposal to administer
"Pancreatine" prepared after the fashion of Pepsine. In point
of fact, it is probable that the activity of the Pancreatine would
be entirely obliterated in its passage through the stomach, unless
it were guarded by the fat with which it is already united. It
is the fat that is wanted, and this is an easily assimilated form
of it.

That the pancreas is an important agent in the digestion
of fat receives powerful support from a class of fatal cases in
which the whole of the pancreas is organically altered in struc-

[1] "Lectures chiefly Clinical," p. 280 of 4th edition.
[2] "Die Verdauungssäfte," pp. 241–259.
[3] Lenz, "De Adipis Concoctione et Absorptione." Dorpati, 1850.
[4] "Archives Générales de Médecine," 1849.

ture, and in which during life a peculiar inaptitude to digest adipose tissue has been observed. I shall revert to this subject in illustrating in a future chapter organic changes in the digestive viscera.

SECTION IV.

Indigestion of Water.

The assimilation of water is the least vital process of the whole of digestion. It would seem capable of entering by the simplest endosmosis from the alimentary canal to the blood-vessels, where it is incorporated with the blood unchanged. The process can be carried on as long as life exists at all, and in obedience to the mechanical laws of diffusion.

Now the chief facts observed with regard to the connection of membranes with liquids, are the following:—

1. If a moist membrane be interposed between aqueous solutions of different densities, two currents will run through it, one from the denser to the rarer liquid, and one from the rarer to the denser, and the latter will be the strongest in direct proportion to the density.

2. The current is increased in the direction of a liquid in motion.

3. The current is increased in the direction from an acid to an alkaline fluid.

4. The activity of osmosis increases with the temperature.

There are, then, in the circumstances under which the blood-vessels and the contents of the bowels are placed, three very marked principal things which promote the passage of fluids into the former from the latter in a greater degree than the reverse. These are:—

1. The comparatively greater density of the blood;

2. Its motion;

3. Its alkalinity.

At the same time the animal warmth keeps up the general activity of the osmosis in both directions probably in the ratio of its degree. Where any of these conditions are diminished (removed or reversed they cannot be during life), then the assimilation of water is retarded, and any excess remains inconveniently in the intestinal canal for longer than usual.

CASE XLIX.—A somewhat corpulent lady had lost much blood by bleed-ing piles before she applied for medical advice, so that she was reduced to a great state of anæmia. She did not come to me about the piles (which were removed), but on account of the flatulent tumidity of the intestines, and a perpetual "*glug-glug*" in them when she moved about. Her appetite was bad, and she therefore washed down her meals with copious draughts of water. She certainly observed that the more liquid she took, the more the "*glug-glug*" was distressing, but still she did not think she drank more than other people. I told her she must drink *less* than other people, and to that end advised the use at meals of weak lemonade without sugar taken in sips, and the sucking of a piece of liquorice when the mouth felt dry at other times.

Here the thin blood of anæmia refused to absorb as quickly as usual the watery fluids from the alimentary canal, exemplify-ing an infringement of the conditions required in the *first* law of osmosis.

Corpulent persons are generally very thirsty souls. In two instances (Cases XXX and XXXI of my Table of Cases of Obese persons, "On Corpulence," page 142) the corpulence was assigned distinctly to this cause. But they are inexplicably touchy about confessing it. I cannot make out why, seeing it is diluted drinks, not alcohol, that is the subject of inquiry. I dare say, therefore, that the patient before us did take more fluid than other people in spite of her denial.

The "glug-glug" of superabundant water may be distin-guished from the noises of flatulence by being caused only by moving the body. Gas generally is loudest when the patient is still after exertion.

CASE L.—Mr. H——, a pork-butcher of healthy appearance, 30 years of age, complained to me of the weight and distension which he always felt after his usual meals; though if he took a chop at a coffee-house or a snack standing he did not feel it. The difference seemed to be that when sitting down com-fortably at his leisure he took a considerable allowance of liquid, which he at other times avoided. He said he had nothing else the matter with him, but observing the breath short, I examined the heart, and found a loud sawing systolic murmur.

Here the *second* law is exemplified. The motion of the blood was retarded by the valvular disease of the heart, and the absorption of fluids in the œsophagus and stomach proportion-ally retarded likewise. It is in burly, otherwise healthy, persons with diseased hearts that this indigestion of water is most generally conspicuous. When the patients are seriously ill and laid up by their structural ailment, it does not so often occur. Perhaps they are not so thirsty, and so do not put the matter to the test.

It is a hint sometimes practically valuable, not to overburden with slops the stomach of cardiac invalids.

Another exemplification of the *second* law of osmosis may be observed in impediments to the motion of the blood from the lungs:—

CASE LI.—Susan B—, a married woman, aged 41, thin, sallow, and hollow-eyed, was admitted under me at St. Mary's, June 15th, 1860. She had been subject to shortness of breath for a long time; but this symptom had been aggravated since the previous March, when she seems to have caught cold. Since then also she had been able to eat her food only very dry, for if she took fluid with it, nausea and vomiting occurred. This was worse when the asthma was worst. The kidneys and heart were healthy. Her appetite was good.

Headaches also were very frequent, but they seemed independent of the gastric ailments, for they are reported when these latter are described as better.

The usual physical signs of pulmonary emphysema were present, and she was treated accordingly with Quinine. Her diet was meat and Pepsine. She was discharged "cured" on July 13th; the "cured" referring to the indigestion for which she was registered, not to the emphysema I presume.

It may be remarked that in this last case the rejection of the water by the digestive organs was gastric, whereas in the two former it was intestinal. There is in this fact no significance of the locality of the impediments to circulation—patients with pulmonary disease quite as often have the gurgling on movement, and cardiac patients will sometimes vomit. Indeed, when you are thinking of them only as *messageries* of the blood, the heart and lungs are one.

An illustration of the *third* law of osmosis may be found in almost all cases of dyspepsia.

Let the reader note first on himself what takes place as a consequence of food in the normal condition of the stomach. Let him take the specific gravity of his urine on rising and look at the color. Then let the stomach be thoroughly roused to acidity by a healthy breakfast at which the usual quantity of fluid is taken. Observe the urine passed during the next two hours. It is paler, of lower specific gravity, neutral, perhaps alkaline. The fluid contents are augmented in greater proportion than the solid, the basic elements in greater proportion than the acid. If pickled fish and light white wines form part of the breakfast, the limpidity is still more decided.

The explanation of this I believe to be that the acid of the stomach being at that period in special excess, the osmosis of water through its walls into the alkaline blood is peculiarly rapid. Water with salts and solids soluble in water enter into the circulation quickly, and fill it. They pass away quickly by the kidneys, carrying off often some of the blood's soda with them, and so increasing the fixed alkali of the urine.

When the stomach gets to rest again and is neutral, the fluids do not pass so fast into the blood, or away by the renal tubes; then the urine resumes its full color, acidity, and average specific gravity.

As the body gets tired with the day's work, the digestion is not so active. And hence after luncheon this physiological variation of the urine is not so marked, and still less after dinner.

Now observe an invalid, the vitality of whose stomach is below par. The physiological variations are much less marked, the urine is never so high and rarely so low as that of a healthy person in healthy condition. Its acidity also alters little during the twenty-four hours. The interpretation of this I take to be the imperfect acidity of the walls of the stomach causing a delay in the absorption of its aqueous contents, which lie there unchanged or decomposing; in either case a burden, giving rise to pain and inconvenience, sometimes to vomiting.

The importance of obtaining for analysis specimens of urine taken at various periods of the day, instead of trusting to one, is too obvious to require much comment. Take for an example the following, which the date allows to be quoted from memory.

CASE LII.—A gentleman came to me yesterday forenoon about a deficiency in his generative powers and other symptoms unnecessary to detail. The specific gravity of some urine he passed was 1.015, and it appeared probable that it was upon this ground that the medical man who sent him to me had been giving him Iron. I told him to come again this morning, and bring a specimen made on rising, as well as another specimen of that after break-fast. The latter was again about 1.015, but the nocturnal collection was 1.026, showing that the vitality of the blood and assimilation were sufficient.

SECTION V.

Treatment of Indigestion Based on the Article of Food not Digested.

In former writings on this subject[1] I attempted a division of the form in which dyspepsia is manifested according to the glands whose secretion may be supposed to be affected. I thought we could assign some to the oral and œsophageal, some to the gastric and some to the intestinal portions of the alimentary canal. I anticipated that we might find a "salivary" dyspepsia of starch, a "gastric" dyspepsia of albumen, an "intestinal" dyspepsia of fat. Further experience has not confirmed my hopes of finding any such trenchant anatomical distinction of the cases which came before us. It seems to me now that a healthy state of stomach and duodenum is as essential as a healthy state of mouth to the digestion of amylaceous matters; that the salivary and intestinal secretions aid powerfully the digestion of albumen; and that even fat is affected by an imperfect activity of the stomach. I have therefore in this chapter based the division solely upon the element of food the digestion of which is most prominently deficient. It must be understood that "prominently" does not mean "solely;" that where one article of diet suffers, the other suffers with it, though perhaps in a minor and masked manner.

This prominence of the indigestion of one or other of the food elements, to what should it lead us in practice? Some, influenced by purely chemical considerations, have answered off-hand that we have only got to omit the objectionable article from the dietary and all is done. That which causes pain is to be left off, and the pain ceases. True, but man is an omnivorous animal, and requires omnigenous food. He can be kept alive perhaps for a time on one food, but not in health. Take an example of the carrying out of such a treatment à l'outrance in respect of vegetable food:—

---

[1] "Digestion and its Derangements."

Case LIII.—E. H—, a Liverpool merchant, aged 30. of muscular build, but rather bloated flabby aspect, came to me in December, 1859. He complained of a foul taste in the mouth like bad fish, low spirits, and want of appetite for breakfast. He had also occasional attacks of headache accompanied by nausea and vomiting, which nevertheless did not relieve his permanent condition. He had been gradually getting into his present state for six years: the administration of bitters and acids had done him temporary good occasionally, but worked no cure. He said that in deference to medical advice he had been most careful in his diet, eating nothing but lean meat and stale bread or biscuit. Vegetables had been forbidden, because they had at first caused flatulence and heartburn, which did not occur at all under the use of the carnivorous dietary. On examination of the mouth, the tongue was seen to be coated with smooth yellow epithelium, especially at the sides; the gums were loosened from the teeth, swelled. red-edged and soft. The patient said they often bled when he cleaned his teeth. I thought at first some of his medical advisers must have given him Mercury, but I could get no history of pills or powders, in which that metal is usually administered, and I am disposed to attribute the whole of his existing symptoms to an exclusively meat diet.

For direct treatment I advised him to eat milk porridge and water-cresses for breakfast, salad and meat and stale bread for luncheon and dinner, and lemonade or fruit-water ice instead of tea. As an indirect aid I prescribed some Bark and Chlorate of Potash. He soon got well.

It is singular how slight a change of diet will bring on minor manifestations of scorbutus.

Case LIV.—During his attendance upon me after a severe operation, one of the leading surgeons in Europe related a bit of personal experience apropos of my complaining of gastralgia after salad or strawberries, I forget which. He said he took a house out of London one summer, and used after his daily work to join his family (who dined early) at tea and mutton chops. After a time he found spots of purpura on his legs, boils, etc. He exchanged the tea for vegetables and beer, and immediately regained his accustomed health.

Patients and doctors both make a great mistake in shunning absolutely all that causes pain or inconvenience. They ought to consider whether the thing shunned is or is not an essential to high health: if it be so, every effort should first be used to get it borne without pain; where that goal cannot be reached, wisdom and duty will often guide us to submit to the pain for the sake of the accompanying advantage.

It may be remarked that the designed attainment of any high degree of voluntary pleasure always involves endurance—

endurance of disagreeable sensations which coming upon us against our will would be real torture. A day's hunting, a match at cricket, an Alpine tour, even a picture-gallery or a ball, success in love, literature or war, are impossible to those who recoil from bearing immediate pain. This ought to be—and (*experto crede*) is—a consolation to many a sensitive sufferer. When it is pointed out that their pains are identical with what they and others have borne without a murmur, nay without notice, in the pursuit of enjoyment, their hopes and aims may be, not so much for the absence of the sensation, as for the vigor which will ignore it.

The avoidance of meat on account of the inconvenience caused by it may bring on an equally undesirable morbid condition, though not so distinctive in its character as the scorbutus exhibited in the last patient.

CASE LV.—In October, 1865, a maiden lady of 34 was placed under my care by Dr. M'Call Anderson, of Glasgow, who quite agreed with me in my view of the case, though circumstances prevented him from attending to her himself. Since girlhood her bowels had been very costive. The presence of the retained feces produced disagreeable sensations, to relieve which for twenty years she had been in the habit of almost daily taking purgatives of her own accord. For the last few years pain in the epigastrium had gradually become habitual with her, and as it was immediately increased by taking solid food, she had entirely ceased to take meat. The consequence was increased costiveness, increased sensitiveness of epigastrium, increased debility. The catamenia diminished in quantity and frequency, making a scanty show for a day or two about three times a year during the height of summer, and never in winter at all. Her complexion was pink and white, and her lips not pale, but she was very thin.

She had an opportunity of breaking out of all her old associations and habits by a visit to some country friends in France, and I urged her to accept it, and to make a complete change in her mode of life. I desired her to let her bowels go unopened for as long as four days if they chose, and then use a simple water enema, if necessary—to bear the pain caused by the first few mouthfuls of each meal, and to eat more and more meat at it each day without flinching—to drink Burgundy. To aid her in this I gave her for a few weeks Quinine and Strychine three times daily.

The last report I have of her is dated March 23d, 1866, and states that she required no enemata, the bowels acting of their own accord every three days; that food scarcely ever causes her pain, though she eats nearly as much as other people; that the catamenia had reappeared in the February even of this backward spring, after being absent since the previous July.

I said just now that as much injury was done by leaving off meat as by leaving off vegetables. But perhaps I ought to say greater, for though it is not so prominent, and has not such a distinctive name, the condition induced by it is longer in getting well, and might just as easily prove fatal.

Remark in Case LV how amenorrhœa was a disease of the stomach; it usually is so. .

The habit of taking purgatives much increases abnormal sensitiveness of the alimentary canal, especially of the stomach. To break through the habit is essential to a cure. I will speak about that in the next chapter.

The injudicious omission of fat from the dietary must doubtless produce similar effects to those which follow the indigestion of fat, in fact the converse of the effects which we aim at producing when we intentionally administer an excess of it as a remedy. As under the use of an easily assimilated oil the skin becomes elastic and firm, the debilitating fluxes cease from the mucous membranes and are replaced by normal secretions, the nerves feel joyous life instead of perpetual pain, and old sores heal up; so we may expect to arise from a deficiency of this article of food a dry wrinkled surface to the body, a persistence of leucorrhœal and other mucous discharges, that feeling of enduring uneasiness which denotes scant life, a deterioration instead of a renewal of all the tissues.

But I am not able to find any good illustrations of the matter. Where fat has been omitted, meat seems to have been omitted also in most of the instances I have referred to, and the symptoms might fairly enough be attributed in a great measure to that; or else there has been a complication of other causes of disease; or else the omission has been intentionally remedial, designed to reduce an over-abundance in the body.

A few years ago, during the prevalence of the attention excited by Mr. Banting's case, I did indeed hear reports of persons having injured themselves by adopting with over-strictness the system by which that famous man tells us he regained the sight of his toes, forgetting that no similar mountain to his had ever impeded their view. But I never saw a real case in point. If the experimenters are really over-corpulent, they feed on their own fat, and submit with case and advantage to the discipline; if they are not so, the instinctive desire becomes so strong that they cannot resist the sight of the forbidden luxury on the table. The possible rectification of their circumference is not worth such stoicism, and they stop in good time.

I do not think, then, that we profit much from those off-hand advisers who suppose they accomplish everything by forbidding the use of the sort of food which produces the symptoms. Neither in the indigestion of vegetable, animal, oleaginous, or watery articles of diet does this restore health. On the contrary, as I have shown by examples which every one may cap out of his own patients, if he will but turn them over in his mind, an actual state of disease may arise from persistence in the remedy.

A short repose for a time, and abstinence from an unnecessary excess in the undigested dishes, is doubtless wise. But that abstinence must not be complete or final. What the patient wants, when he complains he cannot eat so-and-so, is not to have "don't" said to him—his stomach has said so already—but to be enabled to eat it like other people.

The temporary repose may be accomplished often by a change in the mode of preparation of the articles which cause most inconvenience, often by the substitution of something else, not so agreeable perhaps or so common, but which will not be objected to for a time.

The following details may furnish examples and limits.

IN STARCH OR SUGAR INDIGESTION.

The use of *sugar* in such quantity as to cause a sweet taste may be left off. Tea may be taken in the Russian fashion, pouring the hot tea on a slice of lemon with the skin on, thus retaining all the aromatic stimulus of the drink without its indigestibility. And all lozenges and sugar-plums and sweet confectionery will be interdicted. The best substitute is oranges or lemons.

For ordinary *bread* may be substituted biscuit, toast, or Stevens' aërated bread. Baker's bread is usually easier of digestion than home-made.

6

Potatoes may be finely mashed and mixed with meat gravy.

As *vegetables*, stewed lettuces, cabbages, spinach (hot), and golden cress, water-cress, and salad (cold), may be taken. A small quantity only of these is required to keep up the health, and nobody eats so much of them as they do of potatoes.

From green vegetables possible of digestion by weak stomachs must be carefully excepted peas, beans, and, in short, all the papilionaceous plants usually eaten green. They are famous for producing flatulence. M. Chomel attributes this to the evolution of atmospheric air contained in their spongy husks; but I think the cause lies deeper than that—perhaps in the specific action of their empyreuma, arresting the absorption of air in weakly persons. ˙Else why do they not produce equal effects in the healthy?

An example of the mechanical differences made by cookery in the form of starchy food, are the two sorts of crust known as "short" and "puff" paste. In the former, the butter is thoroughly incorporated with the dough, so as to divide the starch-granules one from another, and permeate the gluten; while in the latter the dough forms thin layers, like a quire of buttered paper. If the teeth are imperfect or mastication care-less, those strata of dough are well known to form in the stomach a solid mass, which is difficult of solution in the upper part of the intestines; whilst the friable paste (the "short") is mixed with the rest of the food, and if the butter be fresh, causes no discomfort.

Now, some dyspeptics are such delicate measures of good or bad cookery, that they can take "short" pastry, but not "puff." It is always worth while to make the trial.

There is an advantage in not mixing too much the animal and vegetable food. In a weak stomach they interfere with one another's digestion. A light luncheon of bread and butter, rice pudding, fruit, and vegetables with a little vinegar, can often be borne without inconvenience, which with the addition of meat would have caused flatulence. The dinner after this may be restricted to meat without injury.

Particular care should be taken that vegetables are thoroughly boiled soft all the way through, and dried on a cullender.

A certain quantity of oleaginous matter renders vegetables

in which there is much combined water, less massive in the stomach. Thus, milky rice pudding does not collect into a lump as plain rice is apt to do. In making the latter dish up for baking, eggs should never be used. Baked albumen is one of the most insoluble forms of albumen.

Plain boiled rice should always have a little fresh cold butter mixed up with it. In that way it makes an accompaniment of meat at dinner.

Stewed pears and roast apples are a good substitute for sweets. A little butter improves them also.

But melted butter sauce is an abomination. Nine times out of ten it is rancid, or becomes so five minutes after it is swallowed,—that is to say, directly the flour in it is converted by the saliva into glucose. The best sauces are pepper and vinegar.

IN INDIGESTION OF ANIMAL FOOD, it will be found to be generally the form rather than the chemical constitution of the aliment against which the stomach rebels.

Observe the preparation of food as arranged by nature for the delicate stomachs of the new-born. It is completely fluid ; the various elements are intimately mixed together, and are further aided in their solution by the lactic acid into which it decomposes. Milk is not only a type, but is also itself the most perfect food for extreme weakness. I have never yet met with a stomach which could not bear it either made into whey, or prevented from coagulating by the admixture of lime-water. This fluid meat will pass through the stomach unaltered, the gastric juice will trickle through the pylorus at its leisure after it, and with the intestinal juice will digest the casein in the intestines.

I do not think any one could deserve better of his country than by the establishment of a farm where the milk treatment could be systematically carried out, as at Gäis and elsewhere in Switzerland.

The chief aid offered by art to the conversion of albuminoids into the state of peptone is to increase their softness and permeability by water, so that the converting juice may have access to every particle as soon as possible. The mechanical condition of the nitrogenous aliments is of tenfold more importance than the quantity of nitrogen they contain. If enveloped in an insoluble layer of their own or other substance, they are in fact as useless as gold locked up in a box.

Next to milk, the most digestible form of animal food is properly made beef-tea. The following is the best receipt for dietetic purposes.

Recipe for making Beef-tea nutritious.

Let the cook understand that the virtue of beef-tea is to contain all the contents and flavors of lean beef in a dilute form; and its vices are to be sticky and strong, and to set in too hard a jelly when cold.

When she understands this, let her take half a pound of fresh-killed beef for every pint of tea she wants, and carefully remove all fat, sinew, veins, and bone. Let it be cut up into pieces under an inch square, and set to soak for twelve hours in one third of the water required to be made into tea. Then let it be taken out, and simmered for three hours in the remaining two thirds of the water, the quantity lost by evaporation being replaced from time to time. The boiling liquor is then to be poured on the cold liquor in which the meat was soaked. The solid meat is to be dried, pounded in a mortar, and minced so as to cut up all strings in it, and mixed with the liquid.

When the beef-tea is made daily, it is convenient to use one day's boiled meat for the next day's tea, as thus it has time to dry and is easiest pounded.

Some persons find it more palatable for a clove of garlic being rubbed on the spoon with which the whole is stirred.

The utility of decoctions of animal food depends on several circumstances which modify the advantages accruing from their liquid state. Heat seems to have an effect in some degree proportioned to the period of application, rendering albumen more or less insoluble, at the same time that to a delicate palate there is a decided loss of savor. Thus soups and stews which are "kept hot" are wholesome enough during the first few hours, *may* be digested at a railway refreshment room for some hours after, but on the second or third day give the rash stranger beguiled into a Palais Royal two-franc dinner an infallible diarrhœa. (*Probatum est.*) Though finely divided, the minute fragments of muscular fibre seem to be individually rendered insoluble by continued heat. Good soup is that which is made most like the above-described beef-tea, and is a highly digestible article; bad soup, that which least resembles it, and is to be avoided as poison. Next to good soup in digestibility comes sweetbread.

At a very early stage of the treatment of albuminous indigestion, tender roast leg of mutton can be borne. A chemical view of the process of roasting shows it to fulfil all the indications of perfect cookery. The heat radiated from the open range coagulates the outer layer of albumen, and thus the exit of that still fluid is prevented, and it becomes solidified very slowly, if at all. The areolar tissue which unites the muscular fibres is converted by gradual heat into gelatine,[1] and is retained in the centre of the mass in a form ready for solution. At the same time, the fibrine and albumen take on, according to Dr. Mulder,[2] a form more highly oxidized, and, especially in the case of the former, more capable of solution in water. The fat also is melted out of the fat-cells, and is partially combined with the alkali from the serum of the blood. Thus the external layer of albumen becomes a sort of case, which keeps together the important parts of the dish till they have undergone the desirable modification by slow heat—a case, however, permeable in some degree by the oxygen of the free surrounding air, so that most of the empyreumatic oils and products of dry distillation are carried off. This is no loss either to our stomachs or our palates. If acetic acid be generated it is probably carried off, and if not carried off it is neutralized by the alkaline carbonates, as certainly roast meat is not acid to test-paper if quite fresh. The little that may remain probably renders the muscular fibre more soluble.

Roasting, therefore, is as scientific and wholesome, and *therefore as economical* a process as it is a palatable one, and. is well worth the extra expenditure of fuel which is entailed. Baking can never take its place, especially for invalids; for it concentrates in the meat all the empyreumatic products of slow combustion.

Rapid boiling may effect in some minor degree the case-hardening of the meat above described, but the interior albumen seems after this process also more solid and less digestible.

Slow boiling at a low temperature makes, it is true, a nourishing soup, but converts the muscular fibre into a mass of hard

---

[1] Not, however, the sarcolemma, which an experiment of Professor Kölliker's seems to remove from the class of substances yielding gelatine. See Kölliker's " Mikros. Anat.," vol. ii. p. 250.

[2] Quoted in Moleschott's " Diätetik," p. 450.

strings, which, eaten or not eaten, are in nine cases out of ten equally wasted. The relics to be found in the feces exhibit all the transverse striæ of their original state, quite unaffected by their intestinal journey. The only way to make bouillon digestible is to beat it up in a mortar to a fine pulp and mix it with the soup, as prescribed above in the recipe for beaf-tea.

Of all meats, mutton is the most digestible, because it is, when roast, the closest grained, most friable, and least infiltrated with fat.

Birds, on the other hand, when roasted, still more when baked, are apt to be too much dried up. And, therefore, if you cannot trust thoroughly the goodness of your patient's cook, he had better have his fowls and his partridges boiled. Other birds are either too dry, too oily, or too empyreumatic for invalids.

I have sometimes found toasted cheese borne by a stomach to which meat gave pain. The cheese must be quite new, and toasted very soft. In this form it is both digestible and palatable.

The quantity of albuminous food should be gradually increased from day to day; but the invalid's plate should never be overloaded. The look of more than he can eat sets him against it. Judicious times for pressing food should be selected. A cup of beef-tea on going to sleep can often be borne, when ordinary meals excite nausea.

Ladder of meat-diet for invalids.

Whey.	Turtle soup.
Milk and lime-water.	Sweetbread or Tripe.
Milk and water.	Boiled partridge.
Plain milk.	Boiled chicken.
Milky rice-pudding.	Boiled lamb's head and brains.
Beef-tea.	Mutton chop.
Plain mutton broth.	Roast joint of mutton.
Scotch broth.	

In cases where there is a repugnance to fat, light friable fish, such as boiled sole, or turtle fins, or water soojee, will be tolerated, while red meat excites disgust. It is not nearly so soluble in gastric juice, nor so nutritious, but it makes a variety.

Oysters are best eaten raw, in which state they carry with them their own pepsine for their digestion. Cooking toughens them.

The use of Pepsine, or artificial gastric juice, as a remedy, is especially indicated in the indigestion of flesh food. But I think that since its introduction to general use through the ingenious preparation of Dr. Corvisart, it has caused more disappointment than satisfaction. This is because it has been given in unsuitable cases, and because impossible expectations have been founded upon it.

The cases in which it is really useful are those where a progressive anæmia is accompained by an inability to digest albuminous food. This inability is exhibited in three ways: first, by meals of such diet, even in very small quantities, being followed by a sense of great weight and oppression at the epigastrium, and sometimes by actual vomiting; secondly, by the passage of loose fetid stools containing much unaltered muscular fibre, lumps of fat, and such like remnants of a recent meal; thirdly, by loss of appetite and a nausea roused by the bare idea of flesh food. Often all three phenomena exist together; but each one may be found separately, and is of itself a sufficient indication of the patient's state.

The state of the stomach when these symptoms occur is often an excessive secretion in the upper part of the alimentary canal of alkaline mucus, which envelops the food, and prevents the action of the gastric juice upon it. The consequence is, either its rapid ejection unaltered, or its decomposition, and the evolution of fetid gas. If vegetable food be mixed with the meat, it ferments into acetic acid; and thus you may have sour eructations from the stomach, and diarrhœa. If this excessive secretion of mucus is recent and moderate, the appetite may remain uninjured—nay, may sometimes be morbidly increased; but a long continuance, especially if joined to progressive pulmonary disease, is sure to induce an anæmic condition of the alimentary canal, which results in a disgust for food.

This state of things it is very important to check. If it goes on, the patient cannot take in sufficient quantities the meat which should refresh his degenerating muscles and pale blood; he cannot, if phthisical, take the Cod-liver oil which is to replace his emaciating tissues; he cannot, from weakness, take the exercise which might renew his whole diseased system. And I do not know any remedy which more readily, obviously, and directly does what it can towards checking such a state than Pepsine. It acts immediately and surely.

We must not, however, raise our expectations of the power of Pepsine too high, or we shall be disappointed. I said just now it "does what it can," and I would have understood clearly what position this agent holds in the rational materia medica, and then we shall know what good results may be demanded with reasonable hopes of obtaining them. It is an artificial, and therefore a partial, substitute for a natural process. Gastric juice from a healthy animal is mixed with the food, instead of that which the patient's stomach ought to prepare. And it acts in the body just as it would out of the body under the same circumstances of heat and motion. The chewed meat is dissolved by it just as white of egg suspended in a beaker is dissolved by it; and the putrefactive process is arrested by it in the intestinal canal just as the putrefactive process is arrested by it in the laboratory.

For you may observe that albumen suspended for twelve hours in Pepsine is quite sweet, whereas that soaked for the same time in saliva is most fetid. It is, therefore, a substitute for the natural secretion, and to a certain extent supplies its place.

But like all imitations of nature it is coarse and imperfect. The solvent, instead of being gradually and continuously poured on the outside of the mass of food, is mixed up in the middle part of it, and acts merely chemically, without any of the mechanical and physiological helps belonging to natural digestion, and consequently soon exhausts its energies. The chyme, or albumen prepared for absorption, instead of being wiped off and swept away by the stomach, remains for some time mixed up with the Pepsine, so that the latter is not freed for the solution of a new portion. By this imperfect process only a very small portion of meat can be dissolved. The small quantity of Pepsine in the powder is ridiculously inadequate to the wants of a healthy stomach.

If therefore a patient hopes that by the aid of Pepsine he can get a full and sufficient meal digested at once, he will fail. But let him take about half a mutton chop with the remedy the first day; and if that is digested well, next day a whole chop; but then he has got to the end of his tether, and the digestion of a larger quantity will not be at all assisted by artificial solvents. After a chop has been digested and absorbed twice, or even once, a day by this means for about a week or ten days, the expedient has probably done all the work that can be fairly asked of it, and the stomach has either recovered sufficient to digest alone, or will require different remedies to enable it to do so.

Therefore, for the Pepsine to be completely successful—first, it must be given only to those who cannot digest half a mutton-chop without it; secondly, more than a chop must not be given at once; thirdly, it must not be required to go on alone improving the patient's condition for more than a week or ten days.

But for the time named I advise its being given alone, and the action not interfered with in general by other medicines. Many will really prevent its chemical effect, and all will confuse your judgment of the advantage gained. In this time it will generally be found that the repugnance of the patient to meat has been overcome, and that a small quantity of it at a time can be relished and digested; the morbid fetor of the stools diminishes, and the flatulence and distress arising during their passage through the bowels ceases. A renewed strength and a renewed power of assimilation commences, the sleep becomes more natural, with the diminution of night-sweats and hectic; while, at the same time, the pulmonary symptoms of cough, dyspnœa, &c., relax, and a step at any rate is taken in the right direction towards the cure of the disease. It is remarkable, too, what a slight improvement in the digestive powers will often enable the patient to take Iron and Cod-liver oil. These are acknowledged the mainstays in the treatment of tubercular consumption, and any expedient, however temporary, which will pave the way for their administration, is a great boon.

IN THE INDIGESTION OF FAT a purpose similar to that assigned to Pepsine in the last paragraphs is performed by Pancreatine.

In a classification of curative agents I have put the two together as "Constructive" or "Histotrophic" remedies.[1] But in the form of pancreatic emulsion, as devised by Dr. Dobell, the solvent and substance to be dissolved are united together, and their mutual reaction has already partly taken place. I have already spoken of the necessity for and advantages of this union. It is better than Cod-liver oil, because it carries the agent of its own solution along with it.

THE INDIGESTION OF WATER as a consequence of anæmia is cured by the administration of Iron. Where it results from heart disease or emphysema, it indicates a mercurial purgative, and is temporarily relieved by its emptying the congested portal circulation. An observant patient of mine with emphysema tells me that she finds it a good rule never to drink with her meals.

SECTION VI.

Treatment based on pathological condition.

It cannot but strike any one who reviews either the typical cases I have collated from my notes, or those (not essentially different, I am sure) which have occurred in his own practice, that a general deficiency of the vital powers is more notably exhibited in indigestion than in any other disease. And this is equally apparent in each form of indigestion from whatever cause arising. I always, therefore, look forward to giving tonics as the prime therapeutical aim in all cases. Sometimes that part of the treatment can be commenced forthwith, sometimes it will be necessary to relieve temporarily certain of the prominent symptoms first, but without tonics no cure is effected.

My favorite tonic is Quinine, in two-grain doses in lemon-juice sufficient to dissolve it, and diluted with water to a convenient bulk. Its action seems to be principally on the mucous membrane of the mouth, œsophagus, and stomach, which it astringes and tones up to a healthy state, restraining the secretion of mucus, and making the special secretions more active.

To Quinine I usually add from $\frac{1}{24}$th to $\frac{1}{20}$th of a grain of Hydrochlorate of Strychnia, unless there are some contra-indi-

---

[1] "Lectures, chiefly Clinical," 2d Introductory Lecture in 4th edition.

cations to its use. It relieves flatulence, and that feeling of
sinking when the stomach is empty, which arises from a sluggish state of the involuntary muscular fibres; and in cases of
constipation reinforces the expulsive action of the peristaltic
fibres on the mass of feces. The principal contra-indication
to its use is an over-sensitive state of the nervous system. I
have been obliged to leave it off in several cases of hysterical
women because of the neuralgia which followed it, and in two
instances of men agitated by business I have had want of sleep
and excitement of mind attributed with apparent justice to
Strychnine. In the doses quoted cramps never are produced,
and the slight inconveniences I have named cease immediately
the alkaloid is omitted.

In large quantities Strychnine may sometimes produce
spasmodic action of the muscles. I have had this happen in
hospital, when administering it for other complaints. But even
then not the slightest harm accrues, if the amount is diminished.
Some persons have a fear of its accumulating in the body, and
the effect of successive doses being concentrated into one, which
to me seems impossible in a soluble diffusible salt. The fallacy
has probably arisen thus—in cases of paralysis, for which
Strychnia was originally prescribed, the nervous system is
usually so prostrate as not to respond to even considerable
quantities; after a time the patient becomes more healthy and
more sensitive, and then the dose which had been given day
after day without effect, acts perceptibly, and perhaps vigorously
—acting thus, not because it has accumulated, but because the
nerves have at last become well enough to be conscious of it.
A soluble and diluted salt of Strychnia seems to me one of the
most manageable drugs we have in the Pharmacopœia, because
you can graduate the dose accurately to your requirements.
The extract of Nux Vomica is dangerous, because you never
know the exact strength of the preparation sold.

This treatment of indigestion does not interfere with remedies
addressed to check pain, pyrosis, vomiting, or any of the other
morbid phenomena, which will be discussed in future chapters.
Neither does it interfere with the use of Iodide of Potassium as
a temporary augmentation of the salivary secretion. I have
found it the most universally applicable, and therefore I do not
mention others of less value.

92

CHAPTER III.

HABITS OF SOCIAL LIFE LEADING TO INDIGESTION.

Section 1.—Eating too little. Section 2.—Eating too much. Section 3.—Sedentary habits. Section 4.—Tight lacing. Section 5.—Compression of the epigastrium by shoemakers. Section 6.—Sexual excess. Section 7.—Solitude. Section 8.—Intellectual exertion. Section 9.—Want of employment. Section 10.—Abuse of purgatives. Section 11.—Abuse of alcohol. Section 12.—Tobacco. Section 13.—Tea. Section 14.—Opium.

In the cases cited in the last chapter the causes of the indigestion were, as a rule, out of the power of the patient to modify. Nobody for their own pleasure falls into poverty, catches cholera, is ruined in trade, lives upon potatoes, is worried by clients, nurses the dying, &c.; or at all events they do it with the hope of reward here or hereafter, and it is useless telling them not. The complaint cannot be cured by removing the cause: either it is past and gone; or it is incapable of being removed, as much out of our control, as the changeable weather which in some cases brought on the complaint.

In this chapter I purpose discussing some of the habits of social life which are in a great measure voluntary, which do not promise any sufficient reward, which are persisted in by reasonable persons principally from ignorance, and which therefore we can require our patients to give up, as the principal step towards their cure. "*Sublatâ causâ tollitur effectus*" is a very practical motto when the cause is not too heavy for us to lift.

SECTION I.

Eating too little.

I do not know of anything which more excites our wonder in reading contemporary sketches of the social life of our forefathers than the gross manner in which they indulged their appetites. To pass over the bestialities of the philosophic heathen, as recorded by Juvenal and Petronius, our disgust is equally aroused in civilized Christian times by the mighty emperor warrior and statesman Charles V washing down his six or seven daily meals, his supper at midnight and his heavy breakfast at six in the morning, with great draughts of beer and wine, whether in camp or cloister, or by his saturnine son Philip racking his stomach with enormous loads of pastry till the chronic gastralgia visible in his countenance became a prominent feature in his portraits, or by the genial coarseness of our middle classes as photographed by Chaucer with sympathetic appreciation of all that is human. One is not surprised, in reading such unconscious records of deficient control over the animal propensities, at theologians and physicians both preaching the opposite extreme, and placing abstinence in the niche sacred to temperance. Hence springs the popular notion, too deeply rooted by long growth for increased knowledge to dissipate under several generations, that abstinence is a *sine quâ non* in all medical treatment, and is a cure for all disease.

CASE LVI.—In November, 1856, J. M. V—, a young mercantile man, aged 36, came to me for slight flatulence and constipation. I gave him some Myrrh and Aloës pills to take occasionally, and saw no more of him then. He went on very well till 1860, when he was persuaded by some foolish friend to adopt a system of extreme abstinence, not that his health was to be called bad, but he wanted it to be better than good. The consequence was a relapse into a state much worse than the first. Three or four hours after eating, flatulence bursting upwards from the stomach, rolling about in suppressed thunder among the intestines, or passing off by the rectum, used to cause great inconvenience, especially by night. And the absence of taste and smell in the evacuated air showed it to be the carbonic acid of decomposed amylaceous food. The nervous system was equally deranged. There was great wakefulness in bed, and an inability to apply the mind to anything by day, which steadily increased upon him, and prevented his attending to his business. I gave him some Quinine, and desired him immediately to resume a full flesh diet. A week afterwards he came and said the flatulence and other symptoms were very much better, but the full dose of Quinine made his head ache. He was ordered therefore to take only half a grain twice a day, to keep up his appetite to its work; and then he was able to engage again in business so as to think no more about his health.

Instead of decreasing the dose of Quinine, I have often simply added six or eight minims of Chloric Ether to each draught,

and have found it quite effectual in preventing headache. This
was in cases where I chanced to be very anxious to have the
full dose taken. Here I did not think the magnitude of the
dose important.

That Quinine, however, or some equivalent nerve-renewer is
valuable in such cases I feel persuaded by the rapidity with
which patients get well under its use, and the long convalescence they have without.

CASE LVII.—Miss H. W—, January 28, 1860. The patient is a very thin,
nervous-faced young woman of twenty-three, who complains of a weight at
the pit of the stomach brought on by swallowing any solid. This first began
eighteen months ago at a catamenial period, and she immediately persuaded
the family doctor to interdict all solid food, and she has taken none ever since.
She has lost 21 pounds in weight, though never stout previously, and has
become dreadfully flatulent and hysterical. The heart has become weak and
irregular in strength, and sometimes intermittent.
      She was a long time in recovering even under an improved dietary, so
that I find noted in April, 1861, that though her muscles had become firm
and the general health good, yet there was still some pain at the epigastrium
after dinner, which I attributed to tight lacing.

I attributed it at the time to tight-lacing, but I have seen
reason afterwards to view it as one of the consequences of her
condition, to deficient nerve-power in the stomach.

Even in much more chronically ingrained disease of much
longer duration than in the instances cited a partial improvement may be effected very quickly by the aid of drugs.

CASE LVIII.—The Rev. J. S—, a parish priest and Union chaplain, aged
48, in February, 1866, tells me that when reading hard for his degree at the
University he first became sensible of pain after eating. His theory was
that he ought to eat less; and so he did, less and less; and, with the hope of
working a cure all at once, actually lived a whole year on bread and water
only. In consequence he is troubled with flatulence, debility, and frequent
attacks of palpitation of the heart. The pulse is uneven, and occasionally
intermits. As far as I can ascertain by questioning, he feels more pain now
after eating than he used to when he began this ascetic life nearly a quarter
of a century ago.
      A generous animalized diet, taken frequently, with wine, Quinine, and
Strychnine, while at the same time the oversensitive nerves were deadened by
Opium and Hydrocyanic acid, enabled me to allow him to return home in ten
days; but I of course did not promise that he would ever be the man he
might have been naturally under a rational dietary.

On July 17th he tells me he is able to eat more and more without pain week by week. His pulse is regular, and he has no flatulence. He has left off all medicine except a quarter of a grain of Opium every night. He is more robust than ever I expected to see him.

The above is an instance how ascetism will be persisted in on theoretical grounds in spite of nature's daily warnings to the contrary, and how unobservant even highly educated people are of physical facts. How one whose profession makes him daily conversant with the poor and down-trodden, while his social habits throw him among the rich, can fail to remark the symptoms of deficient diet, is difficult for us physicists to understand.

I used the word "ascetism" in the last observation. Perhaps it was hardly right to do so; for though in familiar conversation applied to abstinence from pleasure with whatever intention, such is not its proper meaning, and it ought strictly to be confined to those self-restraints where the motive is nobler than the mere bodily health, where it is an active devotional exercise, a mode of honoring God.

Where this form of devotion is part of the established worship of any religious community, it is usually made the subject of minute regulations, designed with a view of securing practical results without injury to sanitary condition. The principles of these regulations seem to be that abstinence should not be excessive, and above all not continuous. Moreover, the spiritual patient is never to prescribe for himself.

In the Church of England it certainly does not constitute any portion of the regular religious services demanded of its members. Truly in the homily "On Fasting" a low diet on certain days is urged; but the preacher destroys the force of his advice by inserting the weakening argument that its general adoption would be a great encouragement to our fisheries. The method of asceticism being thus left to individual management, its intention is often mistaken, and its practice abused. Instead of looking upon it as an exercise, as a sacrificial service, in its essence intermittent, occasional, and departing from its essence if not intermittent and occasional, men treat it as a means of destroying the instinctive desires.

CASE LIX.—Last February (1866) an Anglican rector, aged 32, consulted me on account of increasing inability to perform the duties of his ministry. Fits of mental depression more and more frequently came over him, accompanied by a feeling of loss of volition over the limbs. At all times he was weak and incapable of muscular exertion, and was thrown into a cold sweat by any bodily or mental effort. There was loss of appetite, pain at the epigastrium and flatulence after eating, with palpitation of the heart. This local condition of the stomach seemed to have been more prominent a symptom previous to his visit to me, for I remarked that the pit of the stomach had been blistered, by the advice of a former physician I presume. This state of things had been gradually coming on for about two years. He had several times taken short holidays, but with no permanent benefit.

On conversation with him, I found his notion of the relation between soul and body was that of a constant antagonism. It seemed to him that the aim of the former should be to subdue the latter continuously and permanently—not only to knock it down, but to keep it down. He ate merely to enable him to visit and preach and pray; he drank whatever liquid came first; he had married because the world must be peopled, and because he wanted a help-meet in his work. But he rejoiced when his appetite failed, and when he felt no pleasure in his victuals or wish for wine; and as soon as his sweet young wife had borne him two children, they ceased by mutual consent from bodily matrimonial intercourse. The last-named final blow to the flesh had been given four years before.

(*March,* 1867). A year's complete rest, and a constant recurrence to Quinine and Strychnia, have been necessary before he could be pronounced fit for the duties of his profession, as I have certified him now to be.

Surely this is Stoicism or Gnosticism, rather than the religion of the Bible. I am not fond of preaching, especially to clergymen, or of turning texts into traps; but people should not forget the threatenings at the end of Ecclesiastes, where we are told that God will bring us to judgment and make us account for our missed opportunities of enjoyment, for not being cheerful in our youth and loving the beautiful; and where we are urged on those grounds to "remove sorrow from the heart and put away evil from thy flesh." Forgetfulness in youth of the Creator and His creatures, disregard of the Giver as exhibited in His gifts, and neglecting to render Him thanks by using them, always entails a punishment on either mind or body. A joyless man becomes an unhealthy man; in body if they are bodily joys that he has foregone, in mind if they are mental.

SECTION II.

Eating too much.

Habitual gluttony is as rare now as it was common in the days of old. An occasional careless excess in the pleasures of the table may be indulged in, but people feel it to be an excess, promise themselves that it shall be only occasional, and do not go to a doctor for its consequences. It suggests and often spontaneously carries out its own cure, and the shame which accompanies it causes the " remorse of a guilty stomach" usually to be concealed. So that I have no notes on the subject to quote.

I can, however, remember two instances where over-eating really deserved sympathy rather than contempt, and I will transcribe one of them here, it being always pleasanter to reflect the bright than the dark side of human nature. I dare say I shall find some future opportunity of introducing the other also.

CASE LX.—In November, 1859, I was requested to visit a lady past middle life, who, when I entered her library, certainly looked the picture of robust bloom. "Dr. Chambers," said she, "what is a British matron to do who habitually eats too much?" The question suggested the shortest of replies. "Aye, it's very easy for you to say 'Don't;' but, if I didn't, I should be a widow in a week. You know how old and infirm Lord C— is. He has always been used to feed highly, and if I cut the dinner short, or did not encourage him by my example, it would be his death." It seemed that the symptoms of eating too much were a sense of repletion and a want of sleep during the night, feverishness in the morning, a sort of worrying fidget in the bowels, sometimes followed by constipation, sometimes by fetid semi-liquid evacuations, never by natural motions, frequent headaches, and a tendency to depression of spirits. Sometimes she was attacked in the night by what she called "spasms," that is to say, severe pain in the epigastric and umbilical regions. If that ended in vomiting she experienced rapid relief, and was better than usual for several days.

My prescription was an Aloës and Myrrh pill before dinner daily, and a recommendation of a dry diet as mixed and varied as possible, avoiding only soup, slops, butter, and fat. But I doubt if it was quite successful, till the exciting cause of this virtuous intemperance bore his many years and honors to the grave.

I question if my recommendation of a mixed diet was wise. It would have been better for her to have taken a preponder-

7

ance of meat one day and a preponderance of vegetables another, but more generally the latter.

The majority of exceeders have not such a good excuse for their violation of the rules of propriety, and would with reason suppose themselves to be laughed at if asked, " 'Is it for fear to wet a widow's eye' that you eat so much?" They seldom have the discernment shown by the last-named patient in recognizing the habit which is the cause of their ill-health; but they are ready enough to give it up when brought to see that they are committing a contemptible excess.

CASE LXI.—Mrs. L—, aged 32, the wife of a rich manufacturer, came to me in the spring of 1860, complaining of a weight and distension felt at the epigastrium half an hour after meals, and lasting for several hours. It was followed by eructations or returns of small quantities of food, not sour and not accompanied by flatulence. The bowels were loose, the motions never formed, but ragged, and sometimes diarrhœic. There was a nasty taste in the mouth in the morning, feverish and restless nights, and frequent dull headaches, with low spirits and hysteria. The catamenia were irregular and somewhat profuse. She said that these symptoms had commenced nearly two years previously, when her husband had some pecuniary troubles. I questioned her strictly as to keeping up her spirits by indulging in alcohol at that time or since, and believed her not guilty. But she confessed to having become very fond of good eating, and having a great appetite for anything "nice." She was a large-framed woman, and comely, though her outline was growing rather out of drawing.

Remark how in a weaker-minded person the mind becomes affected, while the more robust and educated intellect shown in the previous case bears up and gains strength by resistance.

The error in diet which in a woman produces hysteria, in a man declares itself by melancholy.

CASE LXII.—Richard R—, aged 48, a white-faced and fat clerk, came to me a month ago, November, 1866, persuaded that he had diseased heart by the palpitations of that organ which he experienced, especially in the morning. He had lost interest in life, having succeeded in obtaining a comfortable income more than sufficient for his wants, and having laid by a provision for old age. He was passing a drab-colored existence, taking no pleasure, following no hobbies, and occupied only with the routine of his office and attention to his health. Of the latter he had a bad opinion, and considered that he was delicate and required abstinence from excitement and constant support. Besides his regular meals he was in the habit of taking a slight anticipatory luncheon at 11, an intercalary snack at 4 preparatory to dinner at 6, and a small refresher along with his glass of grog at bed-time. The consequence was sleepless nights, flatulence of stomach, palpitations of heart, returns of small quantities of food by the œsophagus, irregularity of stools, increased obesity, and desponding views concerning time and eternity. To his great terror, I made him go quickly up and down stairs, and examined the heart, the sounds and beat of which were quite natural after this natural excitement. But the stomach was large, and gave a drummy sound on percussion quite up to the apex of the ventricle. A counsel to leave off bacon at breakfast, to eat only at meal-times, and a short course of Hydrochloric Acid, made a new man of him.

How easily such a person as this might be turned into a hypochondriac or a lunatic by coddling and sympathizing!

If the last patient had really got a diseased heart, I should have given him probably a treatment not very different in principle, but I should have especially cautioned him against gorging himself even at meals. For now and then cases occur like the following.

CASE LXIII.—John B—, aged 71, a cheerful old gentleman, came to me in May, 1852. He said he had always taken great care of his health, but had not consulted a medical man since he had rheumatic fever at fifteen years of age. His reason for taking care of his health had been a tendency to shortness of breath, which he said he had experienced so long ago as the beginning of the century, when reading Shakspeare to the young ladies of the period. Examination of the heart showed it to be very weak, irregular in time and strength, with a confusion in its valve sounds, and a dulness on percussion extending four inches in width from the epigastric across the cardiac region. The pulse at the wrist was equally weak. He had always enjoyed his table, but latterly had found that taking the quantity requisite to satisfy him oppressed his chest and made him faint. Nobody could discern better than the patient himself the true pathology of his case, nor give better advice than his own reason suggested. But unfortunately he was not able to follow it, for a few weeks afterwards I had a letter from young Mr. B—, saying that his father had eaten heartily of an indigestible mixed dinner, and lay back in his chair dead.

It very often excites the astonishment of these patients, after having it explained to them that their danger lies in over-eating, to be told to increase the number of their meals. Yet such is in most instances the best way of meeting the case. Small quantities frequently taken are the best device for introducing a full supply of nutriment without overloading the alimentary canal. During the day, four hours is the longest time that an invalid should be allowed to pass without eating something; and for some two hours is a sufficient interval. Very soon the appetite begins to accommodate itself to these habits, and the little meal that is committed to the stomach at once, instead of lying dormant in the paralyzed organ for hours, as was the case under former customs, is enabled to pass away rapidly.

The excess in eating is not uncommonly rather relative than positive. It would not be an excess under normal circumstances, but it is made so by those present. Of this acute examples are given in cases IX, X, XI, in the last chapter, where an ordinary meal was an excess under extraordinary temporary circumstances. The following is a chronic result of a chronic cause.

CASE LXIV.—T. J—, a lawyer, naturally inclined to be corpulent, aged 52, was well till October, 1865, when he sprained his ankle rather severely. He was always used to a good deal of bodily exercise, and of course in his profession equally employed his mind; so that it was not to be wondered at that he habitually fed largely. This did him no harm till the accident to his leg, after which he began to suffer from indigestion. The bowels were costive, and the stools never homogeneous, but consisting of rags of solid matter in much fluid; he had acid risings in the mouth, eructations, wind rolling about at night in the intestines, and breaking off per anum in the morning. What most distressed him and brought him under my care was want of rest at night. He either could not sleep at all, or else woke up after a short nap and could sleep no more. Opiates had made him worse. Worried in this way, he had lost two stone in weight in the six months since his illness began, and appeared to have been striving to replace the loss of flesh by keeping up his usual high feeding. But analysis of the urine showed that there was no lack of active metamorphosis going on, for it was at all times of the day fully acid, clear, and with a constant specific gravity of 1.024 to 1.025, varying singularly little with circumstances. He was nervous and irritable, and, like all nervous people, had a smooth, white tongue. There is small doubt but what a return to active habits would have restored his usual health, but unfortunately some remains of lameness precluded it. He was astonished when told he ate too much, and doubted if that was possible when a man was losing flesh. But experiment proved to him what the symptoms led me to pronounce, namely, that the ingesta were in excess of what was required for the nutrition at the time, though they were not too much for him when he was living more actively.

In this instance the headache which frequently accompanies excess of mixed diet was absent.

The loss of flesh is interesting.

Loss of flesh is rather an exceptional accompaniment of the dyspepsia of excess. The following is a much more common case, causing me a little difficulty in selection, so many are alike.

CASE LXV.—Mrs. H—, a very stout lady of about sixty, came to me in June, 1852, to consult principally about her obesity. But I found her a martyr to gastric dyspepsia, which produced a feeling of emptiness only to be relieved by taking food. This overeating increased her dyspepsia, so that she had a constant diarrhœa, and frequent vomiting. Yet with all this her corpulence increased more and more. Restriction of diet relieved her stomach symptoms considerably, but her bulk was unreduceable. I believe the cause of her death some years afterwards was pneumonia.

Dyspepsia certainly does not prevent corpulence. In thirty-eight cases of obese persons, which I printed in a tabulated form some years ago[1], five of the number suffered in this way. In fact, it is not impossible that one cause of that hypertrophy may be the delay of the victuals, both animal and vegetable, in the stomach, and the setting up in the carbonaceous material of a fatty fermentation instead of digestion. This obesity of persons with weak gastric digestion is peculiarly distressing: the defect in muscular power prevents the use of exercise for a time sufficient to prevent its increase, and hence it becomes a daily growing inconvenience. The encroachment, too, of the adipose upon the other tissues, and the dilute spread of the insufficient blood through an unnaturally large quantity of capillaries, tend to produce atrophy of important parts; and hence we find as consequences of corpulence, dilatations and degenerations of the heart, fatty deposits on the same, Bright's kidneys with dropsy, &c. The addition of many pounds to the body in the shape of fat, requires certainly a very large, although not perhaps a proportionate, addition of blood and bloodvessels to nourish it; yet the same heart has still to undertake this extra labor. The balance then between the systemic and the pulmonary circulation must be destroyed, and the lungs be unequal to the excretions of so much more carbon than they were intended to provide for; hence the blood becomes more venous, more liable to form congestions, and to dilate the yielding walls of the heart by its retarded pace. The effect of diminished circulation in

1 "On Corpulence," p. 139. London, 1850.

also producing degeneration of other parts need not be enlarged upon.

SECTION III.

Sedentary habits.

Among the originators of dyspepsia we commonly find included in books sedentary habits. But when I come to look over my notes, I cannot extract any cases which would exhibit this fact. I do not know by experience if a sedentary life, such as that of a clerk or bookkeeper for example, would induce the defect unless it were joined to some other cause. Alone, with a properly regulated diet, it seems consistent with quite healthy digestive powers. We find it so in the bed-ridden under our care, whose life may be viewed as the type of a sedentary one, yet they do not suffer except from some more than ordinary folly in diet, or from the misuse of some drug.

When therefore those who come before us for indigestion attribute their state to a sedentary life, we must not stop there, but search further for other and more certain causes. For example:—

CASE LXVI.—M. S—, editor of a weekly newspaper, aged about forty, laid on the many hours he spent in the office-chair the blame of enteric dyspepsia, which spoilt his night's rest by waking him in the early morning with flatulence. Charcoal gave him only temporary relief, but dividing his meals more, taking a good luncheon and a light dinner, seems to have set him up completely. This was in 1856, and now he seems quite equal to his official duties, and looks as robust as any leucophlegmatic men ever do.

Let it not be supposed that I underrate the value to health of exercise in the open air. The fresh oxygen, the cheerful occupation, the distraction of the mind from injurious tension, must, however, be taken into account by the physiologist, and not all the benefit set down to muscular motion, which latter element is but a small part of what is usually included under the recommendation of " exercise" by a rational physician. I have come across more brain-laborers whose digestion has been injured by injudicious excess in muscular exertion than by the reverse. Let not those whose avocations are necessarily sedentary, despair of finding by judicious experiment a mode of passing their lives in complete, though not of course blooming health.

The division and arrangement of the meals according to the mode of life is a very important part of the science of digesting them. Much must be left to individual experience, but regular literary men, and others who do routine work at the desk, I generally find are better for taking a meat luncheon and only a light dinner after the day's labor. And if they take a glass of grog, it should be at bed-time. Great late meals washed down with a quantity of alcohol do not suit them.

On the other hand, those who pass a muscular life often suffer from eating in the middle of the day. For instance, I recommended the following to dine late, and to take at most a glass of wine and a biscuit in the middle of the day.

CASE LXVII.—A. W—, a schoolmaster, always dined with his boys at one o'clock, and tried to work off his dinner by playing at cricket with them in the afternoon. But the more he played at cricket the more he suffered from discomfort at the epigastrium followed by intense headache.

CASE LXVIII.—A Welsh country gentleman, aged 57, was under my care in 1862 for weight at the epigastrium, acid eructations, headache. and sleeplessness. He said the beginning of it was over-smoking at Cambridge; but since then he had been to a number of physicians, and taken a great deal of medicine, homœopathic and allopathic. He had been in the habit of much exercise, and always dined at two o'clock. Dining late relieved his symptoms, but he did not seem satisfied without medicine.

Laborers, sportsmen, pedestrians, postmen, are all instances of ready access, from whom it is easy to learn that habitually to eat heavily during the hours of bodily toil produces sooner or later indigestion, and that health and comfort are secured by making supper the principal meal.

SECTION IV.

Tight-lacing.

One wet winter day at Florence I had been spending the morning in the studio of a sculptor of world-wide reputation. We had discussed the perfections of female beauty, and I felt that I was sitting at the feet of a thinker, as well as an "*elegans formarum spectator.*" In the evening we met at a hospitable palazzo, and under cover of the waltz music from a quiet corner of observation saw whirling by us in the flesh much that we had

been thinking of in the marble and the clay; and both our
eyes could not but follow one particular face, famous for the
assistance its great natural beauty received from art. "Face,"
I said, but the mind of Hiram Powers was penetrating deeper,
for he exclaimed, after a short silence, "That is all very well,
but I want to know where Lady —— puts her liver!" Where,
indeed! for calculating the circumference of the waist by the
eye, allowing a minimum thickness for the parietes of the chest,
an area for the spine, œsophagus, vena cava and aorta, the section of the waist seemed to admit of no room for anything else
at all. In such a body the liver must be squeezed down into
the abdomen, elbow like a big bully its hollow neighbors, and
infringe upon their shape. Fortunately for itself it is singularly tolerant of pressure, and may be deformed out of all recog-
nition by the anatomist of external forms, without ceasing to
do its duty as a bile maker, as may be seen well displayed in
Dr. Murchison's graphic woodcuts in the "Medical Times"
(March, 1867). But yet the whole portal circulation must be
carried on under great mechanical difficulties, the due supply
of arterial blood reduced, and its return by the vena cava
resisted. What an inconceivably tough person that must be
who does not become pot-bellied from the downward pressure,
red-nosed from the hepatic obstruction! And must not, therefore, the style of dress which gives birth to such deformities be
an abomination and an eye-sore to the artist?

The organ most deserving pity is the unresisting stomach,
which is dragged and pushed out of all form during the continuance of this packing process. The longer the continuance
the more it suffers. If it is constant, we get cases like the
following:—

Case LXIX.—Emily K—, aged 16, was a full-grown woman in form, and
had been catamenial for three years; but when admitted to St. Mary's, in
March, 1864, she was still wearing an old tough black pair of stays made for
her when a child. The consequence was that she had never been thoroughly
well all that time. The catamenia occurred every three weeks, and, for a
girl of her age, were at first profuse, lasting six days; but latterly they had
lasted only three days. She had constant pain after eating, frequent vomiting, and frequent rising of the food in the throat, on which latter occasions
it was sometimes tinged with blood, especially at the menstrual periods. This
constant ill-health had made her thin and hysterical, but her lungs, heart, and

indeed all the solid organs seemed perfectly normal. When admitted she was vomiting all her meals. At first she had Hydrocyanic Acid, but was no better in any respect for it; but on the 6th of April she was put upon a course of cold showerbaths every morning, with Valerian three times a day. This, with the removal of the obnoxious stays, seems to have been immediately effectual, for on the 12th it is reported she had not vomited for two days, and on the 18th she was discharged "cured."

"Cured"—of her stays. Easy task in such a case as the above, but presenting insuperable difficulties much more often. Women have a very strong won't.

Case LXX.—G.'s "Anonyma,"[1] aged 28, was brought to me in August, 1859, by a gentleman whose mistress she then was. She had borne several children in the course of her career, but still retained a beautiful slim figure which she had when a maiden. This she had accomplished by bandaging very tightly after each confinement, and sternly refusing to have any change made in the shape of her corsets. The consequence was that for several years she never took a meal without throwing some of it up afterwards, and suffered from obstinate constipation, for which she was in the habit of using violent purgatives. She seemed quite as aware as I could make her of the cause of vomiting, but resolutely refused to do anything which might imperil her outline. In fact, she implied she lived by her beauty, and intended to keep it at all hazards.

I do not know how to answer an argument of that sort. Another difficulty lies in the diagnosis of the true cause of the evil. Asking questions is useless; "aucune femme ne se serre," remarks M. Chomel of his countrywomen,[2] and I am sure we may say the same of the confessions of ours. Moreover, if you try to detect them by passing your hand underneath the stays, as M. Chomel used to do, they stinge in, and defend the honor of their corset by a fraudulent kind of gymnastic. So you gain nothing by what is in truth rather a rude proceeding. The best way is to make an excuse to have the clothing taken off, and observe whether it has crumpled and marked the skin by pressure; then to desire the patient to take a full breath, and notice whether the lower ribs are duly ex-

---

[1] I borrow this term from the newspapers in no scoffing spirit, but pitifully and sadly to describe one who has lost her maiden family name by losing maidenhood and family ties, without acquiring a right to any other. It is hard to smile at the loneliness which "no name" expresses.

[2] "Les Dyspepsies," p. 251.

panded, or whether the intercostal muscles and diaphragm have
lost power by misuse.

By that means you can find it out when the tight-lacing is
still continued at the time you see the patient. But in most
cases it has been left off on account of the increasing pain it
causes, and a suspicion that it causes the other symptoms as
well; or perhaps it is temporarily left off for the visit to the
doctor. And I suspect that such is the case with a large proportion of the instances of habitual vomiting, soreness of epigastrium, of hæmatemesis, of ulceration of the mucous membrane,
flatulence, and hysteria, which come before us. These symptoms are most common in the other sex—why? because their
reproductive organs differ from ours? Surely not, or we should
find the same peculiarity universal among females throughout
the animal kingdom, or at least throughout mammals. Yet we
read in veterinarian pathology no hint of a distinction between
the stomachs of our bulls and of our cows. Is it not more
reasonable to conclude that the important difference lies in the
clothes, which we can see, rather than in some mysterious in-
visible influence of the generative viscera over the digestive, of
which there is no evidence?

I should, therefore, in all women where these symptoms appear, suspect at least, for no harm is done by the suspicion,
tight-lacing, though I should not find it still persevered in or
confessed.

As an alteration of form is sometimes diagnostically useful,
it may be mentioned that the prominent abdomen of a tight-
lacer generally sticks out straight from above the pubes, sometimes overhangs it: that of a naturally short-bodied stout
woman slopes up to the umbilicus at an angle of 45°.[1]

In a long-bodied woman, such as in the Phidian proportion,
the abdomen ought to be flat.

In men there is not the same temptation to compress the viscera
for ornamental purposes among those who have the regulation
of their own dress. But it has often struck me that the tight
trowser-bands and buttoned-up uniform jackets, which French
schools delight to enforce, must be very unwholesome, inde-

---

[1] See Albert Durer's " Outlines of Proportion."

pendent of the impediments they offer to cricket and football. One does not wonder at the pale greasy, old looks of the poor lads. They must certainly suffer from indigestion, and probably it is this chronic ill-health which induces certain obscene habits said to be common amongst them.

SECTION V.

Compression of epigastrium by shoemakers.

Indigestions such as I have attributed to the pressure of stays in women, are common in one class of men, namely, cobblers; arising in them from a cause of physiologically exactly the same nature, the compression of the epigastrium by the last on which the boot or shoe is worked, producing on the stomach just the same effects as its compression by the liver in cases of tight-lacing. The following history shows the result in an incipient stage.

CASE LXXI.—Joseph James D—, aged 19, just out of his apprenticeship to a shoemaker, was admitted to St. Mary's Hospital under my care October 13th, 1861. He complained of weakness in the wrists, which became painful after work, and of constipation ; he spoke also of pain in the chest, which induced us to examine his lungs. These, however, were found healthy, and he had no cough. On further inquiry it appeared that the pain he spoke of was in the epigastrium, and was increased by pressure and by taking food. Rest and Quinine improved him rapidly, so that he was made an out-patient within a week.

The loss of power in the wrists, arising from atrophy of the muscles in overworked parts of persons whose stomachs do not take in a sufficient supply of nutriment, in some instances proceeds to a much greater degree ; and there is a case recorded somewhere in my St. Mary's notebooks of a shoemaker in whom the two arms, even to the deltoids, were completely paralyzed by overwork in giving that artistic jerk to the thread which these workmen affect. But I cannot lay my hand on it now. Perhaps I may find it by the time I come to a future chapter on the nervous symptoms produced by digestive defects.

Remark how soon the evil had commenced, on the very threshold of the life the poor lad had chosen!

The next case exhibits a further stage of the same condition.

CASE LXXII.—Philip B—, aged 36, shoemaker, was admitted into St. Mary's under my care November 9th, 1855. He had not been in health for nine years, suffering from what he called "spasms in the chest," that is, pain across the epigastrium, and irrepressible paroxysms of belching. The pain in the epigastrium was always increased immediately after taking food, and was accompanied by a great secretion of gas. When he could get off some of this by eructation, the pain somewhat abated; but the eructations would sometimes continue as long as three hours. During the last nine months he had become emaciated, and felt a good deal of universal debility. The urine was smoky-colored, of the specific gravity only of 1.010, though natural in quantity and free from albumen; the sleep was broken, the appetite good. He stated that unless he took purgatives his bowels would remain unopened for a fortnight together.

Philip's first medicine was Bismuth in Iron. But the Iron did not seem to agree with him; he got into a feverish catarrhal state and had sore throat. During this attack he was kept in bed, had six leeches and afterwards a blister applied on the epigastrium, and took a quarter of an ounce of Castor-oil occasionally. All this time, however, he was gaining flesh; so that between the 27th of November and the 10th of December he had gained four pounds in weight; and the urine was increasing in specific gravity, so that by the 1st of December it was 1.028, but was a little cloudy from lythates. After the acute febrile symptoms had abated he received much comfort from the following draught three times a day, viz.:—

R.—Mixturæ Rhæi co., fl℥j.
    Tincturæ Opii, ℳv.
    Acidi Gallici, gr. v.

He left on December 13th, much improved in health and spirits.

In this instance it will be seen that the evil was much more ingrained by time, and the symptoms were worse and more difficult of relief in proportion to the greater time it has lasted.

The intention of the draught was to soothe the oversensitive nerves with the Opium, at the same time that the Gallic Acid astringed the mucous membrane, and restrained the oversecretion of mucus, which the patient's general catarrhal diathesis otherwise displayed rendered probable to be present in the stomach. The Rhubarb, I think, was designed to prevent constipation arising from the other ingredients. As a rule I like Aloes best for that purpose in gastric cases, and I do not know why I ordered Rhubarb here.

Sometimes when lads begin shoemaking early, before the bones have got quite hard, a peculiar deformity is produced, which acts like a perpetual pair of stays for life.

CASE LXXIII.—William H—, aged 25, bootmaker, was admitted to St. Mary's, June 7th, 1856, for pain at the pit of the stomach which had been almost constant for four years, and was increasing. The pain was accompanied by a local sensation of cold, and what he described as a "dragging." He often felt nausea, but never actually vomited. On examination of the epigastrium there was seen an indentation of considerable depth, and deepest in the middle, which he said was caused by the wooden instrument used in bootmaking, at which he had worked "all his life." The part was painful on pressure. His general health did not seem much broken, and the specific gravity of the urine was 1.020. With rest, Nitrate of Bismuth, and Iron, he lost his symptoms, and was discharged from care June 21st.

But of course it was to be expected that his symptoms would return; for these men spend fourteen hours a day with their heads bent down close to their knees, pressing a hard stick into the stomach; and the injury which was once done could not but be aggravated by time.

The final blow to the stomach given by this trade is exemplified in this next case.

CASE LXXIV.—James P—, a shoemaker, aged 37, was admitted to St. Mary's, May 4th, 1860. He said he had never been well since he was one-and-twenty. His bowels were never moved of their own accord, he occasionally vomited, and he had a perpetual pain in the right side of the epigastrium, which he called his "liver." He continued in this state till 1855, when, as he was vomiting, there came up a sudden gush of blood. Since then the same thing had happened five times, the last time the night before admission. He did not throw up any blood when in the ward, but his statement was confirmed by the passage of a considerable quantity, liquid and clotted, from the bowels. Acetate of Lead stopped the hemorrhage, and by dint of complete rest and Pepsine he was able to take the ordinary diet of meat and vegetables, with the addition of a pint of beef tea at dinner, for a week before he went out on the 25th, taking Quinine three times a day.

The rapid, though probably only temporary, relief of the pain in the epigastrium and the regained power of taking food, shows how much might be done in these cases by rationally removing the original cause of the complaint. No greater blessing to the artisan was ever invented than the Upright Shoemaker's Table, introduced by Mr. Sparkes Hall to the trade. At it the workman stands or sits on a high stool at will, holding his work fixed by a strap and stirrup regulated by the foot. Thus all pressure on the epigastrium is avoided, and Mr. Hall tells me that many of his most skilled hands who used to be off work

from illness nearly half their time, and driven to drink to drown
pain the rest, can now earn daily wages, and are become temperate rich men.

The difficulty lies in the change of method—by no means a
light difficulty. A visit to the Egyptian room at the British
Museum shows that shoemakers have worked in a doubled-up
posture at least since the days of the Pharaohs, and we cannot
expect them to alter in a moment what certainly has some con-
veniences. Moreover all do not suffer. A stomach in a perfectly
robust condition probably can resist even this daily compression.
But when occasionally it is joined to fusty cold workshops, long
abstinence, tippling, accidental illness of any kind, then it tells
chronically, and the injured part is unable to recover itself.
The *dura ilia* make a bad use of their blessings by deterring
the weaker vessel from the trouble of learning a new method,
and are aided by the lazy conservatism natural to the ignorant.
Still I think it is our bounden duty to advise all shoemakers we
come across as patients to adopt the upright bench, and perhaps
in time we may succeed.

I have not found this evil of compressed stomach from the
constrained posture of tailors. They generally suffer from
drinking and bad ventilation.

SECTION VI.

Sexual excesses.

I alluded in the last paragraph of the fourth section to a
perversion of the sexual instinct to be found sometimes accompanying indigestion. I have seen it named as a cause, indeed
it is so named by M. Chomel in the work I quoted. My experience does not enable me to agree in this, though I cannot deny
the possibility of it. Still I believe that more searching inquiry
into those cases where the two morbid phenomena are associated
together, will often enable us to discover a different sequence,
and to call the quasi-voluntary act of lust an effect of feelings
perverted by disease. A perfectly healthy lad never invents
this for himself; and if he has taken it up from imitation, curiosity, or the suggestions of infamous pornographic literature,
disgust and boyish honor soon break him of it. Where it is

continued there is almost always some mental or bodily disease requiring medical care. As for example:—

CASE LXXV.—Augustus T—, aged 24, came to me in October, 1863, saying that for some years till lately he had been in the habit of solitary lust, and that he was suffering from excessive flatulence, and from pain produced at the epigastrium by any quantity of food sufficient to nourish the body. He had broken himself of the habit, but was dreadfully distressed in mind at the degradation of ever having indulged in it, and attributed to it the low state of bodily health he endured. But I found on inquiry that from childhood he had been a greedy boy, morose and weakly, that he had suffered from worms; and that his education was neglected on account of his health, long before the nasty practice he told me of had been adopted.

On the other hand, I can remember in my notes records of at least two cases where the obscenity had been learnt by imitation and practised as often on the average as twice daily for a succession of years without the alimentary canal suffering at all, whatever other functions may have failed.

The natural sexual excess is also said by French writers to produce indigestion. I do not happen ever to have seen an instance. The digestion of prostitutes (whose trade may be considered an excess) has always seemed to me exceptionally good. Their health is less injured by riotous living and spirit-drinking than that of other people who equally indulge. I speak of the class who are patients at the Lock Hospital, where I have been the physician for some years.

SECTION VII.

Solitude.

Eating in a dull heavy kind of way without enjoying it often produces dyspepsia in a moderate form.

CASE LXXVI.—Rev. N. R—, a bachelor of middle age, was my patient in the autumn of 1864, for flatulence of bowels accompanied by confusion of intellect during the second stage of digestion, and sleeplessness. By regulation of the diet, and Quinine with Strychnine, he got well at that time. In November, 1865, he came to me again, saying that when he dined in company he could digest anything, and never suffered, however rash he had been at table. But when he took his meals alone for several days together, his old symptoms of the previous year returned, and no carefulness or abstemiousness prevented them.

I should conjecture the pathological condition to be a partial paralysis of the solar plexus, from attention being directed to it.

Several commercial and literary men have complained to me of attacks of vomiting (that is, temporary paralysis of the stomach[1]) when they took dinner alone, and so were apt to let the mind dwell deeply on some interesting subject; and they have told me in wonder that they could dine out and eat and drink all sorts of rich things with impunity. They did not seem aware of the preservative value of frivolous conversation.

CASE LXXVII.—A famous scientific man of middle age, deeply occupied with his pursuits, and never in the habit of "wasting his time," as he called it, on amusement of any kind, complained to me that when he dined alone, as he usually did on the plainest food, he invariably vomited afterwards. But that in dining out he never suffered even from nausea. At one time he used to read at meals, but that seemed to make no difference at all.

Which is waste of time, work or play? Truly sometimes one and sometimes the other, but each out of their due season, and proportion. The epithet "frivolous" (from the same root as "frio" = what may be easily rubbed out and forgotten) is not necessarily depreciatory. Light thoughts, light occupations that are easily rubbed out and leave no care or impression behind them, are good for mind and body and worldly estate.

SECTION VIII.

Intellectual exertion.

The overuse of the mind sometimes induces indigestion in those previously not very strong.

CASE LXXVIII.—Rev. G. B—, aged 50, after being invalided home from India, got well enough to take the post of secretary to a society. But the brain-fag consequent upon that, without any other change of his habits, brought on nocturnal flatulence, nightmare, and seminal emissions. And during the day his spirits were so depressed that existence was a burden. This was in November, 1862, and a month afterwards he came to report that assistance in his work had been granted him, and that he was quite set to rights, except a little weight at the epigastrium.

It is to be observed that what I am speaking of here is not the original condition of mind which was described as a cause

---

[1] See chapter on Vomiting, in a later part of the volume.

of the indigestion of starchy food especially, in the last chapter (see Case XIV, &c.), but rather the wrong mode of using it. Unavoidable evils were then described, the consequences of which might be alleviated, but the causes were either past or irremediable. In this chapter I am tracing the complaints to habits which are voluntarily taken up, and can be laid down at will.

I do not believe it is the quantity, so much as the quality of intellectual occupation which does harm. Composition, the creation of thoughts, even the putting of old thoughts into new forms, is not, in my experience, injurious. Where it is enjoyed, I believe it a peculiarly healthy occupation. It is the dreary routine work, *invito genio* and against time, which knocks up a man's stomach.

In reality I believe the last two cases are exceptional, and that you will more commonly find some other cause at work in those who accuse intellectual occupation. For example:—

CASE LXXIX.—Joseph W—, an engineer past middle age, with the broad forehead, square jaw, and shrewd eye of a mind like the iron he bent to his will, came to me in March, 1863, complaining of flatulence, with spasmodic pain in the epigastrium, and that he was quite knocked up by the toil of invention, to which he attributed his bodily illness. On inquiry I found that he had been stimulating thought by champagne luncheons, and that it was after these he felt distress.

SECTION IX.

Want of Employment.

The concentration of the mind upon itself we are assured by psychologists will produce mental disease. I confess myself that I have some doubts whether we ought not rather to say that it makes evident and brings into prominence previously existing disease. Because the same class of observers generally also go on to say that the fixing of the mind on any portion of the body will cause morbid phenomena to be therein developed. Now this is an experiment I have often amused myself by trying in a leisure hour; I have looked at, thought about, argued about, and in imagination dissected, my finger tips, nose, toes, epigastrium, knees, &c., till the power of attention was wearied out;

8

but no pain, or redness, or throbbing, or swelling, no stiffness, or coldness, or anæsthesia, has followed. What really happens however in consequence of a concentration of the mind upon the body is this—should there be already existing any slight morbid condition capable of declaring itself to the nervous system, but not in such a way as to draw off from other objects the engaged mind; then, should the attention be unfortunately attracted to this part, the pain is noticed, is in idea multiplied and exaggerated. Anxiety and distress follow attention, and then at last the bodily functions are interfered with (for these passions, as has been illustrated in the second chapter, lower the powers and secretions of the digestive canal), the saliva and gastric juice fail, and the digestion suffers. From thence perhaps, as a tertiary effect, may ensue deteriorated nutrition of the local injury.

CASE LXXX.—An old blind soldier, who lived near the Chelsea Dispensary when I was physician there, used constantly for several years to come to me from time to time complaining of excruciating pain in the abdomen. He had his pension, and was comfortably off in circumstances. No one on looking at him could doubt the reality of his feelings; yet there was never anything in his state of health apparent to account for them. The only cause I could trace them to was his being occasionally left alone by his wife and family; and then his blindness prevented his mind being drawn off to surrounding objects, and he would sit still, allowing any little abdominal discomfort to be depicted in exaggerated colors on his vacant fancy. He had in truth always a little flatulence, but never the "excruciating pains," except on these occasions.

I have not seen much of blind people, but such as come under my notice are always disposed to exaggerate in this way any slight bodily discomforts into real tortures. From want of mental distraction, their internal sensations occupy too prominent a place in their psychical life.

Just in the same way people who voluntarily deprive their minds of occupation, find out the existence of innumerable pains in various parts of their bodies; the anxiety and worry thus occasioned really does deprive them of sleep, injures their digestion, and by the time they are driven to the doctor makes them materially as well as mentally ill. Sometimes these pains arise from actual organic change which had existed for many years unnoticed, and therefore without effect on the general health, ·

and unaffected by it. But when once it is thought about so as to create anxiety, it feels the innutrition hence arising, and grows rapidly worse.

CASE LXXXI.—A paper-maker, utterly uneducated, though very wealthy, aged 70, was brought to me by his wife and doctor in March, 1861. He had had a slight catarrh of the bladder, following an old stricture, many years ; but as long as he was in business he never suffered materially from it. Having made more money than he could possibly want, he thought he would retire and "enjoy himself." But alas, he had nothing to enjoy himself *with*, except, indeed, his money, which is not of much use without tastes to spend it upon. So he took to thinking about his health, considered what was wholesome and what not, what to eat, drink, and avoid, for the sake of his defective urinary organs. The consequence was that his digestion failed, he complained of weight after food, vertigo, flatulence, and "intolerable" pain in the epigastrium. His aspect, as he sat rubbing the pit of his stomach when introduced to me, was one of abject misery. The urine contained a little pus, but he made no complaint about his bladder. He had the white tongue of a nervous man, and his bowels were costive. My next report of him is dated August, 1862, when I saw him in much the same unhappy state of feeling. But the bladder had got a good deal worse; there was more pus and albumen in the urine, and the specific gravity was only 1.015. I do not detail the treatment, for it was various and useless ; and a few weeks after his last visit I received a card from the family announcing his funeral.

As a more cheering illustration *per contra*, I will choose an instance of the same anatomical condition as the last, in order to show that urinary disease is not necessarily depressing to the mind.

CASE LXXXII.—J. B—, a confidential clerk at the India House, getting on for 60 years of age, was sent to me by Mr. Coulson in June, 1856. He had enlarged prostate and vesical catarrh, but managed to avoid all serious inconvenience in that quarter by using a catheter. Now and then his stomach got out of order, but he could generally trace that to a good dinner or some such social imprudence ; and then his bladder discharged more pus. So he went on some years, till I began to observe he was coming to me rather more frequently, and that he had a care-cumbered face, leading me to ask him what he had been doing lately. "Doing? Nothing. I am a gentleman at large now— pensioned off." Poor Charles Lamb! also an India House clerk, I thought of him and his humorous pathos on being pensioned off, and said immediately that it would never answer, it was poison to mind and body. "Ah, there's a good deal in what you say: as the spring comes on I and Mrs. B. will take to gardening : she has a family taste that way. And to gardening they took, and I saw him much seldomer, and heard no complaints of his vesical troubles ; though he dropped in at the end of 1864 to introduce a patient to me, and see

how I was.   I trust they still continue to plant their cabbages and bud their
' roses, and to make wierd skeleton bouquets of dissected leaves for their friends,
and to be as happy and as little ashamed as Adam and Eve in Milton.

Those in whom tastes have been implanted for simple amuse-
ments cannot be too grateful for them.   And I hold it one of
the wisest things we can do in busy middle age to keep up or
acquire such tastes.   When once the inevitable pensioning off
comes, it is usually too late to go through the necessary education.   I have indeed seen a diplomatist, who had held in his
grasp the destiny of nations, commencing at sixty-five the study
of Italian, for the sake of reading Dante; and I thought at the
time it showed more courage even than his old trade of bullying
into reason the masters of armies.   Such courage is rare, and
more generally the mind's mirror gets dimmer and dimmer, till
there arrives with premature haste the state of things so graphically painted in the last chapter of Ecclesiastes.   I am sorry to
say the stock example of this is a member of our own profession,
Sir Astley Cooper, who, when in retirement satiated with wealth
and honors, is described as looking over the trees of his park
with a conviction that some day he should hang himself from
one of them.   He had wasted his life in routine work, and it
was too late to educate the mind to anything else.

The class of patients instanced in the last two cases are such
as have some structural disease, of which I have described the
aggravation by idleness acting through the digestive organs.
More common still are those who have no existing organic
change in any part of the body; and in these the digestive
organs act upon themselves only, and produce distress and
functional derangement.   A state of things arises pithily
sketched by Dr. Markham in a letter introducing a patient to
me a few months ago—"he formerly was poor, worked hard,
had plenty of appetite, little dinner, and little time to eat it;
now he is rich, with lots of time and dinner, but no stomach."
Sometimes the vacancy of mind left by the surrender of involuntary occupation is such that absolute mental aberration
is the result.   There are actual delusions about facts, persua-
sions that they have happened when in reality they have not.
Under those circumstances diagnosis is much impeded by the

difficulty of knowing what is true and what is false of the various symptoms related to you, if they are not in themselves devoid of internal probability.

In these cases I have been much assisted by the observation of a peculiarity in the mental state of the half-insane, which was displayed in the following:—

CASE LXXXIII.—Mr. G—, aged 53, was till the early part of last year engaged in active business. He then gave up his occupation, and supposed he had sufficient mental resources to pass life agreeably. So he settled at a fashionable watering-place, and took to desultory art and literature. But the elegancies of life sit oddly upon him, for he is a grim-featured harsh-mannered man, unlikely to find much favor in that society whose business is amusement.

He had the aspect of strength and health, but complained when I saw him that he had been for several weeks a dreadful sufferer from excruciating pains in the abdomen coming on at night and entirely preventing rest. I had no reason for doubts, till I observed a painful anxiety, which increased as he talked, that I should believe him, joined to an evident suspicion that I did not do so. I however prescribed him some Valerian at night.

When I saw him again in a week the Valerian had evidently made him feverish, and he said the nocturnal pains were worse than ever. I then elicited that last autumn, in fact after he had been trying idleness for some months only, he had been exceedingly low-spirited, and that he used to get nervous and fidgety at night and have paroxysms of causeless terror.

I have since seen reason to conclude that the pains by night and the wakefulness were purely imaginary.

I have always in my lectures on the practice of medicine insisted much on the aid to the diagnosis of mental disease afforded by the peculiar suspiciousness of itself which the mind exhibits. I remember a lady coming into my study saying, " I am *not* one of your nervous patients"—the exordium afforded me immediate evidence that she was so, as the result proved. All lunatics, even in their wildest mood, seem to me to recognize a difference between their delusions and facts, and this makes them often so furiously to insist upon them. Loud talk and shallow faith always run together. This is still more remarkable in the half-insane at an early stage of insanity. Later on, namely in the half-insane stage of recovery, the peculiar suspiciousness is much less marked; indeed an amiable trustfulness often takes its place. With the earlier stages we non-specialists are most concerned, and I am sure what I have named is a valuable aid to diagnosis.

The peculiarity is rarely, or only cursorily, alluded to in monographs on insanity, for the simple reason that specialists do not experience the difficulty, and therefore do not value any means of overcoming it. Nobody is brought to an asylum without there being abundant evidence of mental aberration, and the slighter indications therefore are of no practical moment. Our patients are probably never in a state to render restraint legal, or desirable on any account, remain useful and unnoticed members of society all their lives, and perhaps only manifest a delu-·sion in intercourse with their physician. We hail, therefore, with gratitude any thread to guide us out of the dilemma between a fact and a fancy.

It is not absolutely necessary to have been a hard worker first for idleness to lead the thoughts inwards to the digestion, and put it out of order. Some who have been Lotus-eaters all their lives, still do not get acclimatized.

Case LXXXIV.—Miss M. J—, aged about fifty-five, has as tough a constitution as most people I know of, and had consulted me about catarrhs or some trifling ailments occasionally. When I was away from England in 1865, she took a whim to go and live at an hydropathic establishment. She was not hydropathized, and it is a pity she was not, for it would perhaps have kept her out of mischief. But she used to listen to the inmates talking about their insides, and having very limited mental though plenty of pecuniary resources, she had nothing else to think of. The consequence was she began to suffer from gastralgia, even after the excellent wholesome diet and fine air she was getting at the place; and when she came to London to consult me on my return she was seriously out of health, always feeling a weight at the epigastrium after meals, having acid eructations and sometimes vomiting, and the tongue appearing pale and coated. I made her leave the noxious moral atmosphere, and adopt the physically worse alternative of close London lodgings with their well-known greasy cookery. Then she engaged a companion of her own age and position to talk to, and aided by some Quinine and Strychnine soon got well enough to run over for a trip abroad, with a strict caution to keep clear of spas and invalids.

My main object in this section has been to save these poor sufferers from drugs, which confirm their ailments.

SECTION X.

Abuse of Purgatives.

There is no habit so pernicious to the gastric digestion as systematically taking purgative drugs. And there is none more common.

It is commenced sometimes from mere caprice and imitation.

CASE LXXXV.—I saw last week a fine tall girl of seventeen at home for a few days from school. Her mother noticing how pale and listless she was, inquired into her daily doings, and got out a confession that nearly all the scholars were addicted to drenching themselves with pills; this made them thirsty, and they topped up with another purgative, "Lemon Kali" (an adulterated Bitartrate of Potash) several times a day. As my young friend had never taken physic in her life, except a few homœopathic globules at a former school, and some conventional draughts during the measles, this discipline made her ill; and it opened my eyes to the ease with which bad habits may be acquired. Even in her case it had begun to produce a sensitiveness to the presence of anything in the excretory viscera, which very quickly grows in intensity, and renders the abstinence from purgatives soon a positive deprivation. (September, 1866.)

It is the increase of sensitiveness which does the harm; for shortly this sensitiveness, commencing probably in the intestines, spreads to the stomach, and the presence of food there gives pain and cannot be borne, for the time requisite to normal digestion. The food being undigested, costiveness results; an increased demand for purgatives is made; sometimes even a medical man is induced to order them or to sanction them, and the difficulty of breaking the habit becomes really formidable. I found even a homœopathic physician, who placed his daughter under my care, had been persuaded to allow the growth in her of this living on poison.

The ill-health induced by purgatives is all the more serious in that it affects the most important classes of aliments. In Case XLV an illustration is given of the indigestion of fat, in Case LV of the indigestion of meat arising from this cause.

There is usually great difficulty in eliciting evidence of purgative habits; all the more so the higher in rank and more educated the victims are. Now and then a sensible country girl will make a confession which puts to shame her more refined sisters :—

CASE LXXXVI.—Emma W—, aged 25, a well-built strong country-woman, had to come to London in the summer of 1851 as a nurse to the children of an old friend of mine. Since then she had suffered from pain in the epigastrium (originally excited by tight lacing), waterbrash and debility. Her tongue and face were getting anæmic. For some months her fellow-servants and mistress had been dosing her with purgatives. She said she certainly did feel lighter after she took them, but in spite of that she had sense to remark that she was getting worse and worse, and could not but attribute it to the drugs. Yet she fancied she could not do without them, and feared she should be obliged to leave London and her comfortable place. This was on December 4th that she was sent to me. Before the end of the month, by simply leaving off purgatives gradually, and taking a little Iron, she lost her gastralgia and other stomach symptoms, gained strength and spirits, and remained in London many years a valuable servant, till the junior branches of the family left the nursery.

In the above case it is mentioned that purgatives were left off "*gradually ;*" this I usually accomplish by giving moderate doses of Aloes and Myrrh in pill, and with each change of prescription increasing the proportion of Myrrh and diminishing that of Aloes, then dividing the pill into two, and at last omitting it altogether. Another expedient is to recommend small cold-water enemata which are not really purgative at all, and allow the bowels to act spontaneously, at the same time as they cool the rectum and take off any feeling of congestion and tenesmus, acting in fact as a sort of shower-bath.

I have known the continued use of purgatives kept up by a medical practitioner with a vain hope of making the fecal evacuations of his patient more healthy in aspect.

CASE LXXXVII.—I was summoned in April, 1861, some distance into the country to see a young married woman, whom I found confined to bed with hysterical paralysis of the lower extremities and occasional vomiting. As my coming had been debated and arranged some days, I found prepared for my reception a long row of vessels, set in order of time, containing what had passed from the bowels. Each one was more unnatural, more fetid, more ragged, and with more undigested matter in it than the former. The medical attendant had been purging vigorously, and intended to go on purging vigorously, in spite of the obstinacy with which the patient got worse. When the gray powder, &c., was exchanged for beef-tea enemata, milk, mutton chops, and Pepsine, a rapid improvement followed. In subsequent letters I heard no more of foul stools.

There is a very curious superstition about the use of mercurials. They are supposed to make the alvine excretion normal, though the only visible result is its becoming more abnormal with each dose. They are supposed to do good by "acting on the liver," whether the liver is acting too little or too much. They are supposed to "act on the liver," though it has been shown by Dr. Scott's experiments[1] that the quantity of bile is not increased, nay, is rather diminished when Mercury is taken. All that the metal can be really seen to effect on the hepatic function is a poisoning of the bile, so as to prevent absorption by the ilia, and to cause the secretion to be rejected in a liquid form *per anum;* and that is a very doubtful advantage to most invalids.

The only effect at all desirable following mercurial purgation, and which in fact seems to constitute for patients the attraction to its use, is the relief of certain cerebral symptoms, giddiness, muscæ volitantes, dark globes in the sight, singing in the ears, &c., which result from excess of venous over arterial blood in the brain. It acts in this case as a destructive upon the venous blood, and adjusts the balance by subtraction. Time after time as the rough expedient is resorted to, the strength is lessened by it, and the necessity for its use appears greater more and more subtraction is required. The good and true way of re-storing the circulation to its normal conditions is by addition, by increasing the supply of new-made blood to the arteries.

SECTION XI.

Abuse of Alcohol.

The immediate effect of diluted alcohol on mucous membranes is first to dry them by staying the aqueous exhalation, and shortly to damp them with an abnormal formation of mucus, to retard the capillary circulation, and to deaden the sensibility of the nerves. The last action is its use. Where there is risk to health from undue sensitiveness, alcohol in moderation is an invaluable remedy. It may be considered as an antidote to the condition discussed in the last section; and if a man were condemned to take unnecessary purgatives, he could not do better for his stomach than counteract part of their evil effect by mixing them with alcohol. Experience seems to have led to the

[1] Beale's "Archives," vol. i. p. 209.

same conclusion as science, and we find the most popular drench-ing recipes have either alcohol or some equivalent anæsthetic in their composition.   It is equally antidotal where the sensi-tiveness is the manifestation of weakness in the nervous system, either from exhaustion or imperfection.   And thus it becomes the daily food or daily physic (I care not which it is called) of those whose daily life brings their nerves into this state.

To the health of the bulk of mankind the habitual moderate use of alcohol is probably quite indifferent.   One day they may want a little, and therefore be the better for it; another day they would be in a more perfect condition without it.   So a balance is struck by the habitual users; and their chief argu-ment in favor of fermented liquids remains the unanswerable one *that they are nice.*   No mean argument either, for it weighed with our Divine Master, when He first showed His power by treating the merry-makers of Cana to better wine than they were accustomed to.

The effects of habitual excess (which in some people is taking any alcohol at all, in others is taking what is universally allowed to be "to much") is on the gastric area very similar to that of any other anæsthetic.   A partial paralysis of it is in-duced, it ceases more and more to perform its peculiar functions for the owner; "he cannot eat but little meat, his stomach is not good," though he may still digest vegetables and feel a re-lief from filling the void with them.

If the appetite for food remains large, the weakened walls of the receptacle are liable to yield to the dilatation, as in the fol-lowing instance.

CASE LXXXVIII.—Mr. F——, a burly farmer of middle age, came to me in December, 1856, complaining of a constant sinking at the épigastrium, relieved indeed for a short time by taking food, and partially by a glass of spirits. He ate, however, without appetite, and did not even enjoy his brandy, for it had become a mere matter of supposed necessity with him. Latterly animal food caused disgust and nausea, his bowels, from being cos-tive, had become relaxed, with yeasty fermenting stools, and he had got very down-hearted about himself. The condition had, however, been coming on very gradually he knew not how many years, and he was without difficulty brought to see the connection it had with a habit of taking spirits between meals.

The tongue was coated with patches, showing sharp defined edges, of epithelium on a bright red base.   It was described as being more generally all red, like a beefsteak.   The tympanitic resonance on percussion of the stomach extended right up into the cardiac region and down nearly to the navel, and laterally in proportion; and the abdomen was prominent as well from accumulation of fat in the omentum and parietes.

I put him on a Banting diet, with at first some liquor potassæ to decrease his corpulence, and I ordered fifteen grains of Boudalt's Pepsine powder to be taken with animal food to assist in its digestion.   I persuaded him also to promise that no spirituous liquor should be taken between meals; but he said he had sooner die than surrender a glass of brandy-and-water at supper.

I must confess I had some doubts about the observance of the promise. Yet I was wrong; he did leave off spirits, and he did get much better and more active in business, and continued so for nearly two years.   Then some temptation arose, he resumed his old habits, and was brought up again to London in 1858, in the same state as before.   The same advice was given, but I have no record of the result.

Persons with dilated stomachs are very apt to become obese, though the flesh digested is not sufficient to sustain the muscular strength.   And this sort of obesity is very difficult to manage, from the impediment which the muscular weakness offers to taking exercise.

In women, perhaps, from the bondage of the dress, the stomach does not in my experience become dilated from the paralyzed condition induced by alcohol.   The following case represents the more common injury done to the viscus.

CASE LXXXIX.—Mrs. P—, aged 33, came under my care October 3d, 1864.   She lived in the country in easy circumstances, had no family or society to attend to, and had become lazy, fat, flatulent, and low-spirited. For several years she had been gradually getting into the habit of alleviating her uncomfortable sensations by small doses of brandy, which she took morning, noon, and night, but never in such a quantity as to get into her head.   The reason of her coming to me was the inability, which was growing upon her, of keeping the smallest quantity of food upon her stomach.   It was vomited almost immediately.   She was very hysterical, and the catamenia was irregular.   Leaving off brandy and taking some Valerian and shower-baths stayed the vomiting; but two months afterwards I was obliged to go abroad, and lost sight of her.

The sudden leaving off excess of stimulants will in elderly persons sometimes cause disturbed cardiac action, even when the gastric symptoms are relieved by it.

CASE XC.—Mrs. B—, an elderly lady habitually rather short-winded, came to me on the 26th of October, 1864.   She was suffering from loss of appetite, with frequent nausea and vomiting, which I attributed to a habit recently acquired of taking brandy between meals. The pulse was then regular. I urged her to give up the dangerous habit forthwith, and saw her again on the 2d of November. The nausea and vomiting had ceased, and she felt some return of appetite. But she had a new sensation of sinking at the epigastrium, and was shorter of breath. On examination of the pulse I found it irregular and intermittent. The heart-sounds were normal. I gave her some Valerian, and on the 18th found her still bravely resisting the temptation to brandy, and dismissed her with a prescription for some Quinine and Strychnine.

I am used to quote to such patients as the last *in terrorem* an experience I once had of want of resolution in breaking off dram-drinking—an experience happily rare, and not cited here as illustrative of a class, but still instructive as an extreme warning.

CASE XCI.—In September, 1857, I was called by Dr. Jephson to a consultation in the case of an unfortunate middle-aged woman, who was dying prostrated by uninterrupted vomiting. It is needless to detail the symptoms, which were those of simply retching and sinking, and the nature of the case was made apparent by her desiring her maid to bring her a glass of brandy even while I was speaking to her. Our attempts to feed her with beef-tea enemata and Opium were unavailing, and she died next morning.

She told me the habit had been acquired only the previous year, while staying with some friends in Scotland at their shootings, where a nip of whiskey was the regular preparative for breakfast.

But dram-drinking is by no means confined to uneducated persons, those whose "talk is of bullocks," or to idle women. I am ashamed to say I have been consulted about its consequences by several members of our own profession, who ought to know better and set a better example. *Quis custodiet ipsos custodes?* They tell me the temptation is very great in country practice, sitting in tedious conclave in lone farmhouses during a lingering labor, or watching some long-dying patient with no person that can understand your thoughts within many miles. There is nothing else to do but drink; and then the next day you have to be at work at the usual early hour, and the work can hardly be done without a hair of the dog that bit you.

The last sentence, expressing the necessity for staving off alcoholic reaction, reminds me to mention a test which I am used to apply to discover whether the amount of alcohol taken is such as really to injure the stomach. I ask whether the patient ever is in the habit of taking it in the forenoon. If so, I at once feel sure that the stomach has suffered. When a considerable interval intervenes between the indulgences, and the reaction is allowed to have its way till ordinary digestion is restored, the constitution may very often be still uninjured. But I have not yet met with a forenoon tippler, even though he never got drunk in his life, without a condition of stomach which most infallibly shortened his days. I find it a great advantage in the selection of lives for insurance to substitute a pointed question on this head for the usual aimless inquiry whether the proposer is "sober and temperate." Nobody is anything else, of course; and the answer is a mere declaration of opinion. But "do you take spirits in the forenoon? Is that a habit?" require categorical statements of facts, which if wilfully false would vitiate the policy.

The way in which life is shortened by this stomach affection is generally secondarily through the liver, originating anæmia and ascites: sometimes through the pancreas; when the emaciated gin-drinker, such as Hogarth drew, is produced. More rarely the kidneys break down, and Bright's disease arises. In fact the nearer, physiologically speaking, the organ to the stomach the more likely it is to suffer.

When a patient is persuaded to give up dram-drinking, he often has such a dreadful depression of spirits that his resolution is apt to give way, though he is convinced he is acting right. And sometimes he may have a kind of delirium tremens from the sudden shock, before he can get into the temperate habit of taking stimulants only at dinner, or of giving them up altogether, according to the nature of the case. Still it is best to enforce the absolute rule of no alcohol between meals, and to supply its place temporarily by an Ether and Ammonia draught, then by Ammonia, either alone or with a bitter, and then to stop it altogether.

SECTION XII.

Tobacco.

The more usual toxical effects of the alkaloids absorbable from the Tobacco plant are exemplified in the following typical cases.

Case XCII.—*Smoking.*—Five years ago a young married man of about 32 rushed to me in a great state of alarm, stating that he had suddenly become impotent. This was not strictly true, but still he certainly was less fit for matrimonial privileges than was right in a husband of two years' standing. The next complaint he made was of cardiac palpitation (on examination I found the heart beating unevenly and irregularly), of frequent cold sweats, nervous agitation and causeless fears by night and day too.

I found he had recently returned from sheep-farming in Australia for several years. When there he used to smoke strong Shag in a short cutty-pipe all day and almost all night. He had brought his though dirty companion with him to London, and continued the habit with a certain amount of modification. In the fresh air of the wild downs he had never suffered the slightest illness, but no sooner had he been in London a few weeks than the symptoms detailed had come upon him, and had gradually increased.

He could not at first understand why I should attribute them to the Tobacco, why it should be so bad for him in England, when abroad it seemed to preserve his health. But at last becoming convinced of the difference between British and Australian air, he drew his little black pet from his pocket and broke it in my fire-place. He would never smoke again, rather than risk depriving his wife of her just claims on his attention.

I took the tide at the turn and clenched the promise, which was certainly kept long enough for the palpitations, nervous fears, &c., to be cured without physic.

Case XCIII.—*Snuffing.*—October 22d, 1866. Rev. C. W—, a country clergyman of literary and sedentary habits, has usually enjoyed good health, and in spite of a fondness for his study and dislike of parochial work, visits in his district, and has regularly done two full services every Sunday. It is a difficulty in properly performing the last-named duty which brings him to me. For several months he has noticed that his manner in the pulpit has been getting awkward, and he feels hurried and has an unreasonable desire to get to the end of what he is about. He sometimes cannot help skipping over the latter half of a sentence so as to go on with the next. For some weeks this hurry of manner has been extending itself to his social and professional intercourse on week-days, and to-day in speaking to me he is excessively precipitate and nervous. He can scarce keep his hands still, and clutches at and handles all the little things around him in my study, though evidently ashamed of his solecism in demeanor.

His appetite is good, he has no flatulence, he can eat anything he likes, and drink a bottle of port without feeling any inconvenience, in short he is evidently unaware of having digestive organs. The actions of the bowels and kidney are quite healthy, but he evacuates the bladder more frequently than is needful.

On inquiry I find he is a devoted snuffer, having his pocket box filled up every day, and keeping a second relay on his table as well.

I said at once I would not prescribe for him unless he would make at once at least a step towards giving up this habit. He readily consented to keep a box only on his table, and to have it filled only twice a week. To supply temporarily its place, I allowed him two teaspoonfuls of Tincture of Valerian twice a day.

I saw him again in a week much improved, and in a fortnight after that he seemed quite to have regained his natural dignified manner and to be reconciled to abandon his snuff-taking.

These histories give a pretty full detail of all the important phenomena usually produced by excess of Tobacco, according to my experience. Others are merely a repetition of these, in which it may be observed that the digestive organs seem remarkably free from injury. In fact the only two cases I can find in my note-books where the alimentary viscera have suffered are the following.

CASE XCIV.—Mr. William T—, aged apparently about 50, came to me in March, 1856, complaining of costiveness, pain in epigastrium about three hours after food, flatulence, and dryness of mouth. I could not find any deviation from wholesome habits of life except that he smoked a great deal of strong Tobacco. And the event proved that to be the source of his dyspepsia, for by restricting himself to one cigar after breakfast, and taking some Charcoal and Soda, he came to me towards the end of the month much better.

CASE XCV.—H. C—, a country surgeon, aged 45, complained last year to me that he was really becoming unable to follow his profession from excessive flatulence in the ilia. When he was talking to a patient the bowels would begin rumbling and rolling so that he felt ashamed to stay in the room. He was obliged several times a day to unbutton and lie with his abdomen up in the air. At night sleep was broken, and sometimes rendered impossible by the same nuisance. Curiously enough, when he sat up all night, say with a troublesome midwifery patient, he was not half so bad. While talking with me, I observed he took snuff several times, and on inquiry found he consumed nearly an ounce daily. He, of course, could not be unaware of the cause of his disease, but absolutely refused to give it up. He said life would not be worth having without it.

In all other instances which I have taken notes of, drinking was joined with smoking or snuff-taking as the decided efficient cause of indigestion, so that the cases prove nothing for scientific purposes: or else (as in Case LXVIII, for example) the accusation against Tobacco was shown to be a libel by the symptoms not ceasing when the alleged cause had been long removed.

I must say I am surprised, for several medical writers seem to consider it a matter of course that the pleasures of the pipe

should have a special deleterious effect on the salivary glands
and stomach.

Dr. Prout says, "The severe and peculiar dyspeptic symptoms sometimes produced by inveterate snuff-taking are well
known;"—so well apparently, that he does not enumerate
them, so that perhaps he may mean the nervous weakness
described above—but then he goes on to remark, "I have
more than once seen such cases terminate fatally with malignant disease of the stomach and liver."[1] The insinuation is that
Tobacco causes malignant disease; which is proved false by the
fact of cancer of all the organs being more common among
women than among men; and among men being quite as
common among those who do not smoke as among those who
do. It is very clear that Dr. Prout has misapprehended the
pains of incipient cancer, and ascribed them to the Tobacco
which was taken to solace them.

I must allow that I myself took such ideas as Dr. Prout's
for granted, and supposed that of course the salivary and gastric secretions must be the chief sufferers from Tobacco, till I
came to review my experience and drew out these two solitary
specimens of their being possibly affected by it. They, therefore, must not be considered as the type of a class.

The poison of Tobacco smoke seems to attack more particularly the nervous system. Intermittent pulse, palpitation of
the heart, shaky hands, nervousness, imaginary impotence, and
the like, are produced by it, but not primary affections of the
digestive organs, as a rule. And in snuffing the large quantity of the drug which goes down the œsophagus seems to pass
the mucous membrane with little injury, and to affect the
system only by the absorption of its alkaloids soluble in the
blood.

This last sentence may afford a hint as to the method of
treating our patients. It is avowedly, almost proverbially, difficult to get them to resign the soothing herb. Few of them
take such a wholesome alarm as Case XCII, or, if they did,
would not act upon it. They say the sudden deprivation is too
much for their strength of mind. Now if a Tobacco is pre-

---

[1] "Stomach and Urinary Diseases," page 25.

pared by abstracting the main deleterious agent, Nicotina, a step is set by which the patient may be let down easy, and not run the risk of an abrupt change. For smokers a convenience of this sort is afforded by the Vevay or other Swiss cigars, which are the common leaf fitted for use in cigars by soaking in water till one-third of its substance is abstracted.

I do not know of any kind of snuff manufactured on a similar plan, and consequently there is not the same aid to persuading a victim to surrender the indulgence; but one old snuffer told me he had broken himself of it by the aid of kitchen salt finely pounded, of which he mixed more and more daily with the contents of his box, till it was nearly all salt. Then he took plain salt, and soon gave that up. I have heard also of ginger being employed in the same manner. Another, who had acquired the habit at Cambridge many years ago, and did not like the look of it on leaving the University, used to carry for some time a vinaigrette of Aromatic Vinegar for the same purpose.

SECTION XIII.

Tea.

The following case, illustrative of the pernicious consequences of excessive tea-drinking, is extracted from my Clinical Lectures at St. Mary's Hospital.

CASE XCVI.—Maria D—, a spinster of thirty-two by her own confession, but probably older, has been a general servant in a light place for seven years. She has been happy, and has enjoyed pretty good health, interrupted only by occasional headaches; but for some time lately things have seemed to annoy her more than they ought to do. Three months ago, she had a bad "bilious" headache, which was followed by some paroxysms of laughing and crying. Five weeks back she had an attack of diarrhœa, from which she got better, and went to work again in spite of weakness, for she was loath to let her mistress want her. But exertion was in vain, for she no sooner tried to clean a grate than she fell down speechless, and had a succession of hysterical fits, losing her senses, but not biting her tongue. Then she began vomiting everything she took, and this had been going on for three weeks, and seemed to amount to a complete rejection of all her food immediately it was swallowed. When you saw her, there was excessive flatulence, the air bursting up from the stomach in roaring eructations while one was talking to her.

In this woman, the effect of the wide pupil and sympathetic hemiptosis is

9

not hidden even by the disfigurement of blear edges to the eyelids; and it quite accords with the droll earnestness of her manner, which increases gradually as you let her go on talking about herself, leaving no doubt of her strong hysterical diathesis.

As to cause, that is still more directly traceable to the stomach than even in the last case.[1] It would seem that for some years she has been becoming more and more addicted to tea-drinking. She confesses to caring for little else, so long as she could get her favorite food or physic—or poison—I do not know exactly how to call it. Her mistress was quite angry with her for eating so little meat; and with a far-sighted economy not common in her class of life, took much trouble to keep up the health of a faithful servant. But the weakened stomach refused meat, and she was literally starving in the midst of abundance. (*Nov.* 1, 1861.)

Much ill-health arises among women of the lower orders in this country from the custom of sluicing themselves with tea. (I am not aware if similar results follow in Holland and Portugal, the only other tea-drinking populations in Europe.) Want of appetite for the quantity of coarse albuminous food necessary to working people is induced. In the upper ranks not so much harm is done by the five o'clock kettle-drums and similar sloppy proceedings now so common, because their bill of fare is more attractive to the palate, and they usually get as much flesh food as is good for them in spite of it. Besides which, educated persons have usually the instinct to stop in time a custom which really depends on a mere whim. Still it cannot under any circumstances be a wholesome habit.

Tea seems more injurious to the stomach in the usual form of infusion than otherwise. I remember some years ago being puzzled in viewing lives for insurance by some singularly colored tongues which I saw in those who came before me. On inquiry, I found their occupation was "tea-tasting" for the greater part of the day. Now, tasting tea is performed partly by sipping some of the infusion, but principally by sniffing up the aroma into the nostrils and chewing a few leaves in the mouth. I was given to understand that they sometimes found themselves nervous after a long day's work, that possibly the hand might shake a little in those who worked too hard, and that the tongue acquired this curious smooth orange-tinged coating, but that the digestion and appetite did not suffer from the trade.

[1] A very similar case not necessary to be repeated in this connection.

SECTION XIV.

Opium.

An occasional effect of the Salts of Opium on the stomach is exhibited in the following case :—

CASE XCVII.—Jane B—, a domestic servant, thirty-seven years of age, was under my care at St. Mary's for some painful tumors of the abdomen affecting the uterus and bladder, in March, 1861. On account of the pain, she was ordered a grain of Acetate of Morphia every night. She had never previously had any narcotics. She only took one dose, for that was followed by vomiting, very severe during the night, and recurring at intervals during the next four days.

The possibility of an idiosyncrasy of this sort is no reason for shrinking from the essay of a good and useful medicine, but it is as well to know that it may occur.

The more chronic effects upon the organ are shown in the next :—

CASE XCVIII.—August 12th, 1853.—George N—, an assistant-surgeon, aged 35, states that for eight years he has been in the habit of taking large quantities of Opium. He began the practice in the first instance to prevent his feeling the want of food, when, as a surgeon's assistant, he was obliged to wait many consecutive hours without anything to eat. He at first confined himself to twenty drops of Laudanum a day ; but he gradually increased the amount till he finished a fluidounce of Laudanum daily, and quarter of an ounce of crude Opium in addition weekly. He tried several times to leave it off, but was prevented by the nausea and pain in the epigastrium which he experienced. He had lost much flesh, and got miserably weak ; but he probably would have gone on with his poison had he not been frightened by a numbness and partial paralysis of the left arm, and a loss of memory, which made him think he was going to have a stroke, and caused him to put himself under my care.

I immediately restricted him to one grain of Opium at night, and consequently found him next day in miserable plight, vomiting, with pain in the epigastrium, and with a most melancholy aspect. I gave him strong beef-tea and port wine, but got afraid next day that he would slip through my fingers, and so I added some Chloroform draughts. These relieved the sickness. By the 17th he began to get better, and the Chloroform could then be omitted. On the 20th he felt very sinking for the want of it, but yet fancied he was recovering his appetite. He had, at his own request, a mutton chop and half a pint of porter. On the 22d he remarked his memory was improved, and he got up and dressed. Then his bowels got irregular, and following that lead I was able to restrict the quantity of Opium to what he had in some Chalk and

Opium powders, ordered to be taken when there was diarrhœa. By September 1st he was able to leave it off entirely and take care of himself.

It appears from this to be the digestion of meat and fat which is mainly impeded by Opium. It requires, however, to be taken in great excess for the effect to be produced.

And even then the result is not by any means immediate. That is shown by the case quoted; and I remember also, in 1838 or 9, a sweeper of a lucrative crossing coming to swear an affidavit before my father as a magistrate that the bearer of the said affidavit was in the habit of using two drachms of solid Opium daily. The reason of this measure was that the shop where he was accustomed to deal for the drug had changed hands, and the new-comers refused to serve him with such a dangerous quantity. He was nigh crazy with the restriction, but armed with his legal document he felt safe for the future, and I used to see him at his post many years afterwards.

On the whole, Opium-eating does less harm than is generally supposed—very often much less harm than the pains which it is taken to counteract. The great objection to it seems to be the difficulty of leaving it off, when, as in the case of the surgeon's assistant, it had from its monstrous excess begun to tell on the health. But this difficulty has been very much exaggerated, as well as the temporary pleasures of indulgence, from the description having fallen into the hands of the imaginative De Quincey —a man whose world was in himself, and whose whole biography, when published, let us into the secret of "The English Opium-eater" being really a work of fancy. The same may be said of Coleridge's "Recollections." I find in my notes a special memorandum of the scorn with which the difficulty was treated by a genuine strong-minded man.

CASE XCIX.—During the year 1859 I saw from time to time, for some trifling ailments of which I have no accurate record, Captain B—, a fine, hearty, God-fearing sailor of the old school, seventy-two years of age. He told me that twice in his life he had been a decided Opium-eater, taking as much as a drachm in solid form daily. I expressed my surprise at his having given up to the practice, which surprise he did not at all understand, saying, " Why, I should be ashamed of both my philosophy and my religion, and turn sceptic, if either singly would not strengthen me with resolution enough for that." The occasion for which he took the Opium, some trying mental circumstances, having passed away, he diminished the quantity by five grains daily till he ceased entirely ; and I must say his constitution appeared none the worse.   I hear from his daughter he is still alive and well at 80.

When Opium is given medicinally, that is for the relief of certain bodily or mental pains, and when it succeeds in relieving those pains, it does not seem to produce its special toxical effects: where it is really wanted it rarely does harm.   For instance, in inflammation of the serous membranes, as pericarditis, I have given to young persons who never took it before as much as three grains every three hours, without producing constipation or over-sleepiness till such time as the inflammation had subsided.   (See " Lectures chiefly Clinical," Lect. XV, " On Pericarditis.")   Of course I did not arrive at this quantity all at once, but began with a grain or a grain and a half, and increased rapidly.

I have myself taken Opium for the relief of various inconveniences arising from an amputated limb, but I have never felt the slightest temptation to continue its use beyond the necessary period, or any inconvenience from leaving it off.   The box stands alongside of my razors, and I do not feel one more dangerous than the other.

It is only when taken in great excess, or when persisted in, spite of warning, that Opium seems seriously detrimental to the digestion.

All the habits in this chapter instanced as causes of indigestion are voluntary and capable of being changed.   The cure, therefore, of the indigestion lies first and foremost in that change.   It must be made a *sine quâ non* of the treatment by every honest practitioner.   In aid of that I have given a few hints in passing, but let it be understood that these expedients are to be only temporary : the effect is efficiently to be removed only by removing the cause.

134

CHAPTER IV.

ABDOMINAL PAINS.

Section 1.—Heartburn. Section 2.—Acidity. Section 3.—Waterbrash. Section 4.—Spasms. Section 5.—Gripes. Section 6.—Weight. Section 7.—Wearing pain. Section 8.—Soreness on pressure. Section 9.—Anomalous pains.

In the notes of cases previously used in illustration of my subject certain pains or discomforts are often stated to have been felt in the epigastrium or its immediate neighborhood, without their nature being particularly detailed. Either they were not severe enough to affect the general treatment, and so their form was not noticed; or they could not be clearly made out from the patient's words; or the record was incomplete in this respect, though full enough for the immediate purpose of its citation.

A little care will enable the observer to distinguish considerable differences in these pains—differences which often may modify our diagnosis of the anatomical state of the parts, our prognosis, and our treatment; and I shall devote this chapter to a consideration of them.

The table of contents enumerates the names which I shall use in describing them in detail. I prefer these words to Greek or Latin compounds which profess to include them. The artificially built-up terms have, indeed, a show of science, but are not at all more accurate in reality, and much less graphic than those engendered by daily use.

SECTION I.

Heartburn.

Heartburn is a painful sensation, resembling that produced by swallowing something very hot, which arises at a certain interval after food in the upper part and towards the left side of the pit of the stomach. It runs in paroxysms at the back of the breast bone up the course of the œsophagus, culminating in the pharynx, and each paroxysm often passes off with a feeling as if hot smoke had escaped into the mouth. The pain of pure heartburn is not caused or increased by pressure, and is not felt between the shoulders.

There is in heartburn often a temporary salivation, and the secretion from the glands being voluntarily swallowed somewhat relieves (by its slight alkaline reaction, probably) the discomfort of the cardia; but if it is spat out, no relief follows. An arrest of the passage of this augmented secretion into the stomach will be shortly described under the heading of " Waterbrash," in a future section. Though the sensation is that of a cramp, and the œsophagus is a muscular organ, I do not think there is any real tonic contraction of the fibres. There is no movement in the throat, such as may be readily felt on voluntarily gulping. There is certainly no visible contraction of the back of the fauces. Indeed, when the sensation gets there, it is rather one of relaxation, as if smoke escaped, say the patients. Moreover, if a little fluid be swallowed, its passage is not resisted by any stricture. There appears to me to be a subjective perturbation of sensibility, rather than of contractility, in the milder cases I call " heartburn." Where there is a real spasm, " waterbrash" is produced, as I will explain under that heading.

Though this morbid phenomenon is manifested by the œsophagus, its causes do not lie *in* the œsophagus. Cancer, ulceration, or stricture of that organ, do not originate it in the majority of cases of these lesions, whereas it is a very common consequence of the slighter morbid conditions of the stomach. We may remark that it is easier produced by general than by local states of the viscus, and rather by slight than by severe derangements. We constantly find cancerous tumors and considerable ulcerations in the gastric walls without any such œsophageal symptom at all; whereas a catarrh, a mucous flux, and more commonly still simple atony of the stomach, seldom exist long without it. This would seem to show that a certain amount of health, as well as a certain amount of disease, is necessary to heartburn.

From the effects which alkalies have in allaying temporarily this pain, it may be inferred to arise from the action of the acid contents of the stomach on the cardiac and œsophageal nerves. It is true the gastric mucous membrane itself does not immediately suffer from acid; it secretes acid,[1] and bears acid in contact with its coats without inconvenience. The gullet, too, will do so for a short time; swallowing a mouthful of sour victuals or drink gives a healthy man no immediate discomfort. But we may remark that many influences which, when intermittent and alternated with rest, are indifferent or even pleasant to the sensory nerves, become exquisitely painful, and may even cause material disease of tissue, when long continued. For example, the immersion of a limb in water a few degrees below the temperature of the air is not disagreeable, and may be borne with intermission for any length of time; but it becomes absolute torture if persisted in without an interval of rest or reaction. A moderate degree of pressure, if continued too long, will cause first pain, then gangrene or atrophy. A continual dribbling of feces will make an anus sore—a continual running from the nose excoriate the nares, &c., though we hardly notice it when lasting only an ordinary time. Just in the same way we must look for a quite different class of consequences from the intermittent and from the continued action of acids on the sensory portions of the pneumogastric. But when we trace heartburn to the impression of acid on the œsophageal and cardiac plexus, we do not necessarily imply that the acid (normal and abnormal) is in excess. It very often is not so; and we must refer the symptoms to over-sensibility, that is, to the sensibility of a normally insensitive part, which, I may remark in passing, is always a painful sensibility.

---

[1] There appears no doubt about gastric juice being *secreted* acid, and becoming neutral only from mixture with saliva. See the experiments of Drs. Bidder, Schmidt, Grünewaldt, and Schröder, compared in "Digestion and its Derangements," chap. iv, and "Experiments on Digestion," by Dr. F. G. Smith (Philadelphia, 1856). This last-named renewal of observations on a patient with gastric fistula, formerly servant to Dr. Beaumont, seems to show conclusively that in the human subject the acid secreted is not hydrochloric, but probably lactic. The explanation of finding hydrochloric acid in gastric juice is that lactic acid in a nascent state decomposes the chloride of sodium contained in all animal fluids.

We thus arrive at two immediate causes of the morbid phenomenon in question:—
1st. Too long-continued acidity of the stomach.
2d. Over-sensitiveness of the cardiac and œsophageal nerves.
I will point out the action of these two causes by the citation
of some cases.

CASE C.—Hon. Major C— has attained old age with as little suffering from
illness as most people. What brings him to consult me is a painful sensation
rising up from the epigastrium to the back of the throat at uncertain times
(generally from three to four hours) after food. He is able to prevent it by
eating very little, but he fears that what he takes under this restriction is not
enough to nourish him. He is also able to cure it temporarily by Soda or
Potash, but has heard that is a bad habit. Though he is old, he wants to
be cured. His mouth gets very dry from lack of saliva. I advised him at his
age not to be too solicitous for a second youth, but ordered a Quinine mixture
with a grain and a half of Iodide of Potassium to be taken twice a day for a
few consecutive days occasionally. It seemed to agree with him, and the dryness of mouth was less.

The object of the Iodide of Potassium was to increase the
secretion of saliva; but I did not assign the whole of the indigestion to the deficiency of this fluid, considering it rather as
an effect of the sluggishness natural to old age. At that period
of life less exercise is taken and less food required; so that the
quantity of the diet should be accommodated to the years. If
it be not so, the overladen organ labors.

CASE CI.—Miss K—, aged 40, consulted me in April, 1857, about an intermittent hemicrania which had come on recently through living in an aguish
district. She had a look of chronic invalidism more than was justified by the
recent malarious infection, and on inquiry I found that for many years she
had suffered from what she called "risings in the throat," which came on
about three hours after meals. Dinner was the most painful meal. If nothing
came up, as was usually the case, the "risings" continued two hours or more,
and went away gradually. But if by a semi-voluntary effort she turned the
"rising" into an ejection of a small quantity of food and air, relief followed.
On these occasions what she brought up was very acid to taste and smell. She
had been physicked at various times in previous years for this heartburn without benefit, and had learnt to bear it. She found, indeed, that Soda gave
temporary ease; but fancied the symptoms were aggravated by a persistence
in the remedy, and despaired of being ever better. She came to be cured of
her headache. For this I prescribed, and killed two birds with one stone, for
she was later led to volunteer a confession that the Quinine I gave her to cure
the hemicrania did the heartburn good also.

It is this last observation which leads to my here quoting the case.

By three hours after meals the stomach ought so far to have emptied itself that the cardia should not be distended, and the orifice, relieved of the pressure of acid matters, should be enjoying the change of a trickling flow of alkaline saliva. Though the general contents of the stomach may, and indeed ought, to remain acid longer than that, yet the lower orifice of the œsophagus requires a period of alkalinity, and suffers if it does not get some.

Note—that the throwing up of a small quantity gave relief, because it brought the stomach into a normal condition as to contents.

Note—that what is brought up in heartburn was acid, showing a free communication with the stomach, and therefore that the œsophagus was pervious, not spasmodically contracted, as the patient's sensations led her to believe.

This form of heartburn frequently comes on at night, preventing sleep.

Sometimes the patients will say they have "pain *before* food," which pain on inquiry turns out to be postponed heartburn arising from the last meal.

CASE CII.—Rev. E. M—, aged 26, has worked so hard to raise himself to be fellow and tutor of his college that he has injured his digestion. The false appetite which intellectual exertion brought on, made him overload the stomach at dinner with more than it could part with by next meal. This induced a pain not exactly like that of hunger before each meal, accompanied by a sensation as of something rising up into the fauces. No vomiting or eructation, though the stomach evidently was not empty. He had besides some curious nervous symptoms, for which I gave him Quinine and Strychnine, and he got better of all together.

CASE CIII.—Mr. John H—, aged 42, came to me February 7th, 1866, complaining of pain at the epigastrium towards the left side, rising up in paroxysms to the fauces, and which was shown to be only heartburn by the absence of tenderness on pressure. He declares it does not come on till full four hours after food, and it passes into hunger for the next meal.

It is almost as common in practice, especially among the educated classes, to find heartburn complained of as coming on within the first hour after meals.

CASE CIV.—Henry S—, aged about 40, a solicitor in large country practice, came to me in March, 1856, complaining, among other things, of heartburn commencing within an hour of every meal. He had sometimes made a strong effort at eructation and brought up some of the contents of the stomach, but it gave him no relief. What he brought up did not taste particularly sour, and consisted of whatever he had eaten. I prescribed him a course of Hydrocyanic Acid (♏iv in Infusion of Gentian three times a day). He continued to take that till quite well, and remained well till a hasty journey to Vienna in the autumn of 1860 brought on an attack of diarrhœa and great prostration. After this his old symptoms returned, and were again appeased by Hydrocyanic Acid and a Blister to the scrobiculus cordis.

It is among anxious sensitive persons that we usually find this kind of heartburn, even although they may not be so intellectually and æsthetically endowed as the last-named sufferer.

CASE CV.—Mr. W—, a cheesemonger, aged 30, came to me last January for a feeling of pain rising up from the epigastrium to the back of the throat, as if smoke arose from the stomach, usually under three quarters of an hour after meals. If he ate supper, this would happen in bed and give him sleepless nights.

He was an uneducated soul-less man, but had worried himself a good deal about his trade, and was also anxious about his health, so that his tongue had the quiver and the wet white coat of an overwrought intellectual woman's.

Though doubtless the largeness of the meal contributes seriously to the severity of the heartburn with acidity, it is by no means an essential in its production.

CASE CVI.—During 1861 and 1862 I attended the wife of a retired Anglo-Indian physician, aged about 40, for general sluggishness of the alimentary canal, accompanied by a tendency to mucous discharge per anum. She suffered at first a great deal from "acidity" within the first three hours after meals, but sometimes sooner. She constantly averred, and indeed at my request subjected the matter to the test of experiment, that a small quantity of bread, or any other simple food, brought on the acidity as certainly as a full meal. That this was due to sluggish action of the gastric muscular fibres was evidenced by her deriving benefit from Strychnine; but that it was not wholly so to be debited, its early supervention showed.

Now in the first, second, third, and fourth of these specimen cases we may fairly accuse the stomach of atonic sluggishness by which its normally acid contents are detained too long in their passage, and the nerves exposed too long to that acid. But in the next two the pain cannot be debited to prolonged

exposure, for it would be an exceedingly abnormal thing if the
cardia were not acid at that time.   There must have been an
over-sensitiveness of the gastric plexus.   The last (Case CVI)
is a transition case, which, though it stands alone here, really
represents a larger number than the others, namely, those which
are a transition between the classes, presenting the characteristics of both in various degrees.   We may take the marked
cases as the two ends of the scale, between which the majority
of our patients vibrate.

Although, therefore, it is impossible to make a clear division
of our patients into the two classes, yet has the distinction an
importance, for the nearer the symptom occurs to the meal the
more is it due to hyperæsthesis, and the further off the more
to slow digestion.

And with an eye to this pathology are the patients best
treated.   First, as regards the immediate popular relief by alkalies: when the heartburn does not come on till four hours or
so after a meal, an alkali may be safely taken even habitually;
it is time for the stomach to be losing its acidity, and there is
no harm in assisting nature.   Still it must be remembered that
this temporary antidote, taken in this way, is not a cure.   But
if the heartburn comes on within about an hour of the ingestion of food, an alkaline neutralization of the gastric contents
is positively abnormal and injurious. It prevents the due
digestion of the food, and so deprives the body of nutriment.
It induces anæmia, debility, and possibly some of the further
ills to which these powerful degenerators lead.

In the cases with a preponderance towards the latter class
local anæsthetics come into play.   Opium and its salts, Bismuth,
Zinc, Silver, Henbane, are all of use as a change, but the main
stay of the practitioner is Hydrocyanic Acid.   It does more
good combined with less harm than any other remedy.   Carbonic Acid is also a harmless anæsthetic, and indeed a normal
one, for the natural atmosphere of mucous membranes is Carbonic Acid.   I have sometimes persuaded people to take Car-
bonated (aërated) Water instead of Soda, and they have experienced an equal relief.   I have sometimes combined a Blister
with Hydrocyanic Acid, from my observation of the effect of

epispastics in pleurodynia, and I am inclined to think it anæs-thetic in its action. But it is a disagreeable measure.

Mineral acids are of more use in the heartburn of sluggish indigestion than in the hyperæsthetic; but I think they seldom cure without the aid of other more potent tonics. One great objection to their use is their injurious action on the teeth. Dentists tell me that we physicians make more work for them than any other external enemy to the integrity of the grinders by our administration of mineral acids. In consequence I have made a few trials of Acetic Acid as a substitute, but without much encouragement to persevere.

In hyperæsthetic heartburn the chewing a piece of Liquorice slowly in the mouth will often be of great service, especially when it occurs in pregnant women, to whom one is loath to administer drugs without strong cause. Howsoever, the final reliance of the physician for cure must be in a renewal of the powers of life by tonics, especially by nerve-tonics, especially by Quinine and Strychnine: and then, when a step of progress has been made with these, by the restoration of the blood with Iron. I will not tarry over the shapes in which these remedies may be administered—I am careless of the form so that I get the substance. Some forms may be better than others, but all are good, and all distance competitors so far, that their mutual rivalry is of no moment. It will be seen that in all the cases I quote resort is had to these. They are always at hand, always safe, and within the reach of all purses; which last cannot be said of the tonic I am now coming to.

I have spoken of alkalies as a temporary palliative in heart-burn. There is another way in which they are sometimes em-ployed with advantage, and which may be described fairly as their "restorative" use. I refer to their administration in a continuous course, the dose being taken rather at the times when there is no heartburn than when there is. The effect aimed at is that consequence of alkalies pointed out by Claude Bernard, the augmentation of the acid gastric juice, and so of the normal peptic powers of the stomach.

The test of benefit being derived from a course of alkali is the dose not requiring to be increased as the patient goes on taking it, but on the contrary being capable of being diminished

gradually, while relief from the recurrence of heartburn continues still to be experienced. This shows that the real health
of the stomach is being restored; that a renewed life is developed.

But should the patient be driven by recurring pain to take
larger and larger doses, it is evident that the palliation is simply
a neutralization of the normal acid of the gastric contents. This
induces in the end weakness and over-sensitiveness, and such
patients will in a very short time appear again under medical
care, usually in a worse condition with each recurrence; or they
will become chronic druggers for life, or perhaps be finally
cured by some clever quack who amuses their fancy while he
bids them abstain from active remedies.

Now I think that the due administration of a course of alkalies is best secured by the systems followed at certain Continental
spas. I believe their reputation as panaceas of all bodily ills
to spring from the renewal of the digestion which this mode
of administration of an alkaline course is likely to bring about,
and from the intimate connection which there is between the
digestion and all other parts of the body.

We must not let ourselves be prejudiced against courses of
mineral waters by the little packets of nonsense, brought by
post under various foreign stamps, which set forth in laughable
Anglo-French the omnipotence of their own "Abana and Phar-
par." Doubtless, like the lady in Hamlet's play, they do
" protest too much;" but there is some good in them for all that.
Patients get there something more than the over-puffed and
many-tested springs—Air, Rest, Gentle Exercise, New Diet,
Change of Scene, Freedom from Domestic Nagging, and perhaps from Domestic Physicking, &c. These would do good,
were the waters even moderately poisonous. But the fact is
they are not at all poisonous, and many of them contain not
only Carbonic Acid, which is a gentle normal anæsthetic to
mucous membrane, but also Carbonate of Soda, which rightly
administered has been shown capable of increasing the digestive
powers of the stomach. The right administration consists in
giving it, not when the organ is full, and ought to be acid, but
when it has parted with nearly all its contents, at a time as
distant as possible from the meals.

When, therefore, I wish to prescribe a course of alkali, I think a better plan cannot be devised than sending the patient to a spa containing that ingredient, and desiring them to arrange with the superintendent times for taking the waters in accordance with the above rules.

Of the two most famous alkaline spas, Vichy and Vals, I prefer the latter, because the water contains such a large proportion of Iron (retained in a state of Carbonate by the excess of Carbonic Acid), which assists much in the restoration of strength to the stomach. Moreover, Vals is further off, and in the neighborhood of beautiful and romantic scenery, enough to tempt the patients to a tour, and to help them to shake off the invalidism which associating with sick people at the spa is apt to induce.

In reflecting upon the purely medicinal, exclusive of the psychical, benefits conferred by spas, we must remember not to attribute all even of them to the salts contained in the analysis. Water itself is an important constituent of the gastric juice, and an augment to its quantity and power. It is a direct restorative agent, perhaps the most powerful of all drugs. No minutiæ, therefore, of its administration are frivolous, and the following additional hints as to drinking at the spa are not to be considered impertinent. Let the drinker reflect on the physiological action of temperature on the gastric nerves. Small quantities taken cold act as a tonic shower-bath, and remove local congestion; but if much is drunk at a time, a great depression of vitality follows. Let therefore the cold doses be small. But for exactly the same reason warm or tepid springs may be taken in considerable quantities, and a renewal of vitality be experienced.[1]

The local application to the epigastrium of variations of temperature is often of use to persons with a diathesis to heartburn, provided there is no catarrh of the gastric mucous membrane. It may be applied by the alternation of a cold and hot douche to pit of the stomach—in which case the alternations must be rapid, and the time of application short, say one minute or two

---

[1] I am told by travellers in Central America that the Indian porters preparing for a long, hard journey will drink several quarts of quite hot water. They say it gives them strength.

minutes each; or to the back, in the situation of the dorsal and upper lumbar vertebræ—when a very much longer application is desirable, namely, as long as the patient can bear it without losing the consequent reaction. In spite of the thickness of various tissues which lie between the skin of the back and the stomach, the dorsal douche is very efficient. When a regularly built douche bath is not accessible, a hydropult garden machine is a good substitute, or a slab of ice or ether spray may be ingeniously made to take its place.

I have been led on by the frequency of heartburn in cases of indigestion to pass from its special treatment into that of the general management of the patient. Perhaps the reader will have said that the last few pages would have been more in place at the end of the second chapter. It will be easy to suppose them transferred there if he wishes it. But yet I am not sorry to let it be seen how difficult it is to separate either in fact or fancy the two subjects, and to point out that the cure must not stop short at the arrest of a symptom, but must proceed to the renewal of active life in the organ whose sluggishness and deficient power of resistance to sensation is causing the symptoms.

SECTION II.

Acidity.

What is usually called "acidity" is the ejection from the stomach, with or without heartburn preceding, of a small quantity of sour fluid. There is no objection to the name, if it be remembered that it means merely acidity out of place or too long continued, and be not allowed to lead to our viewing the normally acid state of the gastric contents as an evil to be combated. The fault of doing that has been commented upon in the previous section.

It is not rare to find acidity misinterpreted even by those who ought to know better. I have heard it spoken of as an excess, "excess of gastric juice"—that is to say, too much of a vital act, too much life. Such a mode of speaking, if it leads to anything, must lead to faulty thinking and bad treatment. Instead of being an excess, acidity itself is a proof of deficiency.

CASE CVII.—A medical man complained to me a year or so ago of what he called "over-abundance of gastric juice." "Why do you call it over-abundance of gastric juice?" "Oh, because acid rises up in my mouth, and three or four hours after dinner I sometimes throw up my victuals so sour as to make my throat quite sore." "Well, now," I said, "observe what comes up, look at one of the pieces of meat in it, and you will see it hardly altered from the condition in which it was swallowed. But look at what a healthy person throws up when made to vomit, say by sea-sickness, four hours after a meal; it is all homogeneous, and the lumps of meat are quite broken up. If you really secreted an over-abundance of gastric juice, you would have dissolved your meat more quickly, instead of less quickly than the healthy person."

We know by experiments on artificial digestion, that an in-crease in the quantity of the solvent secretion quickens the solution of albumen. We find, for instance, that the amount of pepsine contained in twenty grains of Boudault's powder will dissolve a piece of hard-boiled white-of-egg much sooner than five grains. The same thing would of course happen in the stomach: were there more gastric juice, there would be quicker digestion. But in acidity such is notoriously not the case; the aliments lie for a long time in the upper part of the digestive canal, and often are passed still undissolved in the feces. It is a chemical act of decomposition directly opposed to the vital act of digestion.

I call a "vital" act any which forms part of the great circle of life, such as is the conversion in the stomach of albumen, previously incapable of solution and absorption, into peptone capable of entering the circulation. Now, when this vital act of conversion is carried on with rapidity by a stomach making abundance of gastric juice strong in pepsine, then chemical decomposition is prevented; nay, it is even arrested after it has commenced, as may be seen by putrid meat not becoming more but less putrid as it passes through the body of a healthy animal. But when the conversion is slowly or imperfectly performed, then the chemical change has time to take place, and does so very soon, being favored by the heat, moisture, and organic matter in a state of change. If the food remain too long without becoming chyme, the protein compounds putrefy with extreme rapidity under such circumstances.

The following simple experiments make the matter very clear

10

to yourself or a class. As far as his own improvement is concerned, the skilled physiologist may skip the next page or two
without loss.

Compare some hard-boiled white-of-egg which has been im-
mersed in saliva at the temperature of 100° Fahr. for a day,
with another portion from the same egg kept the same time in
distilled water. Your nose warns you of the difference directly;
the first is intolerably fetid, the second quite sweet. Exactly
similar is the fate of undigested albuminoid matter, whether
animal or vegetable, in contact with the mucous membrane
inside the body.

But how does that affect the case of acidity? Try another
experiment. Put in one beaker some syrup of grape sugar, and
it remains for hours quite neutral and natural. Set to soak in
some of the same a piece of putrefying albumen for a few hours,
and keep the mixture at the temperature of the body. You
find that a piece of litmus paper put in it is strongly reddened,
showing the copious formation of lactic acid. In another
beaker, the formation of butyric acid from fresh butter by the
same means may be shown.

Just so all the grape sugar and fat swallowed meeting in the
stomach or intestines with decomposing animal food, collected
in a mass or glued to the side by a too sticky mucus, ferments
quickly throughout, and forms lactic and butyric acids in great
quantity.

Remember, the grape sugar swallowed seems something much
more important than merely the grape sugar put in the mouth.
Take some boiled starch, and heat some of it with potassio-
tartrate of copper. There is no change in the blue color of the
salt. Now put some in the mouth, and hold it a few moments.
When it is again heated with potassio-tartrate of copper, the
metal is precipitated, and shows by its brilliant yellow color an
abundant quantity of sugar.

The saliva then begins to convert starch into sugar immediately; very soon it will transform the whole mass. A mouthful
of boiled starch held in the mouth for five minutes will show
afterwards scarce a trace of starch remaining. As, even amongst
wealthy meat-eating nations, from half to five sixths of the solid

food consists of starch,[1] it is evident that one of the most bulky contents of the stomach must be the sugar which has been made by the saliva out of amylaceous food. Here, then, is ample material for the formation of lactic acid to almost any amount. Add to this the oleaginous substances which it is impossible to avoid in any diet, and which from being insoluble in water turn into peculiarly acrid and concentrated acids, and you will have no difficulty in accounting for acidity, without recurring to a theoretical excess of gastric juice. Acidity, then, is an evidence of chemical, and therefore of decreased, vital action—a proof of incomplete digestion, of deficient activity, in the stomach.

There is nothing in acidity to contra-indicate the employment of acids as remedies. They are often most beneficial, especially if taken shortly before a meal. The best to select are those to which the digestive canal is most used, Hydrochloric or Lactic in plain water. The way in which they act is probably by neutralizing the alkline saliva and mucus which the slow digestion has allowed to accumulate in the stomach, and so setting at liberty the pepsine ; for in laboratory experiments it is found that saliva arrests the solvent power of pepsine in close proportion to its amount, and that by acidifying the mixture the action may be restored.

Neither is there anything in the use of acids inconsistent with a contemporaneous course of alkalies, so that, of course, they are not mixed immediately they are swallowed. They may each act separately with benefit on the mucous membrane and nerves, and then the sooner they neutralize one another into a salt and are got rid of the better.

I have indeed heard physiologists deride the idea of taking medicines for "acidity," as if it implied our ignorance of the fact of the stomach being normally acid, especially at its most comfortable times. The best answer to them is that we give medicine for acidity of *mouth*, which they cannot assert to be normal. It is a superficial answer, but the objection is super-ficial also. Of course in reality a man is not a retort or a test-tube.

[1] See the dietaries of soldiers, prisoners, laborers, and others, analyzed by Dr. Hildensheim in " Die Normal-Diät," p. 6. Berlin, 1856.

Some patients will perhaps think that their physician is blowing hot and cold, or rendering inert his own treatment, by ordering acids at one time and alkalies at another; and he will find it a wise plan to give an educated person a short physiological lecture on the subject, explaining the reason of his procedure. He may explain also that the acids given as medicine do something more than in the laboratory; they stimulate the mucous membrane, and so actually increase the quantity of secretion while they intensify its power. There need be no fear, which I have heard some express, that the use of these substitutes for the natural constituents of the gastric juice, or rather the supply of that which ought to exist in the gastric juice, will teach the stomach to be lazy—as doing a servant's work for him makes him less equal to doing it himself. On the contrary, the new vigor put into the system by the healthier and more copious chyme that is formed, renders the organ more active; so that it soon is enabled to go on secreting for itself what is wanted, and to do without the artificial substitute. If patients derive benefit from it, they will be able soon to leave it off.

There is a singular febrile disease mentioned by M. Chomel in "Les Dyspepsies" as an acute acid dyspepsia, in which the whole body, in point of fact, turns sour. As might be ex· pected, it seems to be invariably and rapidly fatal. I confess I do not recognize it from his description. The case he gives seems more like one of pyæmia than anything else in my experience. But its extreme rarity takes away most of the interest which would attach to it, and its incurability the rest.

SECTION III.

Waterbrash

(or Pyrosis) has so far a similar local pathology with heartburn, in that the manifestations of the phenomenon is in the œsophagus. There is, however, this difference, that the spasm which there seems to be subjective only, is here exhibited as a muscular contraction. The tube is closed by it, and the passage into the stomach of the saliva trickling downwards is pre·

vented, so that it collects in considerable quantities, and gushes up into the mouth without any effort of vomiting. The fluid in its pure state is therefore alkaline, and exhibits under the microscope no other formed contents except the buccal, faucial, and œsophageal epithelium.

In his valuable monograph on the diseases of the stomach Dr. Handfield Jones has represented waterbrash as a watery catarrh of the mucous coat of that organ, analogous to bron-chorrhœa, for example. Now, if that were so, it would contain gastric and not salivary elements; and it would also be filled with mucous globules, as the flux of nasal or bronchial catarrh is. It would also always be ejected by a distinct effort of vomiting and nausea, which is the case only when the contents of the stomach are mixed with it.

The alkaline nature of the fluid of waterbrash, contrasted with the acidity of ordinary regurgitations, has been made by some a groundwork for a primary division of indigestions into acid and alkaline. It will be seen by the following pages that I should consider this an arbitrary and artificial division, with-out practical utility or basis in nature. Acid and alkaline regurgitations are often found on the same day in the same person; so, if the stomach were a mere alembic, they would cure one another. But it does not answer to treat of a living body, as if it were a chemical laboratory.

The following case exhibits the principal features of the dis-ease:—

CASE CVIII.—Margaret S—, aged 22, an Irish maid-of-all-work in a small tradesman's family, stupid, ignorant, and bowed down in spirits, applied for advice at St. Mary's, April 13th, 1855, for what she called " sickness at heart" (*Hibernicè dictum*) and " vomiting." She had been ill about two months, during which time what she called "sickness" had occurred daily. She looked in pain, and pressure on the pit of the stomach showed it to be tender when pressed with one finger's point in the cardiac region, though the flat palm laid on the spot caused no uneasiness. This pain was also increased by eat-ing, especially potatoes, bread, and tea, of which her diet chiefly consisted. She was admitted as an in-patient, and then we had the opportunity of ob-serving that what she called vomiting had not really that character. She used to have some eructations two or three hours after meals. But this in-convenience mostly occurred at night, ceasing towards morning with the depth of sleep. When she began to move about for the purpose of rising a sudden gush of fluid would come into the mouth once and again, but seldom or never a third time. There was no retching or effort, and no marked sensation of distress or of relief. The quantity was seldom more than five to six ounces. Preserved in a vessel, it was colorless, slightly opalescent, alkaline, and a little adhesive, like thin saliva. Under the microscope it exhibited large pavement epithelium and a few granular globules. Later in her residence in the hospital it was sometimes found less transparent and acid, as if some remaining contents of the stomach were mixed with it. The tongue was red in the centre, with white coated edges; the catamenia had been irregular for some months.

She had had a similar illness the previous year, but had recovered by rest and medicine. Both attacks she attributed to hard work and bad food.

She was treated with Mustard poultices and Leeches in several relays, at first four being applied, and then three every other evening, for a week or ten days, on the tender spot of the epigastrium, with 15 grains of Bismuth thrice a day for ten days; then with Iron pills and Shower-baths. Her diet was principally broth, with milk and Lime-water.

The Leeches and the Bismuth seemed to relieve the eructations and the pyrosis, but the cardiac pain remained till she got to Shower-baths and Iron.

Perhaps the most conspicuous effect was that which is to be credited to the diet, for she gradually gained twenty-one pounds in weight, advancing from 7 st. 11 lb. to 9 st. 4 lb. between the 27th of April and the 8th of June, when she left the hospital well.

Observe that the loss of blood by leeches did not prevent her gaining blood and flesh by the improvement of her digestive powers. The local benefits to such an important organ as the stomach more than counterbalanced the inevitable abstraction of what, truly enough, she could ill spare.

Some may cry out against such treatment as inconsistent. It is feeding up the patient with one hand and robbing him of his pabulum vitæ with the other. The reproach is just in a certain sense, but that a very limited one, and it may be levelled against half the operations of daily life. We are constantly suffering a small loss for the sake of greater gain— " *necesse est facere sumptum qui quærit lucrum.*" And I reckon the absence of a little blood as of no moment at all compared with the advantage of securing freer circulation or diminution of congestion in the alimentary canal. Do not let us be led away by the superficial notion that blood is blood, and blood is life. That is not true, for blood varies immensely in its composition, some being very valuable, and some worthless. To lose a portion of his imperfect circulating fluid is but little loss to an invalid, and that little loss is amply repaid by the additional nutriment which a more rapid blood-stream will enable him to absorb. The deficiency is soon made up under the restorative plan of treatment.

Where there is localized pain in one spot of the epigastrium produced by pressure at all times, I take the pathological condition to be some local change producing at least sanguineous congestion of the veins, if not tissue-thickening, at that point. Observe the importance in investigating such cases of using the ends of the fingers, and not the flat palm, otherwise the phenomenon may be passed over. It is, perhaps, needful to say that the pressure exerted should be steady and moderate. It is easy to try on yourself what amount a healthy epigastrium ought to endure.

Observe the use of Shower-baths. Their effect is first to drive the blood inwards from the skin, then by nervous reaction to draw it out again. Thus mechanically the capillary circulation is quickened and continues quicker—*vires acquirit enudo.* Compare what was said (page 73) about the increase of osmosis through membranes in the direction of increased current.

It is remarkable how in waterbrash a separation is effected by the sphincter extremity of the œsophagus between that tube and the stomach, and what a barrier it places between the two. There is no particular evidence of this in waterbrash which occurs when the stomach may be empty of food, in the night or early morning, as is most usual. But sometimes it comes on at or immediately after meals, and then much surprise is caused to the uninitiated by seeing nothing of that which is swallowed brought up again.

CASE CIX.—Last August a retired surgeon, aged 64, consulted me for asthma produced by emphysematous lungs. Latterly, also, his digestion had troubled him a great deal; he had, after exertion, pain at the scrobiculus cordis, which he attributed to the diaphragm overstrained by his dyspnœa, but it seemed to me more in the stomach. He had also waterbrash occurring immediately after, sometimes even during meals. He was obliged to leave the room and throw off several ounces, as much as five or six, of frothy, clear, cold-tasting fluid. Although this sometimes made him retch, yet the contents of the gastric cavity were never mixed with it. I gave him Quinine and Strychnine with apparent advantage.

Its being frothy arose from the nearness of the period of its

secretion.   When it has rested in the œsophagus a few hours it
becomes quite bubble-less, as in matutinal waterbrash.

In the next case a few more details of variety in the symptoms
are given, and a pretty good original name for the disease was
invented (I believe) by the patient.

CASE CX.—Mary F—, a widow of 60 years of age, had always enjoyed
good health, and supported herself comfortably as a market-woman till she
broke her arm in crowding to see her son off in a transport for the Crimean
campaign.   This was in January, 1855, and she was thus naturally stricken
down in body and mind, and was almost starved, eating nothing but ill-cooked
vegetables.   In May she heard of her son's death, and this was the final blow
to her health.   The flatulence and pain which she had frequently felt at the
pit of the stomach became more constant, and she experienced a sensation of
coldness there.   She often found her mouth suddenly filled with a "jet" of
watery fluid, a symptom which she called "watery mouth."   She could swallow the fluid by a voluntary effort, but the doing so was often followed by
retching and actual vomiting of the contents of the stomach, smelling sour
and tasting acid, but small in quantity.   Often in the downward passage of
this or of anything else she felt a resistance as of "a ball in the throat."
The greatest quantity of fluid was brought up on getting up in the morning,
when it sometimes amounted to half a pint; but "watery mouth" occurred
at all periods of the day, and sometimes immediately after meals.

The tongue had a white coat on the edges, and was clean in the centre.
The bowels were costive.   The urine was slightly acid, of low specific gravity
—1.012, 1.011, are the numbers recorded in the case-book of the mixed urines.
Her appetite for both food and drink was quite gone, and she felt an especial
aversion to animal food.

She was admitted to St.'Mary's under my care July 6th, and discharged
well on August 17th, 1855.   She was treated with rest, Bismuth, two pints
and a half of milk with a pint of lime-water daily, and a graduated approach
to animal food.   After eating she was to take ℥ss of *Mistura Ferri Com-
posita* (*Pharm. Lond.*).   By the 16th of July she proposed to eat a whole
instead of a half mutton-chop which had been ordered.   On the 20th " no
complaint whatever" is the report, but then she had a relapse, and ejected
some more fluid, and also some rancid oil.   She was then ordered Carbonate
of Potash and Infusion of Gentian, on which she improved slower but
steadier.

Pains in the epigastrium, darting through the chest, are alluded to in my
notes, but are not particularly described.

It has been mentioned that waterbrash is sometimes called
vomiting by the patient—"easy vomiting," or "retching of
spittle."   It is so even when it is evident that the œsophageal

disease is an obstruction of a permanent character preventing the passage downwards of the saliva. As for instance in the following cases.

CASE CXI.—Francis D—, a laborer, aged 57, was admitted to St. Mary's December 28th, 1852. Four months previously he first began to experience discomfort in eating and nausea. Often the first few mouthfuls swallowed would be rejected, after which he would be able to finish his dinner. He complained also of pain in the centre of the sternum, running through to the back, which kept him awake of nights. He stated also that he frequently "vomited," but the matter thrown up was found to be rejected with slight, if any, effort, and to consist of clear alkaline fluid, frothy at top. He stayed under my care three months, now better, now worse, sometimes relieved by Bismuth and gaining a few pounds in weight. But the pain and dyspepsia were not cured, and were considered, probably correctly, to be due to ulceration of the œsophagus.

CASE CXII.—Mary S—, aged 69, was admitted May 25th, 1855, for difficulty of swallowing solids. The dysphagia seemed dependent on two obstructions, one felt at the top of the sternum and the other at the tip of the ensiform cartilage. She complained also of "vomiting," but what she threw up was found to consist of diluted milk in an alkaline condition, that is, diluted with an alkaline fluid, and evidently recently swallowed, or else of saliva. Yet she called it vomiting, and certainly seemed to retch with it. It was never more in quantity than five ounces. Several remedies were tried without effect, and she left on June 8th, discontented at not receiving an immediate cure.

The association of an irritability of diaphragm, exhibited in vomiting, with an irritability[1] of œsophagus, exhibited in the spasm of waterbrash, is again shown in the following case.

CASE CXIII.—Mary Ann F—, a carpenter's wife, aged 52, was admitted to St. Mary's August 4th, 1854. She had been in the habit for some time of taking her meals very hurriedly, but previous to that she seemed to have suffered from various forms of dyspepsia, originally due probably to wearing tight stays as a girl, for her chest is very much contracted by that compression. The last five weeks she had found pain, followed by vomiting, come on about an hour to an hour and a half after taking food. The vomiting relieved her, but if it did not occur, she had for the rest of the day a painful feeling of weight at the epigastrium. At various times of the day also, unless relieved by vomiting, she found clear water rise into her mouth, which was usually tasteless, but sometimes had a bitter flavor. From the frequency of the vomiting she had become much emaciated. The catamenia had ceased naturally two years before.

[1] I use the word, not as explaining anything, but simply to fix the locality of the vital act.

The rest of the hospital and well-prepared food stopped the sickness, so that we saw nothing of it for some time; but she had several attacks of water-brash of clear alkaline fluid. Afterwards the vomiting returned, and sometimes was mixed with the pyrotic fluid, and sometimes contained strings of gastric mucus.

She was treated at first with this pill—

R.—Pil. Rhæi comp. gr. v,
Argenti Nitratis gr. ¼.
Omni nocte et mane.

But in a fortnight she felt very little better. She then took—

R.—Ferri Sesquioxidi gr. x,
Bismuthi Trisnitratis Əj, ter die.

On this she got well and left the ward on September 4th.

In the last three cases it will be observed that a great part of the motive cause of the illness may be fairly assigned to the innutritious nature of the patient's diet. Its innutritiousness arises principally from its insolubility, and that insolubility principally from bad cookery. But yet this cause was not sufficient to produce disease in a healthy body; there was always superadded some depressing influence on the vitality, of either a mental or physical nature.

This is to be noticed even in the waterbrash of the Scotch oatmeal-eaters, where the dietetic cause is so constantly the same, and so powerful as to establish the disease as an endemic.

Dr. Morgan, of Manchester, who formerly practised for a short time extensively in the Western Highlands and Islands, in a letter to me on the subject, says—

" Cases were so similar in their leading characteristics that after seeing some three or four all others were but a simple repetition of symptoms. The history of a typical case was something of this kind. From some cause or other *the vitality of the system in an oatmeal-eater was lowered.* Thereupon the customary diet, whether in the form of cake or porridge, proved a source of irritation ; the patient then lost flesh, and complained of a sense of burning heat in the epigastrium and along the course of the œsophagus. Coincidently with these symptoms considerable quantities of water (a pint or more) "came up" rather than were vomited. . . . Women seem to suffer to a much greater extent than men."

I believe the same is the case still more strongly with the potato-nourished peasantry of Ireland, whose more sensitive nervous system renders depressants of the vitality more common. But I have no written records of the fact.

Dr. Morgan goes on to attribute the innutritiousness of the diet to the form of preparation:—

" I always considered that the eating oatmeal in a semi-cooked state had much to do with it. As a rule, the people do not sufficiently boil the porridge, while in the form of oat-cake the food was still less thoroughly prepared. In using an oat-meal diet I believe that it is very important to carry the cooking sufficiently far to liberate the contents of the starch-granules. Where this is not done, not only does the food fail to nourish, but it proves a source of gastric irritation. . . . If oat-meal is boiled for about half an hour it is, even though coarsely ground, reduced to a gelatinous mass, and in this form it is comparatively innocuous so far as existing pyrosis is concerned —at least such is my experience."

Seeing the importance of cookery, the occupation of the next patient ought to have preserved her at least from the results of bad art. But the occupation may be baneful, as well as its products.

CASE CXIV.—Elizabeth P—, a cook, aged 26, came under my care at St. Mary's July 30th, 1852, for pain in the epigastrium, increased after meals and by pressure. She had also frequent attacks of morning waterbrash. Her tongue was white, but otherwise she seemed in good health; the catamenia and the evacuation of the bowels were regular. She was blistered on the scrobiculus cordis, and took fifteen grains of Bismuth three times a day and broth diet. On the 7th of August she was well enough to eat a mutton-chop. On the 9th the Bismuth was left off, as the local symptoms were relieved; and on the 18th, there being no return of waterbrash, she was discharged.

The exposure to heat involved in the occupation of cook produces general congestion of the portal system, and an after-exposure to cold draughts inclines to a catarrhal condition of the stomach. Hence arise slow digestion, oppression at the epigastrium, and a feeling of faintness, which often leads to dram-drinking. A further stage, more certain if this desire for alcohol be indulged, is pain immediately after food, and then pain on pressure.

Observe the use of external local treatment. That was resorted to in the case of Elizabeth P— because there was pain on pressure as well as after meals. The pain on pressure is an evidence, though not truly an absolute proof, of the existence of anatomical lesion, either continuous congestion or thickening or ulceration. And I find that where it exists local blistering does good, and leeching more good. Even when it is not made out in a clearly defined spot, I am still inclined to suspect in waterbrash such a condition of tissue as is capable of being renewed to a more normal one by the alterative action of counter-irritants. For so many cases occur, like those which follow, where waterbrash is associated with indubitable signs of local lesion.

In this, for instance, there was not only the peculiarity of the pain running backwards to the spine, but also a blood-stain in the mucus to show a solution of continuity in the capillary bloodvessels.

CASE CXV.—John N—, a painter, but without any signs of lead poison exhibited in the gums, aged 35, was admitted to St. Mary's April 10th, 1855, suffering from waterbrash, sometimes of a sour character, and sometimes alternating with vomiting of intensely sour greenish liquid. After he had been in the ward a few days it was observed that the vomit contained tawny mucus like that expectorated in pneumonia, and sometimes streaks of blood. He had also pain running backwards from the pit of the stomach to a spot between the shoulders, which pain was increased by pressure of the finger on the cardiac region. When he was at his worst the waterbrash was least marked, but still it was a feature of the disease. Pepsine gave no relief. Hyposulphite of Soda was tried without benefit. A blister to the epigastrium made him better for a couple of days after it, but he then relapsed. Most advantage seemed gained by the application of a few Leeches to the epigastrium. He became an out-patient on May 13th.

In the next the blood evidently came from an ulcer.

CASE CXVI.—Sarah G—, aged 33, a housemaid, was admitted at St. Mary's, under my care, August 22d, 1857. She had been an out-patient with unaccountable languor and anæmia, which was at last detected to arise from loss of blood by the alimentary canal. After admission it was found she had also waterbrash, and pain on pressure of the pyloric region. And then the locality of the injury in the stomach was fixed by her vomiting blood, both red and brown. The hemorrhage was stayed by means of Acetate of Lead and Opium, and then the waterbrash seems to have got worse. It was considerably relieved by iced milk and by Bismuth, but more by a blister. A grain and a half of Sulphate of Copper daily, which was given for a fortnight, seemed to act as a tonic and enable her to digest better, quicker, and with less pain. She was still taking it when she was made an out-patient October 16th.

It does not appear that the fluid ejected by the brash was ever bloody, thus showing that it does not come from the stomach, as sometimes represented, but from the œsophagus.

I have never tried Sulphate of Copper in simple waterbrash without hæmatemesis, but its beneficial action in this case would seem to offer an encouragement for doing so.

In the history of John N— (CASE CXV) it is mentioned that he was a painter. Though no blue lime in the gums denoted the still presence of lead in the body, I am not sure that we can quite acquit that subtle poison of causing the disease. In the following instance the accusation was brought by the patient.

CASE CXVII.—Mr. Edwin S—, aged 30, a master painter and glazier, came to me July 7th, 1862. He suffered from excessive waterbrash, bringing up sometimes upwards of two pints of clear fluid in the course of the night and early morning. Sometimes this was relieved by vomiting. The matters vomited were acid and frothy, and continued to ferment and swell after being brought up. I had no opportunity of searching them for *sarcina ventriculi.* He had also often heartburn about two hours after eating. His tongue was unnaturally red and clean.

These evils, he said, were always much aggravated by anxiety in business, and it was for such aggravation that he consulted me. But he had suffered in the stomach more or less from boyhood, when he used to work with lead paint.

I gave him half a drachm of Hyposulphite of Soda daily, and fifteen grains of Bismuth every night. In a few days with the medicine and rest he was better, and I prescribed some Iron next with Bismuth. I had no opportunity of seeing more of the case, as his family doctor did not send him again to me.

The Hyposulphite of Soda was administered as an agent to prevent fermentation. I cannot say whether it was effectual or not in this case, as the patient did not vomit afterwards; but I have thought in others that it seemed to effect its intended purpose. That purpose, however, must be well understood to be a very limited one, for it does not cure the cause of the fermentation, namely, the slowness of digestion which retains the contents of the stomach so long as to ferment and communicate their fermentation to new arrivals. This cure must be effected by invigorating the vital energy of the failing organ.

The violent shock to the vitality of the mucous membranes in cholera will sometimes leave behind it a condition of stomach productive of waterbrash.

CASE CXVIII.—Joseph W—, a laborer, aged 42, admitted to St. Mary's October 27th, 1854, had gone through an attack of choleraic diarrhœa in

August, and since that time had not digested his food properly. The epigastrium was tumid and tympanitic on percussion. The tongue was large, flabby, and red, as if flayed. For the last three weeks previous to admission he had suffered from attacks of waterbrash. He was treated with gr. xv of Bismuth three times a day, but was not considered ill enough to remain as an in-patient beyond November 1st, so that I probably saw him only once.

An operating cause of similar nature in dysentery.

CASE CXIX.—Mr. Henry M—, a man of middle age, had several attacks of dysentery in Australia, and has never been quite strong since. He suffers from diarrhœa from the slightest error in diet. It was one of those attacks, brought on by taking a cup of bad coffee at a coffee-shop, that induced him to consult me. I gave him Sulphate of Copper and also Bismuth, which both he said had done him good before. On inquiry I found that he very frequently suffered from waterbrash in the morning and during the night, though very careful of his diet. He traced this to the dysentery, and both to spirit-drinking, which he felt sure predisposed people to dysentery in Australia.

It is also sometimes associated with phthisis pulmonalis, and then the defective nutrition which it implies brings on a condition of general degeneration. This is important from the possibility which exists of staying the degenerative tendency, and so arresting the downward course of the phthisis by attention to the stomach and œsophagus.

CASE CXX.—William J—, aged 21, a carpenter, on admission to St. Mary's, August 21st, 1857, was much emaciated, and presented indubitable signs of solid tubercle in the upper lobes of both lungs, of such duration as to have made the upper ribs flat and immovable. The date of his consumption, from the period of his having "caught cold" and spat blood, was two years. Latterly he had suffered from waterbrash of a morning. It was difficult to make out whether he had pain in the epigastrium, as there was stitch in both sides of the waist, which had its origin in the pulmonary disorganization.

After a few days' Cod-liver oil and Iron the albumen disappeared from the urine, and then the patient began to gain weight. Between the 28th of August and September the 5th he increased 2 lb., and by the 12th 1 lb. more. The extent of further increase is not noted, but he was bettered enough to leave the hospital on October 3d.

The disappearance of the albumen from the urine shows that the derangement of the kidneys was only temporary. But in pulmonary consumption we find such temporary derangements soon end in permanent disorganization, if allowed to become ingrained.

It may be observed that in several previously quoted examples the supervention of waterbrash has been at the period of the normal cessation of the catamenia. It is also apt to follow upon such states of body as cause the arrest of the periodical evacuation in younger women.

In the following case there was also joined an occupation which, as we have seen, tends to produce derangements of the upper organs of digestion.

Case CXXI.—Ellen R—, a cook, aged 82, admitted to St. Mary's November 4th, 1856, had been getting ill gradually for some months, at first suffering from feverishness, headache, and constipation, then finding her monthly periods arrested, though she still had leucorrhœa and pain in the back at the time when they ought to appear. The last-arrived symptoms were a dribbling of saliva from the mouth, and on rising in the morning a gush of clear watery fluid from the œsophagus. This fluid was sometimes made acid by the admixture with it of some of the contents of the stomach ejected by vomiting. She once also, while in hospital, threw up some greenish fluid (? altered blood).

She was treated with Bismuth and ultimately discharged well.

Green vomit may arise from the admixture of bile which has regurgitated through the pylorus. This only happens after violent retching and straining, and the bile may be recognized by its bitter taste. It may also arise from the admixture of blood altered by the gastric juice like the porraceous stools of dysenteric babies; and in such case there is likely to be very little straining and no bitter taste. The notes are not full enough to decide of which nature Ellen R—'s vomiting was, probably the latter, as bilious vomit is rarely joined to waterbrash. Indeed, bile is seldom thrown up in chronic diseases, and appears rather a guarantee of a considerable amount of health.

In nearly all the cases I have quoted waterbrash has occurred in young or middle-aged persons. And perhaps this fairly represents the habits of the disease. Yet it is not unknown in the old, as the following instances will show, and show also that its cure is not to be despaired of even in them.

Case CXXII.—In May, 1848, R—, a farmer, came under my care for waterbrash, from which he had lately begun to suffer. He had also occasional attacks of vomiting. His age was about 70. He got well on Bismuth. I saw him again in 1851, for some pain in the pyloric region of the stomach without waterbrash. However, there was no cancer, for I recollect seeing him several years afterwards in the streets at the time of the cattle-show.

CASE CXXIII.—Mrs. B—, aged 66, was under my care in July, 1861, for waterbrash, accompanying indigestion brought on by anxiety of mind in nursing a consumptive son-in-law.

CASE CXXIV.—Mrs. A—, aged 60 (but older than her age reckoned by annual revolutions of the sun, for the catamenia had ceased eighteen years), consulted me in August, 1863, for indigestion, marked by waterbrash occurring at various times of the day, not confined to the morning.

CASE CXXV.—Mr. Thomas S—, aged 72, was sent to me by Dr. Ellison, of Windsor, January 26th, 1867, respecting severe pyrosis which had on the present occasion afflicted him upwards of six weeks. It occurred daily, and was getting worse; he had several times brought up as much as two and a half pints in twenty-four hours. The sensation was described as of something working in the right side of the epigastrium, and then there was suddenly and forcibly ejected a great jet of liquid. It was almost always quite clear and tasteless, and resembling saliva in appearance. There was often pain at the cardiac extremity of the stomach half an hour before the supervention of the brash, by which it is alleviated.

Mr. S. was a robust man of rubicund healthy appearance, but he said he had been subject for more than a quarter of a century to this waterbrash from time to time; never, however, so bad as during December and the January when he came under my care.

I desired him to come again on the 30th, and to bring with him some of the pyrotic fluid collected. In the mean time I prescribed—

> ℞.—Quiniæ Sulphatis gr. iss,
> Strychniæ Hydrochloratis gr. ₁⁄₁₆,
> Succi Limonum q. s. ad illa solvenda,
> Aqua ad ℥viiss,
> Sps. Rosmarini ℥ss,
> ter die sumendus.

On the appointed day he returned, but without the expected morbid specimen. During the first thirty-six hours he brought up a good deal, and he thought there would be time enough to collect it afterwards. But after the fifth dose of the medicine a great explosion of flatus occurred, and continued for several hours. Since then his disease had quite disappeared.

I hear to-day (April 3d) from his son-in-law that Mr. S— has continued quite well. He has continued taking the medicine irregularly from time to time since I saw him. So I say I think he has had quite enough of it.

The following case, on the other hand, is exceptional from the youth of the patient.

, CASE CXXVI.—Miss S—, aged 15, an undergrown girl, was in my hands in July, 1858, for waterbrash accompanied by a feeling of oppression at the epigastrium occurring when the stomach was empty. and relieved by meals. She was weakly, and retained the insignia of former ill-health in the shape of scrofulous scars in the neck.

In this last example mention is made of the relief which some persons affected by waterbrash experience on taking food. This so frequently occurs in heartburn, and so rarely in ulceration, that I am disposed to view it as an evidence that the waterbrash does not, where it is found, depend on any serious anatomical alteration of tissue.

I have never seen it amount to "bulimia." The patients went to eat often, but they are not often hungry, and they do not want to eat much. I cannot recognize the truth of the ' statement made by some writers, that indigestion leads to bulimia, as I understand the term.

It is a relief which may be prudently allowed to them, so that care be taken that what is eaten be easily digestible. Indeed, a judicious management may turn it to a means of cure by preventing the overloading of the stomach—by "spoiling the meals," as it is technically called.

The treatment of waterbrash has been almost sufficiently detailed in the histories given. It consists of sedative alkalies, and the best are those which lie longest undissolved, such as Nitrate of Bismuth. I give this in doses of from ten grains to half a drachm, either alone in a powder, or in a draught with Carbonate of Soda and Hydrocyanic Acid. The Soda I give where there is much acid rising, the Hydrocyanic Acid where there is local pain on swallowing or on pressure.

In the robust ancient described in Case CXXV it will be seen that Quinine and Strychnine unexpectedly alone effected a cure. They should therefore not be omitted as part of the treatment at least.

Iron is useful in anæmic cases. The red rust goes very well in a powder with the Bismuth, and may perhaps render any other tonic superfluous.

Kino and Opium powder is also a good astringent to the upper part of the primæ viæ, and hardens the over-sensitive nerves of '

11

the œsophagus, but I cannot lay hand on any cases in which I have used it alone.

The local application of Leeches and Blisters must depend on our diagnosis. They are of use in those numerous instances where there is pain on pressure elicited by the finger rather than the palm of the hand—not otherwise, I think. The water-cure by compresses usually does harm; it renders the part more sensitive, and the local application of cold is too depressing.

In food all insoluble matters (such as those consisting chiefly of cellulose, chlorophyll, and raw starch), waxy potatoes, peas and beans, cucumbers, sodden pastry, new bread, half-cooked porridge (according to Dr. Morgan in the letter quoted page 155), and the like, must be avoided. Fresh meat-broth, beef-tea, milk guarded with lime-water, must be the food trusted to. I have found raw oysters well borne; but they must be quite fresh and well chewed.

SECTION IV.

Spasms,

or " *The* Spasms," or "Stomach-ache," as sometimes called, is a peculiar pain, resembling that felt in cramp of the voluntary muscles, extending across the epigastrium. It remits from time to time, but does not intermit like heartburn. Though the pain resembles that of cramp, there is no evidence of any muscular contraction; indeed, examination of the epigastrium shows the stomach distended with a more than ordinary amount of solid matter and air. This pain arises in a stomach rendered atonic, either temporarily by some depressing agent, such as heat or fatigue, or more permanently by general debility, when the organ is filled with some insoluble matters. It is a con-dition analogous to the over-distension of the bladder with retained urine, in which the pain has a similar remittent character. It usually commences from five to six hours after the food which has produced it, becomes gradually more intense, and passes off either by the insoluble mass getting through the pylorus or by vomiting.

The seat of the pain is not easy to fix when it is severe. In the lighter cases, and as it passes away, it seems to tend towards the pyloric sphincter, and to become located there. The very muscular nature of that part may serve to explain the tendency of this, more than any other gastric pain, to be associated with spasms of voluntary muscles.

Other characteristics will be sketched in the cases which follow.

CASE CXXVII.—In the summer of 1842 the writer started, without his breakfast, early in a row boat from the top of the Lago di Como. He was out in the sun without food till noon, when he bought his hatful of hard peaches and little green figs, and finished them at a sitting. In the afternoon pain across the epigastrium gradually came on, but still he ate his dinner. That seemed to ease the pain for a time, but it came on again worse and worse in paroxysms, just like cramp. He travelled on in an open carriage from Como to Milan, but the pain was very bad. On the road he vomited, not his dinner, but the skins of the figs and the peaches in the state in which they were swallowed. After arriving at Milan at midnight he vomited again, this time the dinner eaten in the afternoon. The masses of food had therefore got reversed in their position in the stomach. Soon after the second vomiting the spasmodic pains abated and ceased with sleep.

CASE CXXVIII.—The same party in the following year lunched on a dozen or so of pears at Leipsic after a hot dusty journey from Dresden. Again the pain was relieved by dinner, but returned afterwards. An emetic of mustard-and-water gave relief rather sooner than waiting for the spontaneous evacuation of the stomach.

CASE CXXIX.—The same party, when not in very strong health this spring, committed the imprudence of seeing a troublesome patient before breakfast. At noon pain in the epigastrium came on, was relieved by a mutton-chop at lunch, returned worse an hour afterwards. In the evening vomiting was induced, and the first things that came up were the toast and water-cress eaten at breakfast. With sleep the attack passed off, but the epigastrium still remained abnormally tumid and resonant on percussion.

I promised in a former chapter (page 97) to introduce another case in which gluttony was an act of virtue, if not of heroism— here it is.

CASE CXXX.—Mrs. D—, aged 50, sent for me one afternoon this spring of 1866, to see her a few miles down the country. I found her slowly recovering from an attack of "spasms of the chest" (epigastrium), which had lasted twenty-four hours, leaving the epigastrium tumid and drummy on percussion. She had passed one small light-colored pultaceous stool, so I gave her a Rhubarb and Peppermint draught to elicit another or two.

I directed my principal attention to discover the exciting cause of the stomach-ache, and believed that I rightly fixed on a large cold early dinner, accompanied by a quantity of salad and cucumber. I gave a warning against this, and then went to the village hard by to see an old patient, a poor cousin of the one who had summoned me, making my visit to the wealthy relative an excuse for not taking a fee.

A short time afterwards I received a second summons; found Mrs. D— had another attack of spasms, had vomited, and was better. But what was my surprise to see in the basin a large quantity of cucumber, against which I had given such a strong warning. I found reason for believing that this apparent act of gluttony was committed as an excuse for getting me to see the less fortunate neighbor again. The vomiting made the attack pass over quicker than the former one, and no purgative was required. The contents of the basin were very slightly acid.

Mark the last sentence—it is clearly not acid, but distension, that causes these pains.

The preceding cases have exhibited spasms arising after a heavy meal of insoluble matters taken at an unnatural hour. When the atony or paralysis is induced by dinner, then the attack assumes a different and to the patient a more alarming character. He is woke up by it in the early morning, and I think it is usually more sharp and severe.

CASE CXXXI.—E. N. S—, an energetic but not strong business-man of middle age, has had several of these morning attacks, which he can always trace to a dinner of insoluble matters after an anxious day's work. He is always a good deal alarmed at the time, but they pass off in the course of the forenoon, either by vomiting or pultaceous stools. The matters vomited I have never seen, but he says they retain the taste of food, and are not acid or fermenting.

In whatever parts it may occur, atony, or defect of voluntary and normal action, has a tendency to alternate with involuntary and abnormal action. It is when the legs are tired with over-walking that they are apt to be racked with cramp; it is when the whole body is debilitated and incapable of designed control that it is agitated by chorea.

The atony of purely involuntary parts most generally produces these contractions, not in themselves, but in the neighboring systems of muscles, subservient in most cases to the concatenated acts of the said involuntary parts. Thus, over-distension of the bladder causes stricture in the urethra; over-hard and bulky feces bind up the sphincter ani.

I have already pointed out how chronic slowness of gastric

digestion first is evidenced by heartburn, which appears to be relaxation, and then by waterbrash, dependent on contraction of the œsophagus.

Just so acute atony of the stomach, or stomach-ache, will sometimes produce cramps of the abdominal parietes, even while it is itself distended and palsied.

CASE CXXXII.—At the end of last June I was requested by Mr. Paget to see a young man of nineteen, reported to have cholera. The history I found to be this. He had been working hard at inorganic chemistry, to prepare for an examination, torturing the metals and himself with repeated tests, not shrinking from exposure to Sulphuretted Hydrogen, and, what was still worse, not caring if the laboratory stank of Arsenic. Then came the examination. The evening before it he came home tired and anxious, but ate a good dinner, probably with the false appetite of intellectual toil. He went to bed and to sleep, but was awoke before sunrise by a spasmodic pain in the epigastrium, not increased but rather lightened by pressure. As this got worse he tried to ease it by forcing an action of the bowels, but his success brought no relief to the pain. Breakfast made him somewhat better, and he went to the laboratory. But by eleven o'clock he got so bad that he was driven home and went to bed. In the afternoon cramps came on in the abdominal parietes, and could be alleviated only by the constant rubbing of two sturdy housemaids. These cramps extended from time to time into the legs and arms. The epigastrium was tumid and drummy on percussion. His face was shrunken, pale, and livid, the eyes leaden and anxious, the pulse small and extraordinarily quick (nearly 140 in a minute), the skin cold and clammy. Indeed, I did not wonder at the household calling it cholera. But I was comforted by finding no vomiting or diarrhœa, and by seeing a fair quantity of full-colored urine in the chamber utensil. Towards sundown the pain gradually abated; he had a pultaceous stool, went to sleep, and when I saw him again the next morning was well, though he said his belly was very sore after the cramps. The pulse had sunk to 80.

These cramps have, I suppose, the same pathology as those which so generally accompany the collapse of cholera. In that disease they are developed in consequence of the whole alimentary canal, but more especially the ilia, being devitalized and paralyzed by an extraneous poison; whereas here the stretching of the stomach by the overpowering weight of a mass, which it cannot move on, is the paralyzing agency.

In the case cited above possibly the Sulphur and Arsenic may have had somewhat to do with the illness, but their special poisonous actions were not in any way manifested in the symptoms.

Spasmodic pain will sometimes in weakly and elderly persons be a consequence or an accompaniment of flatulence. At least the pain resembles spasmodic pain, and moves about from one part of the abdomen to another, not being usually fixed in the stomach like that described above. It may be easily relieved by a diffusible treatment at the moment, but care must be taken lest a habit of dram-drinking be thus encouraged, for in good truth it is often among those who are too much inclined that way that periodical spasms occur. I therefore prefer to order an officinal draught to the vague recommendation of a glass of spirits when they feel bad, as medical men often do. The strengthening of the stomach by nerve-tonics is the best cure.

CASE CXXXIII.—An old man-of-war's-man, turned wine merchant, was sent to me by Mr. Way, of Portsea, Feb. 20, 1867, on account of spasmodic pains, which occurred from time to time in the epigastrium. There was no pain to be produced by pressure on the affected part, and indeed, "when the spasms are on," pressure palliates them, he said, and he lies with his stomach firmly driven against the edge of a table or chair with considerable relief for a long time together. They always pass away with a great explosion of wind. Bad attacks of these spasms came on about once a week, but he had often slighter paroxysms at night between times. He had fallen away in flesh a good deal, and the lips had grown paler, and the cheeks patchy in color; his tongue was pale, wet, and smooth; his bowels were open daily; the urine was pale, acid, of the specific gravity 1.015, not albuminous.

His habits have been usually active, as he is fond of shooting and sea-fishing, and has opportunities of indulging his tastes. His worst habit seems to be that of occasional schnaps of spirits in the evening, but he cannot be called an intemperate man.

He was ordered two grains of Quinine and $\frac{1}{16}$ of a grain of Strychnine dissolved in lemon juice twice a day, and told to eat as much flesh meat as he could find an appetite for, and to abstain from spirituous liquors.

I also told his wife to have ready for him in case of an attack of spasms as much Bicarbonate of Soda as will lie on a florin, the same quantity of Magnesia, a teaspoonful of Sal Volatile, and ten drops of Laudanum.

A close resemblance to the pain above described is sometimes found as a manifestation of malarious poison. These cases may usually be distinguished by the entire absence of all other gastric derangement, or the indication of any such derangement in the general health, by the intermission of the pain and its entire absence during the intervals, or by the previous presence of other proofs of ague poison. The following is an example:—

CASE CXXXIV.—In November, 1857, I was consulted by Mr. J. W. W—, a young-looking man of 40, concerning the occurrence at intervals, sometimes regular and sometimes irregular, of a violent "spasmodic pain," as he called it, in the epigastrium. Its usual time of invasion was between three and four o'clock in the morning, after going to bed in perfect health. It would last an hour or two, and then cease with the eruption of considerable sweating. It was worse towards the right side of the epigastrium. His tongue was clean, and he had habitually two natural solid stools a day. He had never had ague, but his house was buried in tall trees in a damp valley in the west of Shropshire, and even he allowed it to be ill-drained. One of his children had died of low fever. He had just had a more than ordinary severe attack of spasms during a night journey by rail. His aspect, however, was that of perfect health, and there was not a trace of tenderness in the abdomen.

I desired him to take two grains of Quinine in a little whiskey twice a day for three weeks at once, and in future to take the same course for a week whenever he returned home from a temporary absence.

Early in 1858 he came to report that the treatment was completely successful, though he had in the meanwhile broken an arm; and later in the year I had a message to the same effect.

It is characteristic of this neuralgic pain that it is not developed by external pressure or by food, and that there is no tenderness of the epigastrium. By this it is distinguished from a kind of spasmodic pain by which the wearing pain of ulcer is sometimes diversified.

I would always give the Quinine without acid in these cases, for it is wanted to act directly on the nerves of the alimentary canal.

SECTION V.

Gripes.

Sometimes, instead of the pain caused by food remaining in the epigastrium, or extending upwards towards the fauces, it descends to the lower bowels and is felt as a twisting sensation about the umbilical and hypogastric regions. This is usually followed and ended by the passage of one or two light, loose, often frothy stools, in which may be not seldom detected articles of diet swallowed scarce half an hour before.

Patients do not call this "diarrhœa," for it is excited only by food, and ceases immediately with the evacuation. They usually describe it as "looseness." One gentleman, who had

been reading the ancients, denominated it "lientery," and I
dare say it is what our forefathers in art meant by that word.

It will be seen by the cases used for illustration that it is
usually dependent upon some morbid condition of the lower
bowels, either the last part of the ilium or the colon. Why a
lesion in that situation should cause the contents of the stomach
to pass through the pylorus too rapidly, when lesions of the
stomach itself, duodenum, liver or jejunum, do not do so,
though much nearer, is not explained.

The following example presents that which is the most unfortunately common shape in which this ailment is found :—

CASE CXXXV.—On September 29th, 1857, I was consulted about a lady,
aged 45, who complained that immediately after taking food a pain came on
in the centre of the epigastrium, which gradually proceeded downwards with
a twisting wavy movement, till within half an hour it, ended in a motion of
the consistence of pea-soup, which varied in appearance according to the
nature of the food it followed, and often smelt of that food in case of its having any characteristic odor. There was no pain on pressure at the epigastrium,
but in the right iliac fossa there was. Ulceration probably existed in that
locality; and the scars of juvenile abscesses in the throat, together with con-
solidation of the two apices of the lungs, made almost certain the conclusion
that they were of a tuberculous character. Sulphate of Copper, Morphia,
Logwood, and Bismuth, were tried in succession, with only the merest temporary advantage. She soon afterwards died.

In mentioning this fatal termination of tubercular ulceration
of the digestive canal, I do not mean to imply that such is the
necessary history of every case. CASE XLII is an instance to
the contrary; but I do not find any mention there of the griping
and emptying of the stomach immediately after the meals, as
in Mrs. B—. And this I have generally found an omen of very
bad import. In point of fact, it is not so much the diarrhœa as
the effect of that diarrhœa upon the upper part of the digestive
canal, especially upon the stomach, which proves so deadly.

CASE XLII shows that it is very wrong, when tubercular
disease of the lung exists, to despair of effecting a cure of the
diarrhœa existing alone; but I must say I have never come
across a case of a favorable termination in consumptive cases
where the upper regions of the primæ viæ were affected by it.

However, in non-consumptive cases much more may be done in spite of the gastric complication.

CASE CXXXVI.—In September, 1858, W. J— put himself under my care. His age was 50. He had lived an active business life without any severe illness. But for the last three years or more he had become affected on the slightest provocation with looseness of bowels. This had gradually become constant, a pain coming in the epigastrium immediately after food and ending in a motion. Examination of the chest detected no lesion of the lungs.

I managed to check this with small doses of Castor-oil and Opium, and extract of Hæmatoxylum. But it recurred again in December, and then I found there was pus and streaks of red blood in the stools, and gave him Sulphate of Copper. This was soon effectual. In February, 1861, it again returned gradually, and I give him Bismuth for a month, but it did not stay the symptoms, and we were obliged to have recourse to his old friend Sulphate of Copper, which set him up again. In 1863 he came to consult me about a cough, but made no further complaint of loose bowels or epigastric griping.

The streaks of red blood in the stools render it most probable that the lesion was in the colon, and the absence of any complaint of pain in the ilio-cæcal region confirms the diagnosis.

I am convinced Sulphate of Copper is the most effectual remedy in these cases. Next to it comes Hæmatoxylum, and next Opium. So far as immediate effects are concerned, perhaps Opium should rank higher, but the good it does is by no means permanent.

By beginning with ¼-grain doses, Sulphate of Copper may be carried to two grains with safety.

The annexed case gives a detailed history of the origin of the disease in non-tubercular persons.

CASE CXXXVII.—J. B. C—, 17 years of age, had a severe acute diarrhœa, brought on by the effluvium from an offensive drain in the house where he was at school. This was in 1858. From that time he became subject to frequent attacks of diarrhœa, brought on by very slight causes, and especially in June, 1861, had one when at college, which was dysenteric, that is, accompanied by sanguineous stools. After this his meals brought on pain in the epigastrium, which was followed almost always by a thick pulpy motion, in which he had looked for blood, but never saw any. In the long vacation he went to an hydropathic establishment, where he said he got worse and was half starved. Whether in consequence of that or the disease, he was very much reduced, perspiring at night and emaciating rapidly, and so weak that I went to visit him at his lodgings several times.

He had never suffered from cough, and was quite sure that there was no hereditary tendency to consumption in his family.

When I first saw him in November of the same year I put him on Hæmatoxylum for five days. It was of no use. I then prescribed—

R.—Cupri Sulphatis, gr. ¼;
Pulv. Ipecacuanhæ co., gr. ij.
In pilulâ ter die.

The employment of this for six days removed the pain in the stomach, and reduced the motions to one after breakfast and one at night, of a solid consistence and greenish brown-color. He then resumed the Hæmatoxylum, which proved sufficient to restore his appetite and strength.

C— continued quite well and went into the army. In 1863, after a long review day at Aldershot, topped up by drinking a quantity of Moselle cup, he got an attack of diarrhœa, and, fearful of a relapse of his old complaint, he came up to see me in London. But it was easily stayed with a little Chalk Mixture and rest.

He called at my house in 1864, when I was ill in bed, to leave a card and say he had got a promotion, so that there is no reason to believe he has continued anything but well.

It is surprising that a state of bowel which has been so long coming on should be so readily and quickly cured. Such cases as these are very wholesome to the mind, strengthening it in faith that efficient treatment is discoverable, if we will only take the trouble to look for it.

Remark the extreme state of weakness indicated by night sweats and emaciation. A mere looseness of bowel would not induce that, but only a looseness which secondarily affects the stomach.

The relaxation of the bowels is not always so immediate in time. Take the following instance.

CASE CXXXVIII.—Miss Louisa P— (age uncertain), in September, 1857, complained to me solely of general languor and pain at the epigastrium of an obscure character, and I put her on Citrate of Iron and Prussic Acid, with milk and meat diet, and directed her to be careful not to press upon the pit of the stomach when sitting at her work of keeping large girls' school. When I saw her again in October I found that the pain at the epigastrium came on about twenty minutes or more after food, that it went downwards to the bowels, and was followed by a soft, sometimes liquid, stool. I put her then on Bismuth and Iron, which she went on with to the end of the month and got well.

Sulphate of Copper would probably have acted quicker.

SECTION VI.

Weight.

This is a feeling like that often locally experienced at the beginning of sore-throat, coryza, influenza on the chest, leucorrhœa, gonorrhœa, or irritable bladder. In those diseases it gives notice that the internal lining membrane of the spot is red, swollen, soft, and beginning to be coated with adhesive mucus. In the more advanced stages, as soon as pus is formed, the sensation ceases.

In all these situations it is sometimes called " oppression," sometimes " tightness," sometimes " distension ;" but I think the word I have chosen is that most commonly applied to the epigastrium. It is so in my notes taken from word of mouth.

Patients will sometimes say they feel as if they had eaten too much, but their account of their meals does not show such to be the fact. And in those whom we know to eat too much we do not find this feeling at all universal, as may be seen by reference to a former chapter (page 97). Besides, if the feeling arose from over-fulness of the stomach, it would be felt most when the stomach is fullest, namely, during a meal ; but such is not their experience.

The first inclination therefore of the medical pathologist is to refer it, when complained of in the epigastrium, to the development inside the stomach of the catarrhal condition alluded to above. And his inclination will be strengthened by the perusal in his note-books of such cases as the following.

CASE CXXXIX.—In the Post-mortem Register of St. Mary's Hospital there is an account of the autopsy of Eliza Ann S—, who died November 25th, 1853, aged 14, of dyspnœa from diseased heart, consequence of rheumatic fever, and towards the end of her life albuminuria with dropsy. It is needless to detail the appearances of the heart and lungs, on which I am not going to comment, except to say that they fully accounted for the illness and death. On opening the stomach its inner surface was found covered throughout with a coat of mucus of extraordinary thickness and toughness. Its transparency was stained by the admixture with it of a good-deal of yellow-brown matter. The microscope showed this not to be bread-crust by proving the absence of starch-granules, and rendered it probable that it was digested blood. The microscope exhibited also the presence of scattered specimens of sarcina ventricula. The membrane itself was stained with many spots of punctate congestion, and the principal contents of the stomach besides mucus was coffee-ground colored fluid of neutral reaction.

She had been under me in the wards for several months, and on referring back to the record of the case during life for symptoms in the epigastrium, I found frequent mention made of " weight," as complained of in that situation, and no other term ever used to describe the sensation. I find also that she very frequently vomited sour matters, and had a sour taste in her mouth: and that the vomiting and the weight embittered the poor little sufferer's last days. Her appetite was good, so that she took a variety of food, sometimes in re-stricted quantities, sometimes not; but neither dietary nor medicine seemed to alleviate the gastric symptoms.

The vomit during life had several times been examined by the house-sur-geon for sarcinæ, and they were not found; nor was it frothy; nor had it ever contained the coffee-ground fluid found after death, but was intensely acid.

The natural conclusion is that the weight and other gastric symptoms were caused by the continually recurring congestion and pouring out of mucus in the parietes of the stomach. And the symptoms and post-mortal appearances were marked enough to make one view this as a typical case.

I must, however, in justice, tell that in twenty-three cases collected by Dr. Handfield Jones, in which an excess of mucus was found after death, no mention is made of weight at the epigastrium among the symptoms during life.[1] Possibly it was not considered of sufficient importance to make a note of. Possibly the diseases of which the patients died, most of them acute and painful diseases, masked even to the sufferers them-selves the minor evil.

I may remark in passing on the difficulty almost universally presented by this last-named factor in the calculation, when an attempt is made to connect the post-mortal appearances with the phenomena recorded during life in all diseases which are not the immediate causes of death. Like other things pain cannot be in two places at once (I speak of course metaphorically), and when you are having a tooth out you fail to notice the operator treading on your toe. The greater ill hides the lesser.

Weight is most commonly felt towards the right side of the epigastrium, and no sensation is conveyed up the œsophagus

[1] Handfield Jones " On the Stomach," p. 74, and " Medico-Chirurgical Transac-tions," vol. xxxvii. p. 109.

towards the fauces. Now it is in the pyloric region, according
to Dr. Handfield Jones, that the catarrhal state of mucous membrane commonly occurs, and I am disposed to attribute to the
nerves of the pylorus, and to a morbid state of that part of the
viscus, this peculiar gastric sensation.

In CASE CXXXIX the lesion of heart was adhesion of the
pericardium with enlargement. In the next instance pure valvular lesion without enlargement would seem to have been
capable of producing a similar state, if one may judge from the
symptoms.

CASE CXL.—Ellen W—, aged 18, a domestic servant, was in St. Mary's
under my care for six weeks, from February 13th, 1852, and again was admitted in January, 1853, for a fortnight. There was a harsh systolic murmur,
heard loudest at the level of the aortic valves, the sound fading away gradually towards the apex of the heart. The pulse was always from 105 to 120,
and she complained of palpitation when asked about it. She was very pale
and weak, unable to do her work, bursting into a perspiration when talked to,
and having a violent hysterical fit when a patient in the ward had an abscess
opened. But her chief complaint was of weight, sometimes amounting to
actual pain, in the epigastrium, and of vomiting.

She was treated at first with small doses of *Hydrargyrum cum Cretâ*, and
saline draughts. She got worse under this treatment, and the pulse remained
quite as quick. She was then put on Decoction of Bark with Quinine, and
the pulse fell. Mustard plasters to the epigastrium seem also to have been
of use. Then the weight at the epigastrium diminished, and the vomiting
ceased; but coincident with that the patient began to have a cough; and as
the expectoration of mucus from the bronchi increased, so the gastric symptoms were alleviated. Then she regained her color, got stronger and heavier
by a few pounds. The pulse went down to 84, and she was made an out-
patient.

However, she was admitted again at the beginning of next year, and gave
us a history of chronic invalidism. She had been allowed to lie in bed and
indulge her feelings of languor. There was no cough, but she said that
another mucous membrane, the vaginal, was affected, and she frequently had
leucorrhœa. She complained of palpitation of the heart, but not of the gastric symptoms so much.

Remark here how the effects were produced now on one
mucous tract, now on another, not on both at once, but in succession. The supervention of the bronchial relieved the gastric
catarrh, and the leucorrhœal, brought on by lying in bed and
coddling, succeeded. Such cases can seldom be cured in the
short time which hospital necessities allow.

Mercurials seem very bad treatment, but just at that epoch somebody had been recommending them in gastric complaints, and I thought that such a one as this, if any, ought to be benefited.

A minor degree or weight at the epigastrium is sometimes produced where the heart is merely excitable, without organic lesion.

CASE CXLI.—G. K. R—, a civil engineer, aged 33. September 26th, 1861. He suffers a good deal from palpitation of the heart, which is brought on by even the slightest mental cause, but not by any ordinary bodily exertion, and there is a feeling as if the heart beat irregularly at times. The stethoscope and percussion detect no abnormality of shape and sound in the organ, except the quickness of beat after a short examination. He is used to the palpitation, and what he would complain of is that when he comes out in the cold after breakfast, to go to his work in London from Greenwich by the steamer, he experiences an oppressive sensation at the pit of the stomach, which continues at least the greater part of the forenoon. The stomach feels as if a weight lay there, or as if it were tumid with wind, which on examination is found not to be the fact.

R— was directed to eat milk-porridge for breakfast, to wear thick flannel over the epigastrium, and to take four minims of Hydrocyanic Acid a quarter of an hour before food.

On the 11th of October he comes to me again, saying that all the local distress has passed away, and that he feels only weakness, for which he is ordered Quinine and Strychnine. He finds milk-porridge a very convenient breakfast.

This action of the cold air is just what one feels in nasal or bronchial catarrh. Flannel is a very good preservative, and acts as a counter-irritant as well, in those who are unaccustomed to it.

The milk-porridge was intended to be a mass of even moderate temperature, in fact an internal poultice, which would at the same time be sufficiently nutritious for a man in hard work.

Hydrocyanic Acid was designed to act on the whole of the pneumogastric nerve, inasmuch as it was through its chronic sensitiveness in the heart that this temporary condition of the stomach was induced.

Doubtless in CASE CXLI the weather had considerable influence in determining the condition. In the next it seemed the sole factor.

CASE CXLII.—Miss D—, aged 50, but old for her age (for the catamenia had ceased four years), requested my advice in November, 1856, for constipation, which was always worst in wet weather. But on inquiry I found that this was not all; she had flatulence and an oppressive sense of weight at the epigastrium, extending to the right hypochondrium, after meals, and it was this which was aggravated by the hygrometric state of the air. She lived in one of the little old-fashioned damp Cinq Ports, and a removal to Ventnor[1] for the winter made all the difference to her.

I am surprised in looking over my notes not to find flatulence more often associated with weight at the epigastrium. The patients so frequently speak about their being "blown out" that one expects it in every case, but manual examination of the abdomen does not detect it. I am led therefore to conclude that this feeling "blown out" must be mainly a subjective sensation. True flatulence is usually associated with more purely neuralgic conditions, and does not, like the subject of the present section, lead to the diagnosis of catarrh. Moreover, the *sensation* of tumidity is by no means a marked feature in real tumidity. Patients often omit to notice it.

There was, however, flatulence in Case I[2] along with weight, and again probably in the following:—

CASE CXLIII.—During the spring and summer of 1857 I had several visits from a thin, withered old gentleman, T. S. S—. His principal complaint was of "weight" at the pit of the stomach, but he must also have suffered from flatulence, as I see that I have prescribed for some time Charcoal and Strychnine in Powders, which I should not have done except for that symptom.

Costiveness is a very usual accompaniment, and in the following case benefit seemed to accrue from a purgative drug.

CASE CXLIV.—Thomas K—, aged 45, an Irish manufacturer, asked my advice July 14th, 1857, for costive bowels. Confusing cause and effect, he attributed to that costiveness a constant "pressure" on the epigastrium, low spirits, want of sleep, and anaphrodisia. I gave him a prescription for five grains of Aloës and Myrrh pill with one-twelfth of a grain of Strychnine every night, and desired him to take them for a week, and return to London to see me again at that time. To his surprise they had not acted as purgatives, but elicited matured stools. He was very much better in every respect, and continued to take small quantities of the two drugs till he was well.

---

[1] There are two climates at Ventnor, the one soft and suitable for sore lungs, the other, which is here prescribed, dry and bracing, though not cold.
[2] Page 31.

The lowness of spirits which usually accompanies weight at the stomach sometimes amounts to thorough hypochondriasis.

CASE CXLV.—Mr. W—, aged 30, was brought to me by Dr. Dunfield, January 19th, 1866, on account of the persistency of a sensation of weight towards the right side of the epigastrium, coming on three-quarters of an hour after meals, by which he had been led to give up business since 1864. He was, however, none the better for giving up business. His nights were restless, and he was often woke up by headache. His spirits were at all times low; he had no actual delusions, but he took the gloomiest view possible of everything, and was inclined to be miserly in the management of his income, which was ample enough for his wants. The tongue was covered with a white fur with transverse cracks. The gums were edged with a pink line, but were not sore. The urine contained floating crystals of oxalate of lime. I advised him to travel abroad.

I suppose it must be from depression of mind being so often associated with discomfort in the pyloric region of the stomach or the right hypochondrium that we derive the term hypochondriasis as descriptive of that mental state—just as irritability of temper is called " the spleen" because it so often accompanies a stitch in the splenic region of females.

The white furred tongue with transverse cracks is very distinctive of an irritable condition of stomach, but it does not always accompany weight.

The pink edges to the gums are also a gastric symptom. They are often found in the dyspepsia of early phthisis; but they are pathognomonic of the dyspepsia, not of the phthisis. As in this instance of Mr. W—, they are often found without any tendency to pulmonary disease.

The deposit of oxalate of lime, instead of urates or uric acid, in the renal excretion, is common in such dyspeptic cases as manifest nervous symptoms. I have sometimes found with it spermatozoa, involuntary seminal emissions being also frequent in the same class of cases, if the patients lie on a soft bed or with the head low or on their backs.

The hypochondriasis is apt to take a form engendered by the situation of the discomfort. The patient will fancy he has something strange and abnormal in the stomach.

CASE CXLVI.—At the end of June, 1857, I saw a few times a Mr. B—, a middle-aged man, who complained of weight at the epigastrium and right hypochondrium. He and I quarrelled because I refused to treat him for tapeworm in the stomach, or to believe that he had one inside him anywhere. The tongue was coated with white, with cracks across it; the complexion was thick and muddy. The patient was excessively nervous and fearful, and complained especially of a "scratching at the back."

What is this sensation which people call "scratching at the back?" I find it used in a letter to me from a girl with hysterical paralysis of the legs, and I have certainly heard it from other nervous patients, but cannot recall the circumstances, nor does it convey any definite meaning to my mind.

As a rule, weight and heartburn do not go together. Patients quite understand the difference, and when skilled by unhappy experience in gastric symptoms treat them as excluding one another. Thus—

CASE CXLVII.—Colonel B—, aged 43, consulted me July 30th. 1866, about a certain loss of power and pain in the legs. Tracing these symptoms to the stomach, I inquired about the habits of that organ. He said it was a weak vessel. Did he suffer from heartburn then? No, he was "remarkably free from heartburn," he said, though he knew what it was very well. What he felt was "a weight at the pit of the stomach and in the liver," better in a bracing climate, worse in a damp relaxing one.

Yet such a thing does happen as the conjunction of weight and heartburn, and when it does the general symptoms are more than commonly severe, even when the catarrh is not bad enough to cause vomiting.

CASE CXLVIII.—J. H. R—, a commercial traveller, aged 42, was sent to me by his family doctor who had watched the case, September 20th, 1858. He had always been a fairly temperate man, and presented a healthy weather-freshened aspect, but with a look of distress or pain in his face. Naturally in the exercise of his calling he had been a good deal exposed to changes of temperature and to wet, and to irregularity of meals. Gradually he began to suffer from indigestion, which grew worse and worse. He had an almost constant weight at the pit of the stomach, especially towards the right side. But he had also decided heartburn and rising of fluid in the mouth of uncertain character, and probably consisting of regurgitated food. This led the way to nervous symptoms, to vertigo and occasional stumbling, and to such confusion of thoughts and difficulty in fixing the attention that he was quite unfitted for business.

He was cupped to a small amount on the back, was blistered, and had gr. xv of Bismuth three times a day, and a small Aloës and Myrrh pill with one-twelfth of a grain of Strychnine every night.

12

His dietary was to be as follows :—

For breakfast.—Stale bread or biscuit, with minimum quantity of fresh butter, milk and soda-water in equal quantities to drink.

For dinner.—Lean meat once cooked, stale bread, one spoonful of mashed potatoes mixed with gravy, weak sherry and water to drink.

Tea.—Same as breakfast.

Supper.—A biscuit and a cup of beef-tea.

In ten days the stomach symptoms quite passed away, and the vertigo was much better.

I have quoted here the dietary, as a specimen of what is required in a case of moderate intensity. It was arranged to relieve the stomach without starving it.

Cupping on the back disperses gastric congestion, and is more convenient than on the epigastrium. At the same time, it may aid in adjusting the disturbed balance of circulation in the brain, which is hinted at by vertigo.

To recapitulate—I think the sensation of weight at the epigastrium is one of the most important evidences of a catarrhal state of the mucous membrane of the stomach. It may exist at all times, but the presence of food intensifies it by increasing the amount of mucus present.

Its spontaneous relief by vomiting when intensified by food indicates directly one of the most important parts of the treatment, namely, that the food should be as liquid, light, soft, and as quickly soluble as is consistent with a full amount of nutrition.

As in all catarrhs, Alcohol is injurious; but in those who habitually take it, dilute wine and water must be conceded in the chronic treatment.

Local treatment of congestion by abstraction of blood and by blistering seems useful. It may be used when the amount of digestion still carried on and the appetite for food justify its employment. Even a considerable amount of anæmia need not contra-indicate it.

The most efficient pharmacopœial agent is Quinine, the conjunction with which also of Strychnine seems likely to assist the peristaltic muscles of the viscus in shaking off their adherent coat of mucus.

Where there is an obvious increase of discomfort very soon

after food Hydrocyanic Acid is useful, not only as a palliative but curative; for allaying the sensitiveness of nerves contributes powerfully to the dispersing of congestion.

SECTION VII.

Wearing pain.

When pain is constant, it assumes what is called a "wearing" character, that is to say, its moral and æsthetic effect is out of proportion to its intensity; though slight, it consumes away all the joy of life. This character is very marked in the case of constant even pain in the stomach. The patient may be known at once by the pitiable worn look of despair engraven on the countenance, *la figure grippée* as the French call it, which Velasquez has made immortal in the portraits of his master Philip, the artist's truth being too strong for the courtier's flattery. Considerable emaciation almost always accompanies it. The reason of this is the destruction of rest by night, for the restoration of rest by opiates checks the emaciation.

There often occur shocks or stabs of sharp agony, darting across the chest or the walls of the belly, and sometimes they flash even into the arms and legs. These have been set down by some as distinctive of cancer. They are not so; I have seen them where simple ulcer was found after death, and I have seen cases of cancer without them.

One characteristic feature rarely absent is its keeping the patients awake at night.

The situation of it points out the locality of the tissue change. Disease of the pylorus is felt in the right hypochondrium, of the posterior wall of the stomach and pancreas between the shoulders, where also disease of the cardiac is usually but not always felt.

If any gastric symptom has preceded wearing pain, the most usual is weight.

Wearing pain is always in my experience increased by pressure, not always immediately, but after an interval.

The prognosis is bad; it is pretty sure to return, for it depends on some organic change of tissue which cannot be restored to its perfect condition. I cannot call to mind ever having seen an example of pain in the stomach wearing and constant, so as to interfere with the nightly rest, in which I have had reason from the future progress of the disease to infer a normal state of the gastric parietes.

In cases of ulcer proved by the fatal event, whenever pain has been noticed, it has been of this character. It is not indeed perennial during all the years that the ulcer has lasted, but it is constant during the time when the degenerating movement is progressive, when the ulceration is marching onwards.

In nearly all cases where bloody vomiting otherwise inexplicable has rendered the diagnosis of ulcer the most probable one, there has been this sort of pain, occurring for considerable periods together.

Cancer of the pylorus causes this pain in the right hypochondrium, even before ulceration, quite as soon as any tumor can be detected by manual examination. Cancer of other parts of the stomach usually causes pain between the shoulders equally early.

I shall cite a series of cases of wearing pain, and make short comments on them afterwards. It will be observed that I treated and prognosticated all as if there was structural lesion of some sort, though of the nature of the lesion there was no absolute evidence.

CASE CXLIX.—Hannah W—, aged 34; a cook, was admitted to St. Mary's April 21st, 1854. She had a worn, unhappy look, and rather sallow complexion. Her body was stated to be a good deal emaciated compared with what it formerly had been. She had enjoyed good health up to about two months previous to her entering the hospital, at which period she was taken with a severe attack of vomiting—"a bilious attack." Vomiting had returned occasionally since, but not with the same severity. Her principal distress was a *continued wearing pain* in the epigastrium, which rendered her miserable throughout the day and broke her rest by night. It was increased by pressure, and to a certain degree by meals, unless the food was very soft and in small quantities at a time. She attributed it to the heat of the kitchen she worked in.

Opiates gave temporary relief, and helped her to sleep at night. So that, after a good dose of Morphia, the tongue, usually coated with a white fur, was noticed to become clean.

Six Leeches were applied to the epigastrium, six grains of Bismuth given three times a day, and the diet restricted for a week to broth without meat in it, and to cold milk and water.

It appeared that previous to admission the patient had been freely treated by means of purgatives, her bowels being very costive. They were entirely left off, and in consequence the bowels opened themselves only once in five days. This rest seemed of great use.

On the 27th she found herself able to eat a bit of beef given her, and the next day some bread, so she was allowed to have it.

On May 2d she had lost her epigastric pain, and on the 6th was able to return home. On the 20th she came to show herself to me, and to report that as yet she had no return of the pain.

CASE CL.—Hannah P—, aged 48, was admitted to St. Mary's August 17th, 1855. She was the wife of a laboring man unable to work by reason of paralysis, and she had for some time supported him by going out to field labor; so that she lived very hard, and, moreover, had lost thirteen teeth, so that even the rough food she did get was improperly chewed. Up to the previous February, however, she had been in strong health. Then she began to suffer pain in the epigastrium at odd times; but it did not prevent her earning her wages till the summer, when it became constant, and she was entirely invalided, partly from the pain and partly from giddiness and a feeling of prostration.

On admission her countenance was worn and sallow; her appetite was good, the pulse small and weak, the tongue cleaner and redder than natural, the bowels costive. The pain at the epigastrium was *constant, and increased by pressure.* She complained of want of sleep.

At first she was treated with Hydrocyanic Acid, but no benefit at all resulted. Then a blister to the epigastrium, on which great relief immediately began. Then she had a grain of Opium every night and the following draught:—

R.—Misturæ Ferri co. ℥j.
Acidi Gallici, gr. iv ter die.

She was able to take a pint and a half of milk with Lime-water in the day, and egg and other diet as well. But she did not lose her pain in the stomach till I cut her down to the milk and Lime-water only, and gave her a drachm of Bismuth three times a day. The latter prescription and the keeping of the blistered surface open for a month was at last successful, so that on September 19th she was able to begin eating half a mutton-chop daily, and on the 28th was discharged.

CASE CLI.—Mary Ann S—, aged 32, admitted to St. Mary's January 23d, 1855, attributed her illness to debility induced by her last confinement. She became subject to pains in the epigastrium, which came on about once a fortnight and *continued without intermission* during the period of the attack. One of these attacks had commenced on the 18th, when the pain was general across the pit of the stomach. On the 19th it passed over to the right side, where it became fixed and constant. She attributed this attack to a meal at which she ate both rice and potatoes. It had much diminished on her admission to the ward.

On examination of the abdomen there was found a circumscribed spot to the outside of the right rectus abdominis muscle, and within two inches of the costal cartilages, which was excessively tender on pressure. This spot appeared also somewhat tumid and tense; the patient said it had been more tumid two days before, and had been reduced by the application of a sinapism. There was resonance on percussion between this spot and the liver, the extent of whose dulness was quite normal.

Six Leeches were put on the epigastrium, followed by the continuous application of a bran poultice. She had gr. xv of Bismuth three times a day, and a diet of milk and Lime-water, with a pint of beef-tea daily. She entirely lost her pain, but eating a bit of meat at supper on the 31st brought back a short relapse, which was immediately checked by the fresh application of half a dozen Leeches. She was made an out-patient on February 2d.

CASE CLII.—Sarah B—, aged 40, was admitted to St. Mary's March 11, 1864. She had suffered seven years from frequent attacks of *continuous pain* in the epigastrium, sometimes accompanied by vomiting; and sometimes the vomiting had contained blood, though it did not do so when in the hospital. She had found by experience that hot food was apt to bring on these attacks, and that the danger was closely proportioned to the degree of temperature. She had consequently acquired the habit of taking everything cold and iced if she could get it. (The notes of this case are imperfect.)

CASE CLIII.—William G—, aged 33, country gentleman, February 1st, 1866. He has suffered for eight months from *almost constant pains* in the right side of the epigastrium, which is increased by pressure and by external cold. His countenance has got sallow, and he has lost more than fourteen pounds in weight. He has no cough, and the chest seems healthy. On manual examination of the painful spot it is resonant on percussion. The bowels are costive. They were regular before he took a quantity of Mercury and purgatives. The pain was increased by riding but not by food. His rest was broken by it. He did not vomit.

I put him on two grains of Quinine dissolved in lemon-juice, with three minims of Hydrocyanic Acid twice a day and sent him to Bath till March 5th.

By that time the constancy of the pain was much abated, and he was enabled to ride without increasing it. He had gained two pounds in weight.

He continued to improve till the middle of May, when he returned to his home in Lincolnshire, a low aguish district, and almost immediately relapsed and returned to me in London. He said he had found several times before that a visit to Lincolnshire made him worse, but he thought the summer weather would make it safe. "To keep him out of harm's way" I have recommended him to travel for a year or two, as it is to be feared these relapses may constantly occur on exposure.

CASE CLIV.—James N—, an upright military-looking country gentleman of 50, was always hearty and strong till he had the smallpox in 1864, after which he became costive in the bowels, and got into the habit of taking so much purgative medicine that without it no action could be secured. In the summer of 1866 he began to suffer from discomfort at the epigastrium, which grew gradually more frequent, so that when I saw him in October it was always produced by food, and often also arose at other times, especially when in bed. I gave him a tonic of Quinine and Strychnine with some pills of Aloës and Myrrh with Strychnine. This at first seemed to afford a little temporary relief, and the patient was anxious to make the best of matters. But he seemed to get weaker and thinner, and then the pain became *constant*, and was observed to be increased by riding on horseback or in a rough carriage. And I thought after a fair trial that the bitter drugs made the pain worse, so I left them off and ordered only some Cod-liver oil. Also one spot in the centre of the epigastrium I found tender on pressure, and at the same point I could feel the pulsation of the aorta with abnormal clearness, not stronger perhaps than natural, but more readily felt by the finger. On these grounds I held it my duty to give a bad prognosis.

After this I heard no more of the case for nearly three months, when Mr. Faithorne, under whose care he then was, wrote to me to say that Mr. N— had persevered in taking medicines for some time, when he found himself no better and left off. He was growing gradually weaker, and derived relief to his epigastric pains only from Morphia. There was no tumour in the epigastrium.

CASE CLV.—Miss B—, a thin, active person of slight muscular development, who looks about five-and-thirty, has been in the habit of walking to church every morning at eight o'clock without taking any food; then swallowing her breakfast and passing the greater part of her day in "parishing," laden with a great pocket full of books, rice, tea, sugar, loaves, &c., fastened round her waist. For some months past she has experienced a pain in the pit of the stomach ten minutes or a quarter of an hour after she begins to eat, and lately this *has become constant.* She has also occasionally vomited after meals, and has noticed mucus and blood in what has been thrown up. This symptom has no connection with the monthly period. Lately she has been disposed to be hysterical, but being of a strong-minded, cheerful temperament of mind, has not given way to it. Catamenia and alvine excretions natural. Feet apt to be cold.

On examination of the bare abdomen, there is pain developed not by gently touching, but by firmly pressing with the finger-tips, the middle of the epigastrium, just below the ensiform cartilage.

Dec. 15th, 1866.—To leave off entirely tea and all viands containing sugar, To lie in bed till half-past eight, and never go out till she has had a good meal. To leave off stays, and the weights suspended round the waist, and when she is strong enough to carry burdens, to carry them in a basket. To take a Quina draught twice a day, and to be provided with a bottle of ammoniated tincture of Valerian, of which she is to take two teaspoonfuls in water whenever she feels inclined to have a cry.

January 27th, 1867,—A strict conformity to rules has been rewarded with considerable improvement, and she has had no occasion to take the Valerian. Still there is pain on pressure of the epigastrium.

To wear a piece of Emplastrum Resinæ six inches square on the epigastrium, and to take four grains of Citrate of Quina and Iron in the stead of the other medicine.

February 28th.—A return to her old home, and, I fear, its associated duties, has caused a relapse into her former state. She appears also nervously excited, to judge by her letter received to-day.

℞.—Tincturæ Valerianæ Amm. fl3ij ter die.

Advised to travel, if she has a chance, in the South of France and Italy for the spring, but to avoid Rome, Pisa, and the like.

April 24th.—Has remained at home, steadily persevering in the tonic. Has gained much flesh, but not entirely lost her pain.

CASE CLVI.—A lady's maid, Sarah S—, aged 33, was admitted to St. Mary's September 10th, 1852, for *constant wearing pain* at the epigastrium, made sharp by pressure or by eating, vomiting and emaciation. She had also suffered from waterbrash of clear fluid, and acidity. She said that in every spring for the three years previous she had had an attack of bloody vomiting. She attributed it to her having worn a long busk to her stays, which consequently she had left off.

CASE CLVII.—James M—, aged 32, a potman, was attacked at the beginning of April, 1860, with *pain of a continuous character* in the pit of the stomach. This continued getting worse till the 16th, when in the act of vomiting, to which he had become subject, he brought up about half a pint of blood black in color. In the afternoon of the next day he brought up as much as three pints of thick black blood in masses so tough as nearly to choke him. The tongue, however, remained clean and moist, and the pulse was only 74; the heart and lungs were healthy, and he had lost the pain in the epigastrium even when it was pressed. All which things considered, it was not thought right to detain him above four days in hospital, especially as he wanted no medicine.

CASE CLVIII.—Sarah G—, aged 33, housemaid, always enjoyed good health to the middle of June, 1857, when she was laid up with sore throat at first. This passed into a *wearing pain* at the epigastrium aggravated by food, and accompanied by several attacks of vomiting, during which she threw up blood. She became an out-patient at St. Mary's under Dr. Markham's care; and he, finding her weakness and paleness increase with alarming rapidity, and seeing the tongue dry and furred as in hemorrhagic fever, recommended her being admitted on August 22d. We then found as Dr. Markham had suspected, but the patient constantly denied, that she passed blood by the bowels whenever they were opened. This required to be done by artificial means, for she was very costive. On one occasion the feces contained a clot of fibrin, washed colorless, as big as an egg. After observing and examining her for a few days, I gave her

℞.—Plumbi Acetatis gr. ij,
Opii gr. ¼. In pilula ter die.

She took this for three days, and then her bowels were open of their own accord, and she passed a dark feculent solid stool containing no blood. The pills were therefore left off, and she was treated with occasional doses of Castor Oil to clear the bowels of the remedy.

But, for some reason or another, perhaps a relapse of the hemorrhage, I began the Acetate of Lead again on September 16th, giving it her only at night, however. On the 21st she passed a quantity of flocculent fibrin without blood. On the 23d, a blue border was observed along the gums, so the Lead was again left off, and she does not seem to have lost any more blood during her residence in the hospital, viz. till October 16th.

CASE CLIX.—Eliza F——, aged 35, was admitted August 21, 1860, having for a fortnight suffered from vomiting of her food, tasting and smelling sour. That morning she had begun to consider her case serious, from having thrown up in addition some clotted blood to the extent of a few ounces. There was pain in the epigastrium, *running through to the back and increased by very slight pressure.*

She was ordered a Blue pill and Castor oil, and then twenty minims each of Sulphuric Acid and Oil of Turpentine in a mixture three times a day; also ice, milk, and cold beef-tea, like all other patients with hæmatemesis: but the next day the treatment was discontinued, as the vomiting had ceased.

There was no more blood thrown up till the 23d; the medicines were resumed, and it ceased. But all along she was passing black stools apparently consisting of digested blood.

Then her bowels became costive, and she took only some Decoction of Cinchona, and was discharged on September 7th.

In CASE CXLIX it may be remarked that the disease is attributed to the high temperature to which her occupation exposed the patient. I question whether this accusation was a just one: but probably the heat caused pain to the injured part, and was on this ground set down as the origin of the injury. For it may be generally observed that hot food, as in CASE CLII, gives rise to distress in tissue lesions of the stomach.

CASE CL is an instance of what must strike every practitioner in a mixed population, that too low living and the low vitality which is its consequence is a more frequent agent in the production of disorganizing lesions than too full living. Almost all patients in whom we can diagnose chronic gastritis are poor people.

Remark in these two cases the use of Opium. It produced no constipation, or it would have been left off; indeed in the first it seemed to take the place of purgatives, with which the patient had been previously treated. This tolerance of Opium without arrest of the excretory functions of the ilia I have found to be the rule wherever the drug is really requisite. When it confines the bowels, it generally is superfluous and meddling practice.

In these two cases I suspected chronic gastritis without ulceration.

In CASE CLI the tumefaction of the localized spot led me to suspect old adhesions of the peritoneum giving rise to the immediate symptoms of congestion. The relapse on the attempt to eat solid food is interesting, as is also the rapid relief by Leeches.

In CASE CLIII local depletion would probably have been desirable. The slowness of relief without it presents a contrast to the other cases. But I was loath to employ it on account of the malarious taint with which the patient's constitution was blighted. Aguish people bear loss of blood very ill.

In both CLIII and CLIV it is to be remarked that the motion of riding was especially noxious, though both were sporting men and more at home a-horseback than on their own legs. This pain from motion is a clear sign of structural lesion. The localized pulsation of the aorta is another pathognomonic phenomenon. The gradual increase of the symptoms is an interesting feature in the case.

CASE CLV, though not poor, yet still took less food than is required by a spare body exercise as hers was. The weight of the bulky pocket pressing on the waist was assigned by her friends as the immediate external cause of her ailment, and I cannot but think they were right.

The renewal of life induced by a trip to a Mediterranean climate, as recommended to Miss B—, for the spring, is very striking to those who have tried it, especially in cases of chronic tissue-change. I made the excuse in CASE CLIII that it was to keep the patient out of harm's way, and so it was partly, but I believe it does more than that, and has really a regenerative power over the degenerated substance of the body. Statistical and self-experimental reasons are given for this in a little monograph " On the Climate of Italy," founded on my own case.

Rome, however, and Pisa, and perhaps a few other places of less note and less tempting, must be shunned by all persons of an hysterical temperament. This caution, though in a less degree, applies equally to young men as to young women, as I' am informed by an intelligent clerical student at the Collegio Pio, who has of his own accord made this remark to me, as to the effect of the climate on young Englishmen who go there to study for holy orders.

CASES CLV and CLVI may both recall the 4th and 5th sections in the last chapter in which somewhat similar pains and effects are attributed to the trade of shoemaker and to tight-lacing. Constant pressure upon the outside of the stomach seems to have a disorganizing power like that of pressure of a foreign body, tumor or the like, inside. The stomach is worn away just as the ribs are, by the gentle but continuous pressure of an aortic aneurism.

In CLV, CLVI, CLVII, CLVIII, CLIX, the diagnosis of ulceration was rendered probable by the hemorrhage being joined to the continuity of wearing pain.

In CLIX the running of the pain through to the back is noticed.

SECTION VIII.

Soreness on pressure.

Soreness on pressure is so generally in all naturally insensitive parts an indication of structural change that we all of us as a matter of course apply this diagnostic sign to the organs of digestion. Where it is constant in any one part, independent of the presence of food, and proportioned in its degree to the amount of pressure, it appears to me pathognomonic, and can hardly arise from any other cause. It is sometimes immediate, and sometimes does not come on till the lapse of a certain interval. In the former case there is a little vagueness in the sign, for some people are so much more sensitive than others, and object so to having the epigastrium pressed, that they cry out without sufficient occasion. Care is requisite not to be deceived by the hyperæsthesia. The pain which comes on after an interval, on the other hand, is a very determinate symptom, and is never simulated or imaginary. It gives very accurate information that an organic change of some kind has taken place in the structure of the stomach.

Whenever, then, tenderness on pressure constantly exists, whether accompained or not by constant or wearing pain, and whatever the other symptoms may be, whether heartburn, waterbrash, or weight, I think we are justified in employing local alternative means, mustard poultices, blisters, leeches, or cupping. Water compresses are not so efficient; I think those who fancy they have found them useful must have fallen in with other forms of gastric pain and mistaken them for tenderness.

An all-important part of the treatment is complete rest.

The action of this may be seen by the rapidity with which the patients get well in the hospital.

Tenderness on pressure does not contra-indicate an analeptic restorative treatment being conjoined with the local. Indeed, it demands it. Numerous instances of this may be seen in cases already cited, perhaps the most striking from the symptoms being capable of being depicted in. number and weight, is CASE CVIII, of a young Irish woman, who gained twenty-one pounds of flesh in twelve days, in spite of being leeched every other night during nearly all the time for waterbrash, with intermittent pain at the epigastrium and tenderness.[1]

Pain felt only on pressure in a part does not require any palliatives except not to press. This is a platitude perhaps; but still both doctors and patients are the better for having the fact brought to mind, since these out of anxiety find it difficult to keep their fingers away, hoping each minute to find the pain gone, and those are tempted by a love of accuracy, hard to blame, into a needless frequency of examination.

SECTION IX.

Anomalous pains.

CASE CLX.—Mrs. S—, aged 40, used to visit my house frequently in 1849 with a daughter, whom I was attending for cutaneous disease. One day, though at the time in perfect health, she desired my advice about a curious pain in the pit of the stomach, which from time to time assailed her. It came on gradually. was not severe enough to lay her up. but constant and worrying while it lasted, namely, for about a week or ten days at the most. The first thing I made out about it was that it usually succeeded to any mental worry or unusual bodily exertion for several days. On further inquiry I found it invariably coincident with the catamenial periods, which, however, were regular, not excessive, and accompanied by even less pain in the loins, uterus, or groins, than most women accuse. It appeared in fact to be a dysmenorrhœic pain, misplaced at the wrong end of the abdomen.

[1] See page 149.

I gave her a course of Quinine and Iron for the nonce, and desired her to take a special dose of Hydrocyanic Acid and Opium if the pain came on again. This seems to have been successful, for though she brought her daughter several times during the next year she said no more about herself.

· Though she appeared in perfect health, the mere fact of being an anomalous pain showed weakness, and constituted the periodical discomfort which is the normal portion of the sex, or *dys-menorrhœa*.

The above is a specimen of the most usual degree in which uterine pains are felt in the stomach, but sometimes they are more serious, as in the following instance :—

CASE CLXI.—Jane R—, aged 25, a housemaid, was admitted under me at the hospital February 16th, 1852. She was a personable robust country-woman, who had lately come up to service in London. Her tongue was clean, her pulse 84, full and strong, her skin normal, her urinary and fecal excretions reported natural. Her mistress said that for three days Jane had complained of pain in the lower part of the chest in front. That it was increased by food, and consequently she had "eaten nothing," that is to say, had taken only liquid food. She got an out-patient's letter to the hospital, but on her way to use it was taken so much worse that she was obliged to be admitted.

She sat up in bed rubbing her epigastrium with her hand, and expressed herself as in great pain. Rubbing, however, gave her no relief, nor did pressure ; but it could be borne without any increase of pain.

The catamenia had been absent two months. A large Linseed poultice was applied to the abdomen, she took a four-grain Calomel powder immediately, and a Senna draught three hours afterwards. The same day the catamenia occurred, not copious (they were never so, she said), but sufficient, the pain instantly ceased, and she was well enough to be discharged on the 18th.

The disgorgal of the portal veins, by a mercurial and a purgative, is a capital way of bringing on the catamenia in a robust, full-blooded person. Remember, however, that I do not recommend it in those more common cases where the amenorrhœa is merely an evidence of the absence of menstrual blood to be discharged.

It has seemed to me that pains in the loins, closely resembling those of the renal calculus, might sometimes be traced to the stomach.

CASE CLXII.—Henry L—, a lithe active Scotchman with golden hair, apparently between 35 and 40, on his return South after grouse-shooting in the autumn of 1866, began to suffer from pain in the loins. This got better without any special medication, but about Christmas, when tossing a child up in the air, he felt a sudden stab, as it were, in the right loin, so sharp that he was within an ace of dropping the child. The pain in this situation, sometimes better, sometimes worse, but never absent, continued up to the time of his coming to me February 8th, 1867. Very often the pain ran sharp down into the front part of the pelvis, there was an ache in the right testicle, and that organ was strongly retracted against the body. He had been treated by a first-rate surgeon with Alkalies, Alkaline Mineral Waters, and with Iodide of Potassium and Turkish baths without success. Indeed he thought he grew worse.

On examination there was pain on pressure with a finger just below the ribs outside the psoas muscle on the right side.

The urine was pale, just acid, of the specific gravity only 1.006, probably from the joint influence of the alkalies and a habit of taking a glass of whiskey toddy at bedtime in lieu of wine at dinner.

Treatment prescribed, February 8.—To wear a belt with loin stiffeners, to take port-wine at dinner, and a nap on the sofa before the meal. At 11 and 4 o'clock daily to take a couple of grains of Quinine.

On the 12th the spec. grav. of the urine examined was 1.025. On the 23d the pain in the back was to be elicited by very hard pressure only. With some Iodide of Potass added to his mixture (I forget why) he was dismissed with orders to continue the treatment for a week. After that he wanted no more doctoring for a time.

But at the beginning of April he dined out twice running, and indulged on each occasion in the fermenting mixture which London dinner-givers are fain to call Champagne, and the consequence was a return of the peculiar lumbar pain, though the urine remained quite natural. I prescribed for him a return to the former prescription. This was in passion week. On Easter Monday he still felt the pain, but like a true patriot joined his corps at the Volunteer Review. The reward has been a complete cure of his pain. Now, had it been rheumatic or renal, as might have been suspected, or anything else but abnormally placed pains of indigestion, it is clear that reviewing on the Dover Downs in English spring weather would have made him worse.

The retention of feces in the bowels is frequently assigned by the public as a cause of pain in the epigastrium. Their fondness for purgatives doubtless often leads to error on this head, but still I do think they are sometimes right, and that the mere retention in the colon and rectum of matters ready for evacuation may give rise to considerable pain in the epigastrium. It is not very easy to hit upon a good illustration of this, for most usually costiveness and even constipation depend on some morbid condition of the stomach or of the whole alimentary canal or of the whole body, and it is difficult to separate the effects of the retention from those of the condition which has engendered it. Thus, for example, you will find that nearly all the chronic cases quoted in the first chapter had confined bowels, but no one would attribute the epigastric pains to that cause, seeing that an obvious indigestion existed in the stomach, the seat of those pains. In the following case, however, the cause of the retention of the feces was quite extraneous, and there was no proof of anything being the matter with the digestive organs.

CASE CLXIII.—Anne M—, aged 23, a domestic servant, was admitted to St. Mary's October 15th, 1856, complaining in various parts of the body of obscure pains, which, however, after admission, seemed to have their definite seat in the epigastrium, and to be worst always after food. She had palpitation of the heart, nausea, and a tendency to faint. Her face was flushed, the skin hot, and the tongue coated; but otherwise her aspect was healthy. After a few days the nurse observed that her linen was stained, and the patient herself stated that she had a vaginal discharge. But actual examination found the organs of generation quite normal, and that the pus came from a small papilla, the remains of an old hæmorrhoid, on the edge of the anus. This was exquisitely sensitive, and the patient confessed that she had voluntarily retained her feces on account of the pain which defecation gave her. Warm baths and softening enemata, with the aid of Valerian draughts, reduced the hyperæsthesia, and with the emptying of copious solid stools from the colon the pain at the epigastrium ceased, and she got good rest at night.

In cases of misplaced pains, I mean pains not in the locality of the parts truly affected, Valerian is a very useful medicine. Its calmative effect on the nervous system is remarkable. That was the reason of its administration to this young woman. It would have been cruel to forcibly open her bowels by purga- -tives, without first deadening the abnormal sensitiveness which had caused her to constipate them.

CHAPTER V.

VOMITING.

SECTION I.

General remarks on the physiology of the process.

In the normal passage downwards of food the involuntary nerves and muscles of the fauces, the gullet, and of the stomach are in vigorous action; whilst the voluntary abdominal muscles and the diaphragm exert no influence over the digestive canal.

In vomiting a converse condition exists—the involuntary œsophagus is wholly or partially paralyzed and relaxed, the involuntary peristaltic wave of the stomach ceases, and at the same time the diaphragm and abdominal muscles are degraded from agents of volition to purely automatic instruments.

The ceasing of the peristaltic wave allows the pylorus to close. It is converted from a portal somewhat stiffly held open by the circular fibres (as if in a sort of erection) into a collapsed valve. The pylorus being closed and the cardia open, it would not require any such very strong muscular effort to empty the stomach.

But the muscles thus abnormally perverted into compressing the stomach are very large and powerful. Hence vomiting is a violent and explosive act.

In spite, however, of the violence and explosiveness, a correct pathology must look upon vomiting as a lowering of the vital powers, as an atony of the digestive tube and its appendages, when the facts are put in the order and light above sketched out.

·Thus it becomes clear why vomiting is an accompaniment of so many states in which there is a diminution, or arrest, or paralysis of muscular action. Unusual or too long-continued bodily exertion, exposure in cold or heat, and such like circumstances peculiarly exhaustive of muscular and nervous power, before eating even a moderate meal, will in some persons cause it to be ejected.

The same result follows in fainting, or when, from excessive mental emotion, the nerves of the gullet experience a temporary paralysis; so that vomiting is produced by disappointment, anxiety, nay sometimes even by sudden joy and pleasure. Still more strikingly is it brought on by structural disease of the stomach, by which the peristaltic wave is arrested, or at least interfered with. Or by a stoppage of the same in the intestines, such as occurs in ileus, hernia, intussusception, and peritonitis.

Vomiting in these latter cases has been sometimes referred to a reversal of that muscular act which carries the alimentary mass onwards—to an *anti*-peristaltic motion. But there seems to me no evidence that such is the case; indeed, an attentive consideration of the phenomena of the act itself would seem to show proof to the contrary. Observe peristaltic motion—it is slow, continuous, and uniform ; possessing indeed strength in its persuasive steadiness, but no irresistible impetus. Compare the two, and note the difference : in vomiting we have a violent explosive power, like a force-pump, throwing the ejected matters out to a considerable distance. Can there be a greater contrast between two acts of the same part? The explanation given above seems much more naturally to suit the phenomena.

In some cases the atony is general, as in vomiting from cerebral diseases of a paralytic character. In others it appears to be more local, as for example in the action of emetics, where the force of the agent falls mainly on the stomach, and secondarily on the limbs; and possibly in some it may be entirely local— an approach to which is made in the quickly acting emetics, such as Sulphate of Zinc, which therefore produces much less depression than most other medicines of the same character. But in all there is sufficient reason to consider the muscular

13

state in vomiting to be one of relaxation or atony, and to view as the main muscular manifestation of atony in the stomach a tendency to vomit.

Vomiting seems less dependent upon the previous or chronic condition of the stomach, and more upon the idiosyncrasy of the individual, than any of the phenomena already discussed. There are dyspeptics who, whatever may be the matter with them, never throw up their food; whilst others do so on the slightest occasion. Even pleasant associations will, in some people, bring on this most unpleasant consequence; an occasional patient of mine, a healthy young lady, has been sometimes taken with retching on entering a ball-room where she expects an agreeable evening, whilst it never happens in going to a stupid party. On the other hand, I have had a patient with cancer of the stomach, and others with various sorts of severe dyspepsia, who could take the most repulsive drugs without inconvenience. The mere fact of vomiting, therefore, affords in itself no clue to the local condition of the stomach. But the time of its occurrence, the circumstances which increase it, and the nature of the matters thrown up, may be most suggestive to the practitioner.

Vomiting when the stomach is empty, or that which, though it accidentally occurs at other times, is most frequent and distressing then, may be safely set down as arising not from any fault of the viscus itself. Such is the morning sickness frequent in pregnant women, and in cases of diseased heart, of abdominal tumor, and sometimes of pulmonary consumption. This has been explained as a reflex action of the vagus nerve excited by the irregular irritation of some of its branches; and on the same principle may be interpreted the more rare cases where it has been caused by foreign bodies in the ear or nose, by tumors in the neck, &c.

When it occurs with a full stomach, we may reckon, as a general rule, that the smaller the quantity of food that produces it, and the sooner it takes place after eating, the nearer to the mouth is the cause. An ulcer of the œsophagus causes rejection of the food before it has got down; of the cardia, or smaller curvature, very soon after it has got down; and a similar lesion of the pylorus or liver, after an interval sometimes of several hours.

When vomiting arises from the paralysis of the œsophagus which is induced by a congestion of the brain, as in apoplexy or drowning, or by poisoned nerve, as in dead drunkenness, it is increased by the horizontal posture; when it arises from deficient supply of blood, as in fainting and anæmia, that same position relieves it. Sea-sickness also is often warded off by lying down with the head low.

The contents of vomit may often afford valuable indications to the practitioner, and will appropriately divide into classes the cases he meets with. They will here serve the purpose of the headings of sections.

SECTION II.

Vomiting of pus.

CASE CLXIV.—Elizabeth S—, aged 25, was admitted at St. Mary's January 23d, 1852. She had suffered for three months from vomiting, at first occasional, but latterly at every meal, so that, in spite of a good appetite and plenty to eat, she had grown pale and thin. After this had continued a month, she began to experience a difficulty in swallowing, which has gradually increased, though the pain caused by it is not so great. The mouthful seemed to lodge somewhere at the back of the manubrium of the sternum, and either to be rejected or retained with great pain, which ran through to a spot between the shoulders. Besides this, she used to have occasional retching and occasional vomiting of glairy and frothy matter, with opaque streaks of pus in it not unlike the sputa of early phthisis.

Gruel, arrowroot, cocoa, raw eggs, and milk were swallowed and kept down, so that she was occasionally not sick for a day or two together. Dry Bismuth powders she also kept down, and thought they relieved the pain. But Sulphate of Copper made her vomit on each occasion that it was tried, as was done several times with the idea of stimulating the ulcerated surface to healing action.

At length she seemed to catch a cold on the chest, and died suddenly after breakfast one morning.

On examination after death its immediate cause was found to be the opening of a fistulous communication between the ulcerated surface of the œsophagus and the pericardium, by which pus and food had made their way into the serous sac.

The stomach, &c., were healthy.

The pus in the vomit here doubtless came from the cellular

tissue around the œsophagus, which was being eaten through by the fistula.

Remark in passing the use of which Bismuth seemed to be to the raw surface. Some persons have found it of equal use in phthisical diarrhœa from ulcerated bowels. I confess I find it in this latter disease less efficient than Sulphate of Copper; but in the upper part of the digestive canal the comparative force of the two drugs would seem to be reversed. Another instance of the use of Bismuth in ulcerated œsophagus will be quoted afterwards (CASE CCVI). I felt considerable satisfaction that this poor woman never had any probang put down her throat. It would have thrown no light upon the diagnosis, and might have gone into the pericardium and been the cause of death. *Imprimis non nocere* is the first commandment in medical morals.

CASE CLXV.—James G—, aged 32, dairyman, admitted to St. Mary's June 27th, 1856, after an illness of a month, during which he had attended as an out-patient. He complained of soreness of throat and of difficulty of swallowing solids. He said he never vomited, but after admission he began to throw up a considerable quantity of pinkish or flesh-colored purulent matter. Sometimes it was ejected by retching, sometimes with less effort. There was nothing abnormal to be seen or felt in the fauces or upper part of the œsophagus. After he had been in for a fortnight a small tumor of carti-laginous density was felt behind the ramus of the jaw, just below the right ear. During the time he remained in hospital he had a Laurel leaf poultice to the neck and Cod-liver Oil; but as no conscientious hope of future amend-ment could be expressed, it was not thought right to occupy the much-wanted space in the ward with an incurable case, and so I lost sight of him.

There could be but one end to what was indubitably a can-cerous ulceration of the œsophagus, and I do not think a hos-pital ward is the happiest place in which to await that end; so neither for the patient nor the public do I think it right to retain such cases in a charitable institution. It is quite different where any doubt exists about the diagnosis.

CASE CLXVI.—Edward J—, aged 56, greengrocer, was admitted into St. Mary's November 16th, 1857, complaining of pain in the left side of epigas-trium immediately after eating. This was relieved by vomiting. His illness had first come on during a voyage to America the previous April. Previous to that he had always enjoyed good health, and weighed 12 stone; but now he was reduced to 9 st. 2 lb. His vomit usually consisted of his food; but on one occasion he ejected a quantity of creamy pus mixed with a strongly acid fluid.

A hard tumor was discovered below the cartilages of the ribs on the left side.

He left the hospital December 14th, probably dissatisfied at the little relief it was possible to give him.

These are the only cases I can find where there was pus in the vomit, viz., common ulceration of the œsophagus, cancer of the œsophagus, cancer of the cardia. It does not appear to be thrown up in common ulceration of the stomach, still less in catarrh of the stomach. The gastric and œsophageal mucous walls are very different from the bladder or urethra. These secrete pus on the slightest irritation; an undue stretching, a hard substance, however smooth, an essential oil, moderate alkalinity of the urine, the infection of a catarrh so weakly poisonous as gonorrhœa, and other equally mild forces, arrest their vitality down to the pyogenic stage. The fauces, gullet, and stomach are much tougher; fortunately indeed, for if stretching, hard substances, spicy oils, alkalies, or acids hurt them, or if moderate doses of morbid or common poisons acted on them locally, who would insure a man's life for a week? To purulent inflammation they are not prone, and therefore we cannot expect to find pus in that which is ejected from them in their usual diseases. When there is pus in vomit, either a malignant tumor has destroyed the walls and taken their place, or there is an ulceration with adhesions into the surrounding cellular tissue.

Care should be taken to ascertain the condition of the lungs, and make sure that the pus does not come from a vomica, the emptying of which will sometimes be accompanied by vomiting.

SECTION III.

Vomiting of mucus.

Mucus is found in the vomit in what is called English cholera, or acute summer gastric disorder.

CASE CLXVII.—Edmund K—, aged 18, was found by a policeman at half-past six in the morning of September 24th, 1851, staggering against some palings, and unable to walk, from a violent pain in the belly, which had suddenly attacked him on the way to work. He was taken to St. Mary's, and kept vomiting mucus and bile all day. The pulse was 90, the tongue dry. Pain on pressure of the epigastrium.

No collapse, cramps, or retention of urine occurred. He had a dose of Calomel immediately, followed by diarrhœa; a dose of Opium at night; and was discharged well next day.

The green matter in the vomit of these acute attacks is shown to be bile by its bitter taste to the patient, and sometimes by its smell to the bystanders. The presence of bile is a proof of previous health, and an assurance that the cause which is disturbing it is a temporary one, however severe it may be, and that the vitality is not deeply smitten. You do not see it in the vomiting of chronic disease, you do not see it in that of fatal epidemic cholera; but you do see it thrown up by the hearty landsman who is roaring over the gunwale of a Channel steamer. and it is hailed as a good sign in a convalescent from cholera collapse. Give an emetic to a healthy man, and you see plenty of bile; give it to a broken invalid, and you most likely will not. Bile, then, is to be looked upon as a bird of good omen. It is regurgitated from the liver into a fairly healthy stomach, and not into an unhealthy one.

Mucus, mixed also with bile, is thrown up in those less severe exhibitions of gastric disorder which are called " bilious attacks."

CASE CLXVIII.—John D—, a retired schoolmaster, aged 55, became my patient February 7th, 1863. For at least ten years he had been subject to " bilious attacks," occurring in the winter, and generally half-a-dozen times each season. He described them as commencing with a hawking up of phlegm; which phlegm did not seem to come from the air-passages, but from the gullet. This usually took place of a morning; and in the evening a severe attack of headache came on, and a vomiting of phlegm and bile. He came to me because he found them getting more frequent and severe, and because he began to doubt if the traditional mode of treating them with purgatives were really the best. I put him on Quinine and Strychnine, and saw him again on the 23d. He said that in the meanwhile he had been threatened with a bilious attack, but it had been warded off, he thought, by the medicine. I heard of him in 1866 from a relative as a much heartier man than he used to be.

The summer gastric disorders, which I first exampled, are probably brought on by the absorption into the body of some poison diffused through the air or water, and which, when widespread and intense, constitutes the terrible epidemic cholera. They fall on the robust equally often with the weakly. These winter bilious attacks are more like what we called "catching cold," and are certainly induced by changes of temperature in damp climates. Like colds in the head or chest, they affect delicate-formed and delicate-constitutioned persons principally. Much may be done, therefore, to ward off the attacks by strengthening the constitution. Quinine and Strychnine is the best treatment; purgatives do harm. I say purgatives do harm, because an unprofessional friend of mine, who used formerly to be treated *secundum artem antiquam*, finds even homœopathic treatment better than purgatives; and restorative treatment would be still better than homœopathic, would she but try it.

Another thing which seems to me a broad hint against purgatives for "bilious attacks" (by which I mean attacks of gastric catarrh in a body healthy enough to eject bile by vomiting) is, that where there is purging arising without the aid of drugs, the sickness lasts much longer than if there is none. He must be indeed a devoted admirer of the pharmacopœia who imagines that the artificial diarrhœas excited by its help can do good where nature's diarrhœas do harm. Compare the more usual forms of bilious attack, of which I have quoted examples, with the following:—

CASE CLXIX.—Elizabeth J—, a domestic servant, aged 32, came into St. Mary's October 18th, 1855. Since the first week of the month she had complained of headache and weight at the epigastrium, and on the 10th was seized with diarrhœa, which still continued, though less severe than at first. On the 17th she was, in addition, attacked with pain in the epigastrium and vomiting, which was very frequent on admission. The pain was much relieved by mustard cataplasms. The tongue was red and clean, the pulse weak and quick. The motions were green, and the vomit was green too, with shreds of mucus stained with port wine that had been administered to her.

The vomiting was somewhat appeased by Hydrocyanic Acid in effervescing draughts; and she also took some Chalk and Opium, with a little gray Oxide of Mercury in it. But the vomiting did not cease till the 20th, after which she began to amend, asked for food, and was able to get up on the 22d, and to leave for home on the 26th.

CASE CLXX.—Anne G—, aged 49, was admitted July 31st, 1857. She had been in her usual good health up to a month previously, when she caught cold from exposure, and distressing nausea and vomiting ensued, which in a few hours was followed by purging; the stools being watery, frothy, and free

from offensive odor. The skin then was cool, her tongue clean, the pulse 85.
The purging continued at the rate of four motions daily. She vomited after
each attempt to eat; the ejecta being green and yellow, mixed with clots of
mucus and undigested food.

The patient had also considerable anasarca of the lower extremities, and
she suffered from palpitations. The force of the heart was weak, but the
sounds natural, and the lungs healthy.

She was treated without drugs, but had five grains of Pepsine powder three
times daily, half a pint of beef-tea, and milk diet guarded with lime-water,
food being administered in small quantities every three hours.

By the 8th of August she was so much better in all respects, that she was
able to eat mutton chop, and to take a grain of Quinine dissolved in Tartaric
Acid three times a day.

On the 12th she had the full hospital diet. The bowels having become
costive, an Aloës and Myrrh pill was ordered every night.

On the 15th, there being still some pain at the epigastrium complained of,
the solid diet was reduced to half, and a pint of beef-tea given for supper.

There was also, on the 17th, a threatening of return of diarrhœa; but it
was promptly stayed with Chalk mixture, and she was discharged well on
the 19th.

Remark how in the first of these two cases the preceding
diarrhœa did not prevent the occurrence of stomach symptoms,
and how long they endured in the second case where the diarrhœa had come on at the same time.

I abstained from perturbative practice during the height of
the disorder, not out of scepticism in the pharmacopœia, but
the contrary; there was no physical condition capable of establishment by its means which I would induce.

The costiveness of the bowels after a natural diarrhœa is a
very usual reaction. Had astringents been given, one might
have attributed it to them.

The weakness of this patient was shown by the anasarca; but
we had no reason to suppose that there was any chronic degeneration of the viscera, or she would have been retained in
hospital.

In the next case the character of the disease is much more
chronic.

CASE CLXXI.—Helena F—, a domestic servant, aged 40, February 25th,
1858. Since the beginning of December she had complained of uneasiness'
at the epigastrium after meals, accompanied by nausea. Six weeks before
admission she began to relieve the feeling of weight by vomiting her food
several times a day, and in the morning to vomit frothy and stringy matter

occasionally streaked with blood. She never threw up any *clots* of blood, and it was always of a dark color. Since her illness the catamenia had ceased, having been previously copious and painful; she had got very much emaciated, and in good truth she presented an aspect closely resembling that of pulmonary consumption. But stethoscopic examination of the chest showed it to be quite healthy, and she had no cough.

The absence, however, of cough is no proof of the absence of pulmonary lesions in the sort of case which at first blush hers seemed to resemble. For instance—

CASE CLXXII.—Mr. J. P— came to me May 30th, 1860, complaining of flatulence, and of uneasiness without pain in the epigastrium, and of having become muscularly weak and nervous. He frequently threw up, as he said, from the stomach, stringy mucus, especially after eating. But it was not mixed with food in general. He had no cough. He remained under my care till the 10th of June, when I examined his chest and was obliged to tell him that I found bronchophony, crackling râles, and dulness on percussion at the apex of one lung, He seemed dissatisfied with the diagnosis, and I did not have another visit.

I should like to have made out for sure whether the mucus really came from the stomach, or whether it was the contents of the bronchi thrown up by a nauseating effort instead of a cough. Vomiting certainly does occur in phthisis, and the vomit contains mucus. But that mucus is also often purulent, which gastric mucus rarely is, so that it is probably swallowed or ejected from the bronchi by the emetic strain.

Such cases as the last should remind us always to examine the chest, cough or no cough, in any forms of disease which are ever associated with consumption. And the only way to guard oneself from the imputation of mistaking the disease is to declare the diagnosis to somebody at once, for the patient will often break down very suddenly while you are casting about for an opportunity of letting his state be known.

SECTION IV.

Vomiting of blood.

The immediate symptoms differing from ordinary vomiting which precede and accompany the vomiting of blood are described in the annexed cases.

CASE CLXXIII.—May 13th, 1862. Mrs. H—, aged 42, awoke at two in the morning, feeling very hot and restless: a sudden faintness and dread came over her; she felt sick; the sickness felt somewhat better, and she got out of bed. Almost immediately, if not in the very act of rising, a flood of blood or bloody fluid gushed up from the stomach. She had not previously considered herself an invalid, but had for several months had irregular catamenia, and for three days before her attack had experienced a dull pain at the epigastrium and right shoulder. On examination of the epigastrium, it was painful on pressure, but not one spot more than another. A feeling of nausea was excited by the examination, and also by taking food. The tongue was somewhat dry. She did not bring up her food; not even when the swallowing a cup of hot tea suddenly (on the 16th) had caused her to retch violently.

A pill was administered on the 13th containing four grains of Acetate of Lead, and on the 14th she was ordered—

      R.—Acidi Sulphurici diluti ℳxx,
           Olei Terebinthinæ ℳx,
           Infusi Hæmatoxyli fl℥j, ter die—

and had no return of vomiting.

CASE CLXXIV.—James P—, aged 37, was admitted into St. Mary's Hospital May 4th. 1860. He had been long subject to vomiting, and had five times vomited blood, for the last time the night before admission. The blood, he said, always came up with a sudden gush, and was dark in color. He described the symptoms preceding the hæmatemesis as commencing with headache and pain in the right side; after this he felt heavy and drowsy; then he got giddy; and then the blood came up.

The tongue was dry and furred, the pulse 92 and bounding. The pulsation of the aorta in the epigastrium was very distinctly felt.

(Further particulars of the history of this man may be found at page 109, where his case is repeated on another account.)

The above are the ordinary symptoms which occur without prognosticating any immediate danger. When a fatal result is to be feared, they are more severe.

CASE CLXXV.—Hannah H—, aged 48, a cook, was admitted to St. Mary's June 24th, 1852, for hæmatemesis. She said she had brought up a great quantity of fluid blood the day before, and that while she was throwing it up she felt complete inability to move her limbs.

While in the hospital, between that date and July 3d. she several times vomited blood, the vomiting always coming on quite suddenly without previous warning. but being followed by deadly faintness and by an increased pallor of face. On the last-named day, as the house-surgeon was going his morning rounds, he saw her suddenly turn paler, and so he laid her back on the pillow. In ten minutes after she threw up a quantity of liquid florid blood, mixed with clots. She did the same an hour afterwards, and again in the evening. She continued hiccuping, lay with her brows knit, but in no pain. The pulse rose to 136. the skin grew burning hot, the tongue coated with a white gelatinous fur. The voice was reduced to a whisper. In this state of hemorrhagic fever she continued, without becoming comatose, till she died, about fifty-five hours after her last vomiting of blood.

Post-mortem.—It was found that an ulcer had eaten into the coronary artery of the stomach.

The symptoms may be nearly as severe in cases which do not ultimately prove fatal.

CASE CLXXVI.—May 24th, 1848, I was summoned to Mrs. M—, aged 82, who had fallen down that morning in a sort of fainting fit. On recovering herself, she threw up, as was alleged, from the stomach about a wineglassful of blood. The tongue was dry, and in the centre brownish. The epigastrium was painful on pressure. The pulse exhibited the largeness and loose sharp stroke distinctive both of the hardened arteries of old age and also of hemorrhage, both which factors were probably united in this patient.

The next day the tongue was quite dry and brown, and the abdomen more painful on pressure. She had vomited a great quantity of red and black blood, and passed a number of black stools.

On the 26th the tongue got moister. On the 27th she again fainted, and her face became anxious, and the tongue dry and brown. I thought she would vomit blood again, but no. she only passed it in a black stool. On the 28th she was better again, and with one more relapse on the 29th she finally recovered, and had no return of bloody vomiting, though she lived several years afterwards.

The treatment had been lemon ice, Bark, Alum, Opium, and Sulphuric Acid.

She must have had a very vigorous constitution to have survived such a serious illness at such a time of life.

Apropos of age the following case may have an interest, for I think it is the next oldest patient with this complaint that I have had under my charge.

CASE CLXXVII.—Elizabeth A—, aged 60, was admitted into St. Mary's August 21st, 1855. She called herself a strong, healthy woman, though subject to occasional "bilious attacks," and ruptured on one side. After feeling a weight in the belly for a week, she had on the 11th vomited a small quantity of blood. On the 18th she again vomited, and this time nearly a quart of blood, bright-colored, and with clots in it, but not frothy. There was slight pain on pressure of the epigastrium, but this, she said, was much more severe at the time of the hæmatemesis. She also passed black blood by stool. The hepatic dulness was normal.

She was but little pulled down by her attack. So, after a fortnight's rest in the hospital, she went out.

Vomiting of blood may occur again and again without risking life.

Sometimes it is an annual affair, as in CASE CLVI, and again in the following.

CASE CLXXVIII.—Henry H— came under my care at St. Mary's in November, 1862, having in October, on two occasions, in going home from his work, thrown up what seemed to him near a pint of blood. The same patient had, in October, 1861, also been in St. Mary's for hæmatemesis. It is right, however, to remark that he had emphysema of the lungs as well; so there might perhaps be a question about the certainty of the diagnosis of the blood coming from the stomach.

In the following cases there were intervals of three and four years.

CASE CLXXIX.—In the middle of September, 1855, Selina Y—, a widow of 48, was woke up at two in the morning by a single attack of profuse hæmatemesis. She was my patient at St. Mary's for the weakness thence arising, but she was not alarmed at the occurrence, as she had been ill in the same way three years previously.

This woman experienced again another gush of blood in October, 1859, and was under my care at St. Mary's afterwards. Whilst in the house she had waterbrash one morning, and ejected a cupful of clear aqueous fluid with streaks of red in it. She reported that in the interval of the two admissions she had lost no blood.

It is remarkable that in this case there was no tenderness of the epigastrium.

CASE CLXXX.—Catherine C—, a servant, aged 28, was admitted to St. Mary's October 19th, 1858, for an attack of hæmatemesis which had just occurred. She described herself as a person of good constitution and strong body, but she acknowledged to having had a similar illness three years previously, which had reduced her more than the present one. The ejecta had consisted of what seemed to her a pint of blood, at about 4 P. M. on three successive afternoons, with very slight antecedent symptoms.

Again a longer interval:—

CASE CLXXXI.—Elizabeth F—, a servant, aged 23, retched up a tumblerful of blood a few days before her admission to St. Mary's, May 30th, 1862. She said the same thing happened eight years before to a greater extent. and that ever since she had, besides irregularity of the catamenia and debility, suffered from time to time with sickness after meals, but had not seen any red in it till the present occasion, a few days before admission. She had again several attacks of vomiting whilst in the hospital, but threw up no sanguineous fluid.

And again a longer:—

Case CLXXXII.—At the beginning of February, 1862, I was requested by Sir Ranald Martin to meet in consultation on the case of a gentleman about 60 years of age, who was gradually dying of excessive vomiting. He was an old Indian, and he described the first beginning of his gastric ailments, leastwise the first thing which drew his attention to the stomach, to be an attack of hæmatemesis thirty years previously. Twelve years after that he had another, and two or three at shorter intervals which I forget. But during the final attack of vomiting, of which he died in the course of the spring, he lost no blood, the cause of death being the excessive exhaustion and emaciation only. It was diagnosed all along, and proved by autopsy to be due to gastric ulcer.

It would probably not be difficult to fill up all the intermediate years with instances, but those quoted are enough to show that blood in the vomit is not by any means a sign of immediately impending danger. It really would seem, unless those bad symptoms detailed in the two consecutive cases (CLXXV and CLXXVI) should be present, to afford in itself a good omen for some time to come.

The appearance of the blood vomited is very various.

Sometimes it is seen in *streaks* among the mixed matters ejected. See case of Helena F—, Case CLXXI, &c.

More commonly it comes in *a gush*, as in James P—, Case CLXXIV ; Selina Y—, Case CLXXIX.

Sometimes it has remained long enough in the stomach to become *coagulated into large masses*, and then it is somewhat hard of ejection.

Case CLXXXIII.—James M—, aged 32, a potman, was attacked at the beginning of April, 1860, with pain of a continuous character in the pit of the stomach. This continued getting worse till the 16th, when in the act of vomiting, to which he had become subject, he brought up about half a pint of blood black in color. In the afternoon of the next day he brought up as much as three pints of thick black blood in masses so tough as nearly to choke him. The tongue, however, remained clean and moist, and the pulse was only 74 ; the heart and lungs were healthy, and he had lost the pain in the epigastrium even when it was pressed. All which things considered, it was not thought right to detain him above four days in hospital, especially as he wanted no medicine.

Sometimes the sanguineous effusion has remained long enough not only to be coagulated, but to be *partially digested, or rather cooked*, by the gastric acid; and then it assumes the reddish-brown color that it does in black puddings or German sausages from a similar partial cookery.   It is more like that than coffee-grounds.

CASE CLXXXIV.—Henry C—, aged 50, was admitted under Dr. Nairne into St. George's Hospital in March, 1842, and died in about a month with an enormous cancerous mass in the liver ; part of this had ulcerated the wall of the stomach by pressure, leaving some bloodvessels with open mouths, which must have continually been pouring out their contents.   There was no cancer of the stomach itself.   The vomit during life consisted of " coffee-grounds" (as technically called), with only an occasional admixture of red blood.

From this instance it is evident that " coffee-ground" vomit is not exuded in the state in which it is seen (for here, of course, it must have exuded red), but has remained for a certain period, I cannot say how long, in the cavity.   The brown stains found in the walls of mucous canals after death are in fact ecchymoses, which have probably existed a long time.

Sometimes the color is still more changed—it turns *green*.

CASE CLXXXV.—John N—, aged 35, a painter, admitted to St. Mary's Hospital April 10th, 1858.   He had constant pain at the cardiac end of the stomach (increased by pressure), waterbrash, and frequent inability to keep down his food.   On the 15th he vomited half a pint of grass-green matter, was intensely acid to test paper, complaining at the same time of pain between the shoulders and of acid rising in the mouth.   On the 16th, before breakfast, he vomited some of the brown matter usually described as coffee-grounds, after which the acid rising in the throat was alleviated for a few days.   On the 28th it was found that there was blood in the vomit ; but he does not seem to have had any gushes of it.   On May 8th he is reported as having vomited the green fluid and blood also.   I have no further notes of the alterations in appearance of his vomiting; but it was relieved by leeches, and he was made an out-patient.

One cannot doubt that the various colors visible in the vomit were due to one and the same cause, namely, blood.   The great acidity of the fluid forbade the idea that it consisted of bile, or even that there was bile enough to cover it.

Another mode in which hemorrhage of the stomach manifests itself, and by no means an uncommon mode, is by the stools being stained black or blackish-red.

CASE CLXXXVI.—Esther R—, aged 34, admitted to St. Mary's October 14th, 1853.   She had been in St. George's Hospital six years before for hæmatemesis, but what she complained of on admission was the passage of blood

for the last two months by the bowels; and truly enough we found the stools
sometimes with inky matter intimately mixed up with them, sometimes exhibiting clean masses of red blood. She was much weakened and blanched by the
loss. Desirous of assigning this to its apparently most probable source, I
treated the patient first with purgative enemata, and then with terebinthinate
and astringent (Iron Alum) enemata, and gave her also Decoction of Bael,
which is said to act most on the lower bowel. Nothing stopped it till she
took m̃xx of Sulphuric Acid with m̃ iij of Battley's Liquor Sedativus three
times a day. A few days after commencing this she had a natural fecal
evacuation, and then improved rapidly under the use of Quinine.

It is clear that the last-used course of treatment must have
acted upon the stomach principally, for it certainly does not stop
bleeding from the colon. And from the stomach came the
hemorrhage on a former occasion; so that I presume it did so
on this also.

Why had she bloody vomiting in one illness, and bloody stools
only in the other? One may lawfully conjecture that the lesion
which was the fountain of the hemorrhage has extended gradually nearer to the pyloric orifice, and was at last so near that the
sphincter did not block the passage through—that is to say, it
was in the pylorus itself.

This escape by the ordinary course of the alimentary canal
is a very dangerous course for the hemorrhage to take; for
instead of causing extraordinary, even unwarrantable alarm, as
hæmatemesis does, it is liable to evade notice till the patient
drains to death. It was nearly doing so in the case quoted.
The same risk was run in the next case.

CASE CLXXXVII.—Sarah G—, aged 33, a housemaid, always enjoyed
good health till the middle of June, 1857, when she was laid up with sore
throat at first. This passed into a wearing pain at the epigastrium, aggravated by food, and accompanied by several attacks of vomiting, during which
she threw up blood. She became an out-patient at St. Mary's under Dr.
Markham's care; and he finding her weakness and paleness increase with
alarming rapidity, and seeing the tongue dry and furred as in hemorrhagic
fever, recommended her being admitted on August 22d. We then found, as
Dr. Markham had suspected, but the patient constantly denied, that she passed
blood by the bowels whenever they were opened. This required to be done
by artificial means, for she was very costive. On one occasion the feces contained a clot of fibrin, washed colorless, as big as an egg. After observing
and examining her for a few days, I gave her

R.—Plumbi Acetatis gr. ij,
Opii gr. ¼. In pilula ter die.

She took this for three days, and then her bowels were open of their own

accord, and she passed a dark feculent solid stool containing no blood. The
pills were therefore left off, and she was treated with occasional doses of
Castor-oil to clear the bowels of the remedy.

But, for some reason or another, perhaps a relapse of the hemorrhage, I
began the Acetate of Lead again on September 16th, giving it her only at
night however. On the 21st she passed a quantity of flocculent fibrin without
blood. On the 23d a blue border was observed along the gums, so the Lead
was again left off, and she does not seem to have lost any more blood during
her residence in the hospital, viz. till October 16th.

This is the worst of Acetate of Lead—you are so likely to
have the chronic poisoning peculiar to the metal induced by it.
It seems to occur in direct proportion to the length of time the
salt is taken, and not to the dose. It is better on this score to
give a very few large doses, even to run the risk of griping
your patient, than many small ones. A couple of doses of ten
grains each will likely enough be sufficient. I have so administered it in hæmoptysis with great satisfaction.

More usually the bloody vomiting and black stools occur at
the same time, and then there is no difficulty in discovering the
true cause of the latter. The following is the most familiar
history :—

CASE CLXXXVIII.—Eliza F—, aged 35, was admitted to St. Mary's
August 21st, 1860, having for a fortnight suffered from vomiting of her
food, tasting and smelling sour. That morning she had begun to consider
her case serious, from having thrown up in addition some clotted blood to
the extent of a few ounces. There was pain in the epigastrium, running
through to the back, and increasing by very slight pressure.

She was ordered a Blue pill and Castor-oil, and then twenty minims each of
Sulphuric Acid and Oil of Turpentine in a mixture three times a day; also
ice, milk, and cold beef-tea, like all other patients with hæmatemesis: but the
next day the treatment was discontinued, as the vomiting had ceased.

There was no more blood thrown up till the 23d; the medicines were resumed, and it ceased. But all along she was passing black stools, as of
digested blood.

Then her bowels became costive, and she took only some Decoction of Cinchona, and was discharged on September 7th.

(These two last cases have been related before, as instances of "wearing
pain.")

To find the remains of blood in the stools is very satisfactory
in hospital practice, in order to confirm the statements of the
patients, which are not always to be trusted. They will talk

about throwing up blood to excite attention, when in reality it is only simple vomiting.

Besides designed imposition, we have also to guard against the mistake of confounding blood coming from another source with that from the stomach. This is easy enough to run into. An example of the doubt has been given in CASE CLXXVIII.

Waterbrash is sometimes found along with vomiting of blood. It is remarked in CASE CLXXIX, Selina Y—, and also in CASE CLVI, Sarah S—.
It was observed again in the following.

CASE CLXXXIX.—Mary S—, a cook, aged 23, was admitted to St. Mary's July 27th, 1853. She stated that her health had always been excellent till six months previously, when she began to experience pain in the chest, and frequently to vomit after her meals. She brought up her food mixed with a yellow (? sanguineous) fluid. She was under medical treatment and got cured. But six weeks before admission the pain returned. Frequently instead of vomiting she used to eject a quantity of clear watery fluid (waterbrash). But what brought her to the hospital as an in-patient, was her having three times lately thrown up blood by vomiting. Previous to the hæmatemesis there had been felt darting pains in the epigastrium. Rest for ten days in the hospital, and half-a-dozen leeches to the epigastrium, put a stop to all her symptoms, and at her own desire she returned to her situation; so there was evidently no sham in the case.

I suspect in these cases, where waterbrash is joined to bloody vomiting, that the lesion which occasions the latter is near the cardiac orifice. You do not have here the pains in the right shoulder which point to pyloric lesions.

I have spoken without hesitation of hæmatemesis as arising from some lesion of the mucous membrane, by which a more or less number of larger or smaller bloodvessels have been broken. The precise mode in which the rupture is effected is not easy to ascertain. But it does not require any very great violence. A blow on the epigastrium not hard enough to bruise the outside skin may, for example, cause it; as in the following.

CASE CXC.—Susan L—, aged 45, the dissipated drunken wife of a laboring man, had a fight with her husband, and got her eye blacked, her back kicked, and a punch on the belly. The two former were bruised, but not the latter. She was brought to the hospital April 3d, 1856, because on going out into the

14

Edgware Road the day after the fight she felt very faint, and threw up a good deal of blood. It was at first considered to have come from the thoracic viscera, seeing that she had no bruises or pain ou pressure in the abdomen. But examination with test-paper, and what she had thrown up on her clothes, showed it to be acid, and to have come from the stomach therefore. The pulse was hardly to be felt; she was delirious, the skin clammy, and the feet cold; so that we had to rouse up her ebbing life with hot water and mustard. · She had also ℥xx of Oil of Turpentine every hour; but as at night she still continued vomiting blood, a slab of ice was laid on the epigastrium, and Alum and Tannic Acid administered by the mouth.

Next day the bleeding had ceased, the pulse became more perceptible, and the mind clear.

There was no return of the symptoms, but I kept her in till the 21st for safety.

In the last history it was mentioned that there being no pain in the epigastrium made the diagnosis doubtful at first blush, though the injury was proved so indubitably to be in the stomach by after events and observations. In the case of Selina Y— (CASE CLXXIX) there was also no pain in the epigastrium. Perhaps a clue to the condition existing under such circumstances may be afforded by the following.

CASE CXCI.—Elizabeth A—, a cook, aged 35, but unmarried, was admitted to hospital under me June 10th, 1861. She had the appearance of good health, and said she had always enjoyed it till five days previously, when she felt so nauseated and giddy that she thought a bilious attack must be impending. A disagreeable rising in the throat lasted all day, and at 10 P.M. she vomited violently and threw up blood. There was perhaps also bile mixed with it, as she said it tasted bitter. This was on Wednesday, and on Friday she again vomited blood and passed black motions, and on the Monday came to St. Mary's. In the mean time she had been taking pills bought at a small chemist's shop, and therefore probably containing Mercury, the usual panacea in counter practice. At all events, she was salivated on admission.

In this case there was no pain in the epigastrium without or with pressure. The liver on admission was found much enlarged laterally and vertically, yet neither was it painful on pressure. She had no medicine.

On the 12th the liver was much smaller. The black color had disappeared from the motions, and she had no more vomiting of blood, and went out on the 28th.

I suppose the cause of hæmatemesis here was congestion of the liver—a condition which is said often to occur in practice, though I confess I can seldom make it out by percussion.

From the above cases I would conclude that the vomiting of blood denotes, if not perhaps an open bloodvessel, yet such a losed pathological state of the mucous membrane as requires a completely alterative renewal, and that such alterative renewal is best brought about by general analeptic remedies, by the local removal of congestion, and by the restoration of capillary circulation through local depletion. I should infer also that the quickest arresters of the immediate hemorrhage are Turpentine and Acetate of Lead internally, and ice externally.

SECTION V.

Acid fermentation of vomit.

The contents of a healthy stomach ejected by any accidental cause have a certain amount of acid reaction from the presence of the acid gastric juice, and indeed this is necessary to their solution. And a degree of sourness in the viands consumed seems to favor their digestion. So that acidity in itself is not a morbid phenomenon in vomit.

Let it be understood, then, that I do not refer in this section to the ordinary normal acidity of the gastric contents, but to the fermentation, principally into acetic acid, of the whole mass, to a decomposition of undigesting food.

For truly in some cases of vomiting the excessive acidity of the mass is a very marked feature. The throat is burned by it, the teeth roughened and the eyes made to smart, just as by taking into the mouth a strong solution of acetic acid. And the sour smell of an acid volatile at a low temperature is diffused through the air. In fact, the whole mass of the ejecta has become acid, instead of merely having acid mixed with it.

The cause is the retention in the stomach of the remains of the meal so long that they have had leisure to ferment throughout, instead of being digested as they became soured by the gastric juice. The cause of the retention most generally is the coating of the lining membrane with adhesive mucus, which impedes the peristaltic movements and prevents the gastric solvents from penetrating the mass. The acidity does not cause indigestion, but the indigestion causes the acidity.

Besides retaining the mass so long that any internal decom-

position to which it may be from its nature apt is aided by
time, the mucus also in itself is an encouragement to chemical
action. A familiar instance of this is the rapid decay of the
urine in a catarrhal bladder. The mucus is probably itself in a
state of chemical change which is thus propagated to the mass.

The decomposition of the mucus is shown by the frequency
with which different sorts of low parasitic growths, or moulds,
are developed in it. The well-defined species *Sarcina Ventriculi*
is the most distinctly marked of these, which, though detected
occasionally elsewhere, certainly finds its most congenial home
in the stomach. In other places it has been found in completely
dead matter (as by Virchow in gangrenous lung) or else a floating wanderer in excreted fluids,[1] but on the lining membrane
of the stomach it may be seen fixed and growing in the mucus.
It is not often that an opportunity occurs of proving to the eye
that such is the habitat of the *Sarcina*—we frequently find it
vomited, but the patients seldom die during their illness, the
complaint not being a fatal one. One indeed has offered itself
to me in CASE CXXXIX, a girl of fourteen, who died in St.
Mary's Hospital of enlarged heart.[2] She had frequent attacks
of sour, but not frothy vomiting, before death, and at the autopsy we found the great curvature of the stomach thickly clothed
with a stringy mucus, very difficult to detach, in the outer layer
of which a considerable quantity of *Carcinæ* were imbedded.

It is pretty clear from this that the mucus, and not the
stomach's contents, is the root-soil of the *Sarcina*.

Being fixed then in a permanent home, and rapidly replacing
with new growths those which are wiped away by the food, the
Sarcina is probably not inert. A great number, perhaps all,
of those cryptogamous plants whose nature is to grow upon
decomposing organic matter, have the property of promoting
decomposition, so that they are not only consequences, but
causes also, of decay. It is found, for example, that the gutta-
percha covering to electric-telegraph wires, when laid down
near the roots of oaks, becomes rapidly rotten from the presence
of a fungus peculiar to that tree. Put your jam in a new cup-

[1] Parkes, "On the Urine," p. 213.
[2] See page 171.

board, and it will keep much longer than in one where mould has previously grown. Saving housewives used formerly to keep what they called a "vinegar-plant;" it is a simple-celled cryptogam found in old casks. If placed in sugar and water, it makes the whole undergo the acetous fermentation in two or three weeks, instead of the process occupying several months. The mould found in yeast (the *Torula Cerevisiæ*), though not essential to alcoholic fermentation, certainly augments the rapidity of its induction ; so that it is entirely in accordance with known physical laws if the presence of sarcinæ, or of the yeast-plant, on the mucus of the stomach should bring on fermentation in the food before the obstructed absorbents have time to take it up. Both have been found in the contents of the stomach ejected; and it is shown by the case I quoted, that sarcinæ at least may exist adherent to the mucus without being thrown up, at least in quantities sufficient to be discovered. Probably oftener than we fancy these moulds are unseen promoters of the rapid fermentation which takes place so mysteriously in the stomach of invalids.

The chief factors in this fermentation, then, I take to be mucus adherent to the walls of the stomach. With this mucus there gets intimately mixed up some dead animal matter which decomposes and moulds and so encourages the fermentation. The dead animal matter often is blood exuding from the gastric parietes; for the mucus is so tough that the food taken into the stomach has much difficulty in blending with it.

The following cases are typically illustrative.

CASE CXCII.—Cornelius K—, a laborer, was admitted to St. Mary's June 27th, 1856. For the last ten years he had been in the habit of occasionally vomiting blood, on the average about three times a year. Of late he had vomited more frequently, but there was not always blood in what he threw up. Sometimes the vomit was very fluid and sour, sometimes it contained yellow matter, and when blood was thrown up it was dark and clotted. He had constant pain in the epigastrium, but that was much aggravated by pressure, and also before and after ejection. His most usual time for vomiting was about four in the morning ; if it recurred again in the twenty-four hours, it was usually in the evening. He was much emaciated by his illness. The tongue was very clean.

After admission the vomiting was found to occur with regular periodicity morning and evening. The matter thrown up was copious, brown, and frothy. It diffused a strong smell of acetic acid. Often, when left to stand, it went on bubbling and frothing, so as to flow over the edge of a small vessel. Once only were *Sarcinæ* detected in it.

He was treated for a week with a drachm of Hyposulphite of Soda three times a day, but it did not seem to check the symptoms at all. He then had eight grains of Quinine with twenty drops of Laudanum every night, and for nearly a fortnight he did not throw up. However the trouble then returned, though not so periodically. He complained of loss of appetite and pain after swallowing fluids. He then had ten drops of Oil of Turpentine three times a day without benefit, and with some increase of pain at the epigastrium. Then he had six Leeches on the epigastrium. After this the vomit, though intensely acid. seems to have contained no more of the brown frothy matter. He left the hospital August 4th, having gained so much flesh that he thought himself able to work.

CASE CXCIII.—Eliza T—, aged 35, a married woman, was admitted to St. Mary's January 14th, 1853. She had a child nine weeks old, and during her pregnancy she related that she had suffered much from sickness. She also frequently had a pain come between the shoulders, which extended round to the abdomen, and lasted about four hours. Since her lying-in the sickness had continued, and on admission she had pain at the epigastrium on pressure.

After admission we found that she had constant uneasiness following her meals, and that she was never at ease till either the food returned spontaneously, or she ejected it by exciting vomiting. On examination of the matters vomited spontaneously, they were found frothy, and containing a considerable quantity of *Sarcina Ventriculi* in each specimen.

She was treated at first with Leeches to the pit of the stomach and Hydrocyanic Acid internally. She got better at first, but then relapsed; when she was put upon two drachms of Hyposulphite of Soda thrice a day. She had no more vomiting at all after this, and went out in eleven days in good health.

CASE CXCIV.—Alfred F—, aged 25, died August 9th, 1859. At the post-mortem examination there was found an ulcer the size of a crown piece in the duodenum, about an inch below the mouth of the gall-duct. The ulcer had penetrated the coats, but the gut was at this point adherent to the pancreas, which had prevented perforation. All the intestines were filled with partially digested blood, and this hemorrhage seemed to have been the cause of death, for the lungs and liver were completely blanched with bloodlessness. He had died fainting from loss of blood by the bowels.

This man had previous to death been my patient at St. Mary's, with jaundice and vomiting of brown, sour, fermenting matter, in which, however, no *Sarcinæ* were found. The tongue had been throughout his illness remarkably clean.

CASE CXCV.—Mr. Edmund L—, aged 27, December 19th, 1861, for the last six months has been in the habit of throwing up an hour after many of his meals. especially dinner, a quantity of sour-scented matter, "frothing like yeast," according to his description. He has no constant pain at the epigastrium, and very little on pressure. His previous illnesses have been an attack similar to this seven years ago, and a sharp pain, like pleurisy, last year.

I prescribed for him—

R.—Sodæ Hyposulphitis ℈j.
Acidi Hydrocyanici diluti ℳv,
Misturæ Camphoræ fl℥j. Ter die.

Dietary.

For dinner.—A mutton chop; stale bread; water.

For other meals.—Milk, with one-quarter of its bulk of lime-water; stale bread or captain's biscuit.

He vomited just after leaving my room, but only once again after commencing the use of the medicine. After a week he was troubled with some intestinal flatulence, which was entirely obviated by fifteen grains of Charcoal every night and some Pepsine at dinner. He had also some Strychnine as a general remedy for his indigestion.

SECTION VI.

Fecal vomiting.

To quote instances and discuss this subject in detail would be to travel out of the province of "The Indigestions" too far; yet a formal notice of it can scarcely be omitted from an enumeration of the morbid matters ejected in vomiting.

Feces, or more strictly speaking matters having a feculent smell, are found in vomit only where a mechanical impediment has completely arrested the onward movement of the peristaltic wave in a lower part of the intestinal canal. It lasts as long as the impediment lasts, and ceases with its ceasing. The cure lies solely in the direct removal of the cause.

Fecal vomiting is popularly ascribed to a reversal of the peristaltic motion; but I do not think it desirable to resort to such a strained explanation. When we reflect that about twelve quarts of secretion, bile, and intestinal juices together, not counting food, are daily poured into the intestines,[1] it is easy to see that the onward wave and absorption have only to be arrested for the ilia to be overfilled, and for their contents to overflow upwards into the stomach. There they naturally produce vomiting, just as they would if swallowed. No reversal is necessary.

No such an arrest takes place most notably and obviously in strangulated hernia, in which without any inflammatory action

---

[1] Bidder and Schmidt.

having arisen we have vomiting, which does not endure long
without becoming feculent. And a like paralysis falls upon
the muscles and absorbents of the bowels in peritonitis, also
inducing vomiting.

It is true that this vomiting in either case, though it tends to
become feculent, does not always arrive at the point of being
so. There may be too little feces already prepared in the canal
to odorize the great mass of liquid; or the arrest of movement
may take place too high in the ilia; or it may be just complete
enough to fill the ilia while yet some feces drain off at the
lower end. These circumstances do not alter the essential nature
of the act.

I think the smell is derived from the contents of the lower
ilia. I doubt much if liquids can overcome the ilio-cæcal valve,
even when paralytic. It is a valve, not a sphincter, and offers
a resistance even in the dead body.

SECTION VII.

Vomiting of unchanged food, and hysterical vomiting.

By far the most common cases of vomiting are those in which
the ejecta consist of food scarcely if at all changed from the state
in which it is swallowed. Sometimes it is moderately acid from
the admixture of a small quantity of gastric juice; sometimes it
is neutral.

It is not my intention here to discuss accidental or occasional
vomiting from external causes, which may be considered rather
the business of the physiologist, but such as having a deleterious
influence on the general health comes under the care of the
physician.

This sort of vomiting happens soon after food has been taken,
and is always preceded by a feeling of discomfort at the epigastrium, often by nausea; indeed, it seems often to be a sort of
semi-voluntary movement to relieve that discomfort.

I do not know but what in all vomiting there is something of
an exertion of the volition; but in some cases there is a much
more marked feature, and the voluntary character may be made
use of in the treatment. It is an important point to observe,

and I shall therefore cite first some typical examples of its being under the control of the will.

CASE CXCVI.—Miss Ellen B—, aged 14 or 15, was under my care in the spring of 1863, for general ill-health and emaciation. There were some glandular swellings in the abdomen and groin, but hardly enough to account for her extreme degree of emaciation, dry skin, and depression of spirits. On further inquiry it appeared that for four years she had experienced discomfort around the waist after eating, and had been in the habit of going away secretly soon after meals and vomiting up what she had taken. She said she could not help it, but yet it appeared that when circumstances prevented her retirement, she was able to restrain herself for a time. Acting on this hint, I desired her parents to exert their authority and forbid the ejection of food. I gave her (with the Iodine ordered for the glandular swellings) some Cod-liver Oil, and sent her to be amused at the sea-side. In a fortnight I heard from her father that she had become convinced of the importance of keeping off vomiting, but that still from habit the food would rise, on which she swallowed it again, according to her own very appropriate phrase "chewing the cud." The best evidence he could give of the success of treatment was, that she had gained in weight four pounds the first week and four pounds the second. This girl, though neither hysterical nor insane, was yet very original in her notions, and had apparently out of her own head devised the vomiting as a relief to epigastric discomfort.

The vomiting may at first have been wholly intentional, but latterly it seemingly assumed a more involuntary and reflex character, as shown by the rising of the contents of the stomach into the fauces in spite of the efforts of the patient to keep them down.

In the following case the vomiting was at first involuntary, and then when the patient got better and was really able to prevent it, she designedly induced it as a relief to her discomfort.

CASE CXCVII.—Emily G—, aged 20, maid-servant, presented herself at St. Mary's September 24th, 1858. She was reported subject to hysteric fits, for which she had already been an in-patient in 1857. She was very pale and leuco-phlegmatic, and the catamenia were irregular. She had an hysteric fit on the 25th. On the 28th she complained much of headache, and began vomiting after all food. The next morning the catamenia appeared. The vomiting continued very obstinate, in spite of Valerian in decoction and tincture, and Bromide of Potassium. Showerbaths at last stopped it, and then she designedly brought it back by putting her fingers down her throat.

In the last case the catamenial period seemed to bring on the gastric symptoms. In the next it relieved them.

CASE CXCVIII.—Mary H—, aged 16, was admitted into St. Mary's Decem. ber 16th, 1853. She was complaining of flatulence in the bowels, eructations, and vomiting of food. She had been wearing a large wooden busk to her stays. The catamenia had been regular since the age of fourteen, except the last period, which was overdue ten days. She continued vomiting everything she tried to swallow all that day, the next, and the next after that. On the 20th she vomited part of her breakfast, and then the catamenia appeared, and she vomited no more, though kept in a few days to see if the symptoms returned.

The frequent connection of vomiting in the female sex with that same state of constitution which induces hysteria and also irregularity of the catamenial periods, leads one to employ Valerian even when the menses are regular; and it is often successful.

CASE CXCIX.—Mary Ann T—, aged 18, was an in-patient at St. Mary's December 3d, 1855, for an attack of continuous vomiting of all food, which had lasted six weeks. She said she had been subject to attacks of this sort since her childhood; but they had not prevented her arriving at puberty at fourteen, and menstruating regularly ever since, having a good appetite, and growing up a plump, cherry-cheeked girl. She was given simple diet with milk and lime-water, with a mixture of Rhubarb and Gallic Acid three times a day.

The sickness continued as bad as ever on the 5th, the bread and the milk taken being rejected exactly as swallowed. Then she was ordered

R.—Infusi Valerianæ fl℥j,
Tincturæ Valerianæ co. fl℥j. Ter die.

An immediate good effect followed. She did not eject the medicine, and the next day she was able to retain the milk. She had a little relapse of sickness on the 10th, but after that continued well, and left on the 20th.

Functional vomiting is sometimes so bad that no remedies can be kept on the stomach, and then a very good expedient is to give that organ a complete rest.

CASE CC.—Esther D—, a stout young woman of 21, was admitted to St. Mary's August 23d, 1859. She had been ailing for a fortnight with headache and general malaise, and pain in the left hypochondrium. On the 21st she had an hysterical fit, and afterwards commenced vomiting very violently. She had great pain across the pit of the stomach, and the vomiting and this pain were immediately induced by an attempt to swallow.

She lay on her back, with the knees drawn up like a person with peritonitis. But, very unlike a person with peritonitis, the abdominal muscles were violently exerted in breathing. Her skin was hot and dry, her pulse 120, her tongue coated with a yellowish fur. Altogether, she was extremely ill, but a good deal of her febrile state seemed due to her being partially under the influence of Mercury, which had been assiduously given up to her admission. The gums were ulcerated, and blood oozed from some part of the fauces staining the vomit with streaks of blood.

Ten Leeches were put on the epigastrium, but they did not seem to relieve the pain.

She was ordered to have no food or medicine by the mouth, but half a pint of beef-tea in an enema, with five drops of Laudanum every three hours.

She was fed in this way for ten days, when some warm beef-tea was given her; that she threw up, but was able to retain it when quite cold. After this she was able to retain her food for a week or so. But then the vomiting returned, though not so bad as before. She was treated with Valerian, with Strychnine, and with Blisters; but the success of each remedy was very temporary. On October 10th, a cold shower-bath was ordered to be taken every morning, and an immediate stop was put to her vomiting. The symptoms did not occur at all again, though she was kept in till the 28th to be watched and to have baths.

Entire rest given to the stomach for a few days will put a stop, final or not I cannot say, to vomiting of a much more chronic character, and even where the souring of the mass seems to point to something more than the functional nervous paralysis which has caused it in the hysterical cases already quoted.

Case CCI.—Charlotte S—, a dusky, tough-looking spinster of 28, admitted to St. Mary's March 26th, 1860. Eighteen months previously she had caught cold, and after three days was taken with vomiting very soon after eating. The matters vomited are the food she has been taking, often accompanied by a considerable quantity of fluid tasting sour. This has made her weak and diminished the catamenia, which are scanty, though regular, and accompanied by a good deal of pain. The last six weeks she had got worse, and could keep no food on her stomach at all.

On admission her pulse was 96, full and strong enough, the tongue was furred, the bowels were costive, the urine was slightly alkaline, not albuminous.

For two days she remained in hospital, vomiting all her food, but taking no medicine ; for, either by accident or intention, I had written no prescription. On the 28th she was ordered to have no food by the mouth at all, but half a pint of beef-tea with five drops of Laudanum as an enema every three hours. She retched no more.

On the 31st some milk and Lime-water, in small quantities at a time was given her to drink, and she kept it down. Still, however, the enema was trusted to as the chief nutriment.

On April 4th she tried a mutton-chop, and succeeded in retaining it. On the 13th she left well.

It was observed that when she took to meat again the urine was acid, deposited urates, and contained a little albumen.

In the following case the habit was still more ingrained by time, and also the color of the vomit induced a suspicion that there was hemorrhage of the mucous canal in some part, either œsophagus or stomach, yet it was cured by a temporary rest.

CASE CCII.—Mrs. S—, a small, swarthy, bright-eyed woman of 22, was brought under my care March 9th, 1861, for constant vomiting of three years' duration, which she attributed to having caught cold during a monthly period, having her courses checked for several months at nineteen years of age, when a virgin. Her food was always returned by the mouth within ten minutes after swallowing, and was unchanged in appearance. Besides this, she also vomited at other times, when the stomach was empty, if her mind was excited. Indeed she did so in my own room, ejecting some reddish-brown granular and flocculent matter, which looked exceedingly like semi-digested blood.

She was not much emaciated ; her catamenia had returned ; she had married six months before I saw her, and had a miscarriage at an early term of fœtal life—four months after marriage. All which proofs of vigor seemed to show that a good deal of nourishment must escape the rejection by vomiting. She said she felt constantly hungry, and was evidently of a hysteric temperament.

I advised that she should be kept entirely without food and nourished by enemata of beef-tea and Laudanum, for a week, whilst at the same time the stomach was further quieted by the application of a few Leeches to the epigastrium, and some Bismuth.

On the 17th I heard from Dr. Woodhouse, of Hertford, who had undertaken to watch the case, that they had not arrived at continuing the treatment a full week, but that for two days the patient had taken food and kept it down. He reported well again on the 18th. But on the 20th he said the sickness had returned with great pain in the right groin. It was again stopped by a recurrence to the treatment for a week. The whole number of days' rest was thirteen or fourteen. On the 28th Mrs. S— was able to take four meals a day, and began Iron and Quinine, which on April 29th she was going on with, having had no return of her sickness. In the spring of 1866 I heard from her sister that she had continued well ever since.

In a former case (Esther D—, CASE CC), the agency of complete rest to the stomach and of shower-baths may have been compared. The first seems more calculated to work a powerful and immediate effect, but that of the latter was more permanent. In the next case I trusted to shower-baths at once, and with apparent success.

CASE CCIII.—Miss Frances C—, aged 21, a younger sister of the last patient, is a very plump girl, with a pink and white doll's complexion; but when she first came to me, on the 4th of April 1866, she and her mother positively affirmed that she hardly ever, for five years, passed a meal without vomiting. She seems a calm, sensible person, impressionable perhaps, but not hysterical. She says that by a violent effort she can keep things down; but that effort produces violent pain at the upper part of the sternum. The vomiting had been worse and her efforts to restrain it more ineffectual since a violent purgative course which had been administered by an occulist to reduce an inflammation of the tarsi. Since then, also, her bowels had been very costive.

I ordered her a cold shower-bath at twelve o'clock every day, and the following draught :—

R.—Acidi Hydrocyanici dil. ♏iv,
Tinct. Valerianæ comp. ℥j,
Infusi Valerianæ ℥j.
Bis die semi-horam ante cibum.

On the 12th this medicine was changed for four drops of the Prussic Acid before meals, and

R.—Zinci Valerianatis gr. iij,
Opii gr. ⅓.
Omni nocte et mane.

On the 28th she called to show herself as quite well ; but she purposed to continue the shower-baths every morning as a substitute for the British tubbing.

The hereditary nature of the constitution tending to this disease is shown by the two sisters being afflicted in a similar way. Seeing their mother one day I took the opportunity of cross-examining her, and found that though she had never been subject to vomiting, yet she used to have regular hysterical fits when a maiden.

Strong mental impressions sometimes have a singular effect both in bringing on and stopping chronic vomiting of this sort. An upsetting shock will induce a relapse; a calmative control, or the idea thereof, will arrest it.

CASE CCIV.—Miss Hannah M—, aged 19, was sent to me by Mr. Ayres, of Ramsgate, in January, 1858. After a preface of hysteria, she had suffered daily from rejection of food for six months, sometimes throwing up everything eaten, but always unchanged in appearance. She had also frequently difficulty in swallowing, so immediate was the rising of the gorge at food.

I gave her some Valerian, and she was soon well. Then she went home to

Ramsgate; and, being soon afterwards frightened by a chimney catching fire, was attacked by vomiting again.

She returned to London and sent for me. Immediately on my visit—without any remedy—the vomiting ceased, and she swallowed everything easily. It was the most "*veni, vidi, vici*" cure I ever saw.

The cure here was purely moral. And of shower-baths, too, I think we may class a great part of the strength among psychical agents. To take a cold shower-bath demands a certain control over the will, even when you are driven into it by a stern nurse, and the bracing up the mind to the resistance to the instinctive shrinking against the shock is the best possible lesson which the physician is at liberty to prescribe. Strength of will is gained by willing.

I have already said that I looked upon the temporary paralysis of the œsophagus as the most essential pathological condition in vomiting,. A confirmatory evidence of this is found in some cases where temporary paralysis of other parts is exhibited along with the vomiting. I extract the following case out of my clinical lectures for 1863 :—

CASE CCV.—I will call your attention to a case of vomiting, namely that of Hannah P—, aged 18. She is a respectable farmer's daughter, and seems to have been much petted at home. She has large black pupils to the eyes, and puffy eyelids, and allows that before her present illness she used to have hysterical fits, but not since she has suffered from what she came here for, namely, chronic vomiting. I should rather call it a rejection of food, for it occurs even while food is being taken, almost always before it is swallowed. This happens at every meal, and has lasted a year and a half, and during that time she has been for a short time in her county hospital with relief but not cure. She has also pains in the back and in the splenic region. She declared she was unable to walk or even to stand without assistance, and when placed upright in the middle of the room she fell down at first. Nevertheless, after a scolding and a decided command to exert her will vigorously, she at last began to put one leg before the other, and progressed a few steps even on the first day. The catamenia had been absent three months, and indeed had never been established at regular periods.

This girl, after retaining mutton-chops and porter for a fortnight, and exhibiting her muscular powers by a walk to Oxford Street and back, went home well July 13th. While in hospital (convalescent) she was employed about the wards; and being thus brought in contact with a young woman recovering from rheumatic fever, she infected her also with a desire to vomit, which, however, was checked in the bud. And I afterwards heard, from one of my pupils, that she next winter not only relapsed into her former condition, but again communicated it to a neighbor of her own age.

There is a peculiarity about hysterical paralysis which in a great many cases guides to its nature—and guiding to its nature is here more than anywhere a most important step in the cure. When you set the patient up on the floor, assisting her with one or two hands, or with your hands under the axillæ, according to the degree of paralysis and the amount of aid wanted, the body is immediately thrown forwards, and all your strength is called for to prevent her falling on the face. Other paralytics fall to one side or the other, or backwards, and do not stumble forwards in this way. The peculiarity was well marked in the above instance, and aided the diagnosis.

So also in the vomiting which is associated with it in its nature there is a peculiarity which is a diagnostic guide. It can generally be controlled by a violent effort of the volition.

And thus to exert the volition is a help not only to the diagnosis, but to the cure, as has been shown by several instances of a typical sort.

Mention was made in a parenthetic addition to the history of this last case of the communicability of this kind of complaint. It is an instance of the mysterious power of sympathy which influences so much of our outward life from the cradle to the grave. In this instinct of imitation there are indeed degrees, but no essential differences between that which helps the infant to acquire the power of speech, and that·morbid condition in which the mind and body are slavishly enchained to reflect the acts engendered by the feelings of another. It is innate in all, but is weak or strong in proportion as the mind is capable of going· alone, or is necessarily in the habit of depending on others. This is the reason why it prevails so much among the female sex. I have had so many instances of hysteria, chorea, and allied diseases whose pathology lies between mind and matter, being caught by lookers-on, that I cannot hesitate to call their transmission an infection by the eye.

Care must be taken to distinguish from this class of cases those in which from some mechanical impediment or lesion the food cannot be swallowed, such as that cited in illustration of another part of the subject at page 194, or the following, where the result being happier, more doubts might have been expressed as to the diagnosis.

CASE CCVI.—A respectable cabman's wife, Ann A—, aged 32, was admitted to St. Mary's July 22d, 1853. She was exceedingly emaciated, weak, anæmic, and had a loud murmur, probably from anæmia, with the first sound of the heart. For a month she had been obliged to reject her victuals after chewing them, from inability to pass them further than the back of the throat. They seemed to stick at the level of the os hyoides. From this point a pain ran to the back of the neck, between the shoulders. Quite at the posterior part of the fauces the mucous membrane looked redder than elsewhere, and was redder still lower down.

She was ordered rations of beef-tea and milk, and the following electuary:—

R.—Bismuthi Trisnitratis ʒj,
Sacchari fæcis ʒiss.
Fiat electuarium quotidianum, cujus lambat panxillum subinde.

When able to swallow better, she had some Bark and a blister on the throat. She lost the pain, was able to swallow, and left much relieved on August 7th.

It will be seen here that the food is not swallowed at all, and there is no emetic effort. It is simply rejected.

I am glad of the opportunity in citing this case of again recording the good effects of Bismuth, alluded to under Case CLXIV.

To sum up, I would deduce from the very common class of cases of which I have cited typical examples:—

1st. That the chronic vomiting of matters swallowed unchanged immediately after swallowing is almost peculiar to women.

2d. That it is allied to hysteria.

3d. That, like hysteria, it is now more a mental, now more a bodily affection; now more under the patient's control, now less.

4th. That the efficient employment of drugs being in a manner barred by their rapid ejection, other means are more imperatively called for in this disease than in most others.

5th. That the weakening of the patient's will being the marked feature of this morbid constitution, the strengthening of the will is the best antidote.

6th. Rational persuasion is available in some few, extremely voluntary, cases.

7th. The most powerful remedy is the cold shower-bath, for the reasons given above.

8th. When physic can be retained, the most efficient is Valerian.

9th. A forcible change of habit by resting the stomach, and giving it nothing to bring up, is a valuable aid; but it is doubtful how far it would answer without other remedies.

SECTION VIII.

Vomiting in pulmonary consumption.

CASE CCVII.—Cyrus K—, aged 22, came in July, 1855, with a complaint of languor, sleepiness of an afternoon, weight at the epigastrium an hour or two after meals, and occasional vomiting in the morning. He had had a good deal of hard work latterly, and attributed his indigestion to that. But his mentioning a cough induced me to examine his chest, where I found marked deficiency of respiration and dulness in the apex of the left lung, and crackling in the lower lobe of the same side. He was also a good deal emaciated, and he owned to having spat blood before he was ill. I thought there was tubercle just beginning to soften; gave him for a time Lime-water and milk, Cod-liver Oil, and Steel Wine after meals. And then I urged him to go to the West Indies, where he had connections, for the winter.

In October, 1861, he came again, telling me that he had gone to Bermuda and stayed, not only for the winter, but ever since. In 1858 he had spitting of blood, and he had yellow fever in 1859, but had grown fat in spite of them; and he had continued well till he was now come to England, where, after a few months' holiday, he found his old dyspepsia returning, and was wisely resolved to go back to his more appropriate home. There was crackling in the apex previously dull, but I do not think the lungs had got materially worse.

This is the slightest degree of stomach derangement, for it did not even prevent the taking of Cod-liver Oil.

CASE CCVIII.—An unmarried lady of 32 was sent up to town for my opinion by Mr. Gardner, of Painswick, May 22d, 1863. She had been ill since the previous October with vomiting after meals. The food returned at short intervals in mouthfuls in an undigested state. The matters rejected were almost always free from acidity. Sometimes this would begin in the morning and continue all day, sometimes would not come on until the evening.

She had a slight feeling of weight or oppression at the epigastrium, but there was no distension or tenderness.

Her general health had not suffered much, the menses remaining regular, and at these periods she thought she was better. Though she had some cough, it was not a marked feature in the case.

15

On examining the chest, I found evidence of crude latent tubercle in the lungs. There was deficient respiration in the right apex, and a bronchial interrupted expiration in the left.

The degree of dyspepsia and the degree of tuberculosis are not proportioned to one another. There was in this case much more vomiting and less tubercle, or less advanced tubercle, than in the last. Perhaps it was because of the patient's sex.

CASE CCIX.—William J—, aged 21, was admitted to St. Mary's August 21, 1857, for pulmonary consumption of two years' duration. (The upper part of the chest was much flattened, and the shoulders drawn forwards; there was bronchophony and bronchial breathing, and various creaking rûles in the apices of both lungs, most in the right.) He suffered from several dyspeptic symptoms, and among them from vomiting. He stated, however, that this latter only occurred if he attempted to move about and take bodily exercise after meals.

He was able to keep down Cod-liver Oil if he remained quite quiet afterwards; and upon that, and Iron after meals, he gained two pounds in weight between August 28th and September 5th. He was then treated for a week with Hyposulphite of Lime (eight grains three times a day), but gained only one pound in that time. His sickness never troubled him as long as he kept quiet and rested in the hospital.

The sickness only on exertion looked as if it depended on general weakness, rather than on any morbid condition of the stomach.

I introduced the last clause in the history, not as especially bearing on consumptive vomiting, but to take an opportunity of saying that I have not found Hyposulphite of Lime such a good renewer of life as Cod-liver Oil. Some consumptive patients did not gain any weight at all under its use, in spite of the improved diet of the hospital.

The next two cases exhibit the coming on of vomiting in consumption coincident with the first hæmoptysis.

CASE CCX.—Emma K—, aged 25, was sent from a penitentiary to St. Mary's Hospital, on account of her failing health, July 22d, 1853. She described herself as having been weakly for a couple of years, but had no marked symptoms till a fortnight before, when she began coughing up blood. At the same time she commenced vomiting, and continued to throw up everything she took. She was rapidly losing flesh.

On stethoscopic examination the apex of the right lung was found dull and very painful when pressed.

ľ Gallie Acid (for the hæmoptysis), Hydrocyanic Acid, Morphia, Quinine, were severally given, without any advantage to the sickness. Chloroform in eight minim doses was of temporary use, but the most effectual remedy was Opium in grain doses. Under this her sickness ceased, and she immediately began to gain flesh and strength, and left the hospital in fair condition August 26th.

With the vomit there was at first a good deal of light green fluid, probably blood swallowed and digested.

Case CCXI. Bridget S—, a domestic servant, was admitted to St. Mary's January 26th, 1857, with pulmonary consumption of eighteen months' duration. It had begun with hæmoptysis and vomiting. The vomit usually was merely the contents of the stomach, but sometimes she brought up clots of blood at the same time.

The good effects of Opium in checking the vomiting of consumptives was alluded to in Case CCX; the following illustrates it still more strongly.

Case CCXII.—B.'s Anonyma, aged about five-and-twenty, was placed under my care in March, 1861. She had a large vomica in the upper lobe of the left lung, and the greater part of the lower lobe impervious with tubercles; but she had suffered very little from pulmonary symptoms, would not hear of her being in a consumption, and talked about going to dances in a low dress as soon as she could get about again. But she was utterly prostrated to her bed by the constant vomiting of all she ate, and retching when she ate nothing. The bowels were obstinately costive, and she had taken as much as twelve grains of Extract of Colocynth without effect.

I gave her Opium, beginning with a grain and augmenting it to six grains daily. Then the vomiting ceased, and she recovered her appetite and fondness for luxurious living. She ate twelve shillings' worth of strawberries (in April) daily, and an immeasurable quantity of brown bread ice. Her bowels recovered their functions, and she passed naturally colored and formed stools in spite of the Opium. She slept naturally and easily without excess or stupor.

She died in the summer, but was able to keep off her vomiting to the last by the help of the Opium. I think, however, she increased the dose. So that her ending was made much more easy, and probably postponed by it.

A different form of phthisical vomiting, sadly common, is that which occurs in an advanced stage of large vomicæ, from the nauseousness of the sputa. It is very distressing to the patient; and almost equally so to the physician, for his remedies afford little or no relief.

This vomiting is more frequently found when the vomica occupies the lower or middle parts of the lung than when it is at the apex, for the reason that in these first-named situations the cavity is more apt to eat itself into the neighborhood of the ribs, and it is the contact of bone which makes the pus grow so horribly fetid.

CASE CCXIII.—Captain H—, a man of fine build and healthy family, aged about 36, came under me in August, 1862. He had long been subject to cough, but had never spat blood. His complaint was of considerable pain in the right side, which, as an old Indian, he attributed to what they call "liver." In the lower lobe of that lung there were dispersed cracklings to be heard with the ear, and there was slight general dulness on percussion diffused through the lobe. This was in front; behind the sounds were healthy. The sputa consisted of transparent mucus.

Leeches and Chloroform considerably relieved the pain for the time, so I suppose it was dependent partly on pleural inflammation.

In October of the same year his pain in the side was less sharp. The expectoration and cough were worse. There was very marked dulness with absence of respiration in the right infra-mammary and infra-scapular regions. He went to the South of France for the winter.

In the May of 1863 I saw him again. He suddenly, during a violent fit of coughing, had thrown up a pint of pus, and continued coughing it up. If the cough ceased for a little time, the pus would collect, and then, on being expectorated, tasted and smelt so intolerably nauseous that vomiting invariably was produced. This took place always every morning, when the matter had collected during the night.

I one day examined some of this fetid sputa under a microscope, and found pus-globules of various sizes, some regular and normally granular, some swollen and exhibiting their nucleus, fat in globules, granular masses (? rotten fat), tabular crystals of cholesterine, and spicular crystals which my microscopic lore was not sufficient to enable me to identify. It was anything but "laudable."

Poor Captain H— was very patient, but a more distressing case I have rarely seen, so excessive was the disgust from the constant vomiting and fetor of the expectorata.

He got a little better for a time at Malvern in the summer, and was kept from sinking by Cod-liver Oil and Quinine. But the abscess or vomica never healed up, and continued to secrete fetid pus. The dulness on percussion also increased in extent, so that there was scarce any breathing over the whole lung; I supposed that a fresh deposit had taken place of tubercle, or whatever other matter solidified the pulmonary tissue. He died at Lisbon the next winter in an extreme state of consumption. However, he never had any diarrhœa.

The vomiting and fetid expectoration never seemed bettered by any medicine, except perhaps Quinine, and that he could take very little of, such a headache it gave him.

It is impossible to bring to bear on a pulmonary lesion any of the usual applications that surgeons make to fetid abscesses; or else in such a case as the above one would be glad to use them. To bore an opening through the thoracic walls would probably be a great comfort to the patient and a prolongation of life; and I should be glad to find the operation consented to. But it is impossible conscientiously to speak of it as likely to effect a cure, and naturally the knife is shrunk from—I do not expect ever to try it.

SECTION IX.

Occasional causes of vomiting.

CASE CCXIV.—Wilson M—, aged 29, a coach-painter, was taken in by me at St. Mary's Hospital the morrow of Christmas Day, 1862. He had always been a strong man till the middle of November, when he was aware of a pain across the loins and down the thighs, a tightness across the belly, and head-ache. At the same time he noticed first his face, then his body, swollen of a morning before going to work. He got himself cupped in the loins, but thought it did the pain no good.

On admission, the whole person was anasarcous; the pain in back and belly remained; the bowels were regularly open once a day; his appetite for food was sufficient; his thirst more than natural; the urine was albuminous, pale, scanty, of the specific gravity 1.012.

Hot baths, a dose of Jalap, and Nitre draughts three times a day, were prescribed. He was kept in bed.

On the 31st of December the legs were natural, and on the 3d of January the general anasarca was nearly gone. He was ordered—

   ℞.—Tincturæ Ferri Sesquichloridi m̥xv,
    Tincturæ Digitalis m̥x,
    Misturæ Camphoræ fl℥j—Ter die.

During the use of this his feverish thirst diminished. He was made an out-patient on January 16th, his urine remaining albuminous.

On the 18th of February he was readmitted as too ill to be an out-patient. He had become very feverish and thirsty again, his tongue was white, and he had dizziness of head and obscuration of eyesight. At the same time he had been attacked with vomiting, and thrown up as many as twelve times in the day. He had no return at all of the dropsy, though the urine remained albuminous.

A day's rest in bed reduced the frequency of the vomiting to once a day. But that and the feverishness continued for several days longer. Hydrocyanic Acid seemed to do him good.

This case shows that it is the albuminuria, and not the dropsy (as some have suggested), which causes vomiting in Bright's disease. I think it very likely that the gastric glands may in the more obstinate of these cases be degenerated after the same fashion as the kidneys. In a continuous series of a hundred post-mortem examinations recorded by Dr. Handfield Jones in the "Medico-Chirurgical Transactions" for 1854,[1] out of twenty-four cases of real degeneration only seven had the glandular structure of the stomach completely healthy.

CASE CCXV.—Ann F—, aged 52, married, was admitted under me at St. Mary's March 18th, 1853, complaining of a general throbbing, faintness, and what are recorded in the book as "general dyspeptic symptoms," of which the most marked were vomiting and tightness across the chest.

On auscultatory examination, the ribs were found rounded and immovable, and the cardiac region overlapped by emphysematous lung, so as to be, with all the rest of the chest, unnaturally resonant.

Hydrocyanic Acid and Chloroform relieved the dyspeptic symptoms somewhat. The remainder of the history has no bearing on my present subject.

CASE CCXVI.—Jane K—, aged 27, having had a distorted spine from childhood, it was impossible to ascertain precisely the anatomical condition of the lungs; but, as the heart was healthy, the probability is that the shortness of breath she suffered from arose from pulmonary emphysema. The reason of her coming into St. Mary's in June, 1856, was frequent vomiting, which exhausted her very much. This did not occur in any relation to meals, but at night. She was benefited by Hydrocyanic Acid, a Jalap purge, and a fortnight's rest; after which she went out without complaint.

It is worthy of remark how the worst time of the twenty-four hours for the lungs of the broken-winded is also the worst for their stomach. It is at night that their paroxysms of dyspnœa come on, and at night this woman had hers of vomiting.

Dr. Hyde Salter, in his useful monograph, remarks: "It is very rare to see an asthmatic with a perfectly sound, strong stomach, about which he has never to think, and in the history of whose case dyspepsia has no place. Sometimes the dyspeptic symptoms exist in a very aggravated form, and they are frequently such as to imply that the stomach disturbance is one of deranged innervation—that its sensibility, or its movements, or the nervous superintendence of its secretion is perverted. In these cases the stomach and lung symptoms are part of one morbid condition; the whole thing is deranged pneumogastric innervation, the dyspeptic symptoms being the manifestation of the gastric portion of this deranged innervation, and the asthma of the pulmonary portion of it." He gives then a good example of the alternation of the diseases, asthma and vomiting.[1]

It has been observed already, in CASE LI (page 75), that liquids often disagree more than solids with emphysematous and cardiac asthmatics.

Some cases of intermittent vomiting seem connected with ague poison.

CASE CCXVII.—Stephen A—, aged 54, an active, well-to-do farmer from the marshy neighborhood of Colchester, came to town to consult me May 24th, 1860. He stated that he had always been a temperate man, and appeared to speak the truth. He had suffered from weight at the pit of the stomach, especially in wet weather, for near upon ten years, and at various times has occasionally thrown up some stringy phlegm from the stomach. (*Gastric catarrh.* See "Weight.") In the summer of 1859 he had rather a severe touch of ague, which pulled him down a good deal, and he had never been quite the same man since. The stools were sometimes "yeasty," sometimes dark, rarely natural. Since his ague he had vomited every other day, and at the time of the vomiting had a spasmodic pain just beneath the ensiform cartilage. He occasionally had vomiting and occasionally had pain at other times, but seemed pretty clear as to their general tertian character. I ordered him five grains of Quinine every night and morning, and as I did not hear of him again I presume it was sufficient to effect a cure.

In the following case, a living irritant seemed the cause of vomiting.

CASE CCXVIII.—Bridget W—, aged 20, spinster, was admitted to St. Mary's January 11th, 1861. She had very obstinate vomiting, especially in the morning, which resisted Oxalate of Cerium, Bismuth, and Pepsine, which were severally tried. Of the three, Pepsine seemed of most use. Then she had a diarrhœa, and passed two worms (probably the ordinary Round-worm, but I did not see them). It was found that she had been very subject to worms since the age of fourteen, and was of a very mucous diathesis, having leucorrhœa and frequent catarrhal coughs. She was ordered Turpentine, but I have no note of the result.

Hæmatemesis has been spoken of as the result of violence. Chronic vomiting also may be produced by the same cause.

[1] Salter, "On Asthma," chapter xii. section *a*, page 216 (Edit. of 1860).

Case CCXIX.—George S—, aged 21, a porter, was admitted to St. Mary's September 25th, 1858. He had had a fall six months previously, cutting his head and otherwise knocking him about. He was very giddy afterwards, and felt a violent pain near the navel, to which Leeches were applied with relief. The pain extended backwards to between the shoulders. The next day vomiting of nearly all ingesta commenced, and continued more or less all the six months.

On admission, there was dulness on percussion, and tenderness to the right of the epigastrium; but this proved afterwards to be due to feces impacted in the colon.

The vomiting was very constant. He was obliged to be fed on a couple of mutton-chops very slightly done and pounded up, of which a teaspoonful was given every two hours with a little milk. He had fifteen grains of Pepsine every four hours. But he kept on vomiting, and lost 2½ lbs. in weight.

On October 9th he was put upon Liquor Calcis, and milk and beef-tea, continuing the Pepsine. Then he did not vomit for eighteen days, and got back to meat; but had to leave it off after a few days and resume the liquid. He gained at one time six pounds in weight while free from vomiting, but lost some of it during a relapse.

A good deal of hard feces was brought away by clysters, apparently with relief.

He complained of much pain in the epigastrium, which was relieved by a Blister dressed with Acetate of Morphia sprinkled on the raw surface.

He got gradually better, with occasional relapses; due perchance to imprudences, and was discharged November 24th.

The pathological explanation of this case I take to be partial paralysis of the intestinal canal by a sudden shock to the solar plexus, very much as the voluntary nerves are paralyzed in a concussion without lesion of the brain. This would account for the loss of vitality in the colon and stomach and œsophagus at the same time. Remark how gradual and slow was this man's restoration.

In all physical lesions of the nervous tissue, the main elements in the treatment are time and repose. With these the foolishness of prescriptions scarce impedes the cure; without these the most judiciously selected means fail. The slowness of renewal is very distinctive of the nervous system, and is explained in a great measure by the difficulty exhibited in that tissue of parting with its substance by vital metamorphosis. In his experiments on the effects of inanition, M. Chossat, comparing the losses of substance in different tissues, found that the nervous suffered least; and indeed it retained its full weight after several weeks' starvation.[1] It is the true *ultimum moriens* of physiological interstitial decay, and of course it is the last and slowest renewed.

I have already, in a chapter on the social habits leading to indigestion, given examples of alcohol as an occasional cause of chronic vomiting;[2] but I omitted to mention a drug which I have found useful in that complaint. It was first used in this way by Dr. Marcet.

CASE CCXX.—Jonathan B—, a middle-aged gentleman, asked my advice in May, 1861, for nervous trembling, indigestion of food, and vomiting, arising from indulgence in spirit-drinking between meals and in the forenoon. I gave him

> R.—Zinci Oxidi,
>   Pilulæ Aloës cum Myrrhâ, āā ʒiss.
>   M. fiant pilulæ xx, quarum sumat unam ter die.

After taking these for ten days, as he afterwards informed me, he was quite well. Of course he had left off the excess of alcohol. Still I think the Zinc was useful.

CASE CCXXI.—Charles W—, a patient with tubercular lungs, who used to consult me in the spring of 1862, had lodgings at Greenwich in an open situation for the sake of the air. He seemed to get all the worse, and took to vomiting in the morning, and having pain in the epigastrium. He always felt so much better during the day, and got so much worse during the night, that I was led to inquire more particularly into the peculiarities of his lodgings. I heard his bedroom was colored green, and on his bringing by my desire a piece of the wall-paper, I found it tinted with a light powdery arsenite of copper. He lost the dyspeptic symptoms when he changed his apartments.

I feel no doubt that here the arsenicated water color was the cause of the vomiting, though that is not its universal effect. When made into an oil paint, Scheele's green is not nearly so dangerous.

Mechanical compression of the epigastrium by tight lacing, and by handicrafts where that part is exposed to injury, has already been spoken of as a cause of emetic indigestion. Another mechanical cause is umbilical hernia, though I cannot now lay my finger on a case in point—I have made an error in tran-

---

[1] "Recherches expérimentales sur l'Inanition," Paris, 1843, page 91.
[2] Page 123.

scribing the reference. I do not, of course, refer to the acute vomiting of strangulated hernia, but to chronic vomiting.

Allied to these mechanical causes is cancer of the stomach and parts adjoining, which often causes vomiting. Cases of this and of ulcer, however, I will postpone to a chapter on the morbid anatomy of the stomach, for they illustrate that much better than they do the phenomena of indigestion. And the same may be said of gouty inflammation of the stomach.

Whether the vomiting of pregnant women will be capable of explanation on mechanical principles, or whether we are to look to increased knowledge of the nervous functions to interpret it to us, is doubtful. Its occurrence in the morning might favor either view, for there is both a change of mechanical relations in a change of posture, and also a marked weakness of nerve force, at that hour. The vomiting of pregnancy may often be stopped by directing the patient to leave off alcoholic beverages, of which the feelings of weariness from having to drag about an extra weight and general lowness of spirits often induce women to consume an extra quantity during their breeding. Instead of taking more, they ought to take less alcohol than usual at that time; instead of blunting their vitality, they ought to leave it free, for it has its fullest work to do. A simple milk diet, guarded by alkalies, for a few days, will frequently quite check the vomiting of pregnancy.

Vomiting the food first taken seems sometimes to arise from simple nausea consequent on taking food with repugnance, and is then curable by remedies which awaken a natural appetite. For instance—

CASE CCXXII.—Amelia D—, aged 20, was admitted to St. Mary's June 19th, 1857. On admission, her general condition was as follows: *stature* small; *weight* 84½ lb.; *complexion* fair; *skin* healthy; *pulse* 92, even, feeble; *tongue* clean, flabby; *bowels* daily; *urine* normal; *catamenia* monthly.

She was well fed, and not overworked; but her employment necessarily confined her a good deal to the house. The thorax was healthy, though she told a tale of having had cough and hæmoptysis.

She complained of pain in the left side, and sickness in the morning, especially after breakfast. Her appetite was very bad, and the sight of food made her gorge rise at it; but still she forced herself to eat.

She was at first dieted on milk guarded with Lime-water, rice-pudding and ice, and took a grain of Opium every night.

But after five days she was no better, so the Opium was left off, and ten grains of Boudault's Pepsine powder three times a day substituted for it.

In three days her appetite had returned, the vomiting and nausea had ceased, and she spontaneously asked for meat. She continued taking that with relish and without vomiting.

It would be easy to cite cases where drugs had effected the same purpose, but I chose rather to select an instance of the simplest form of restorative treatment (namely, the direct replacement of a deficient digestive solvent, so as to aid formative nutrition[1]), in order to direct the reader's thoughts to the true theory of healing.

SECTION X.

Sea-sickness.

The principal cause of the vomiting produced in those unused to it by the motion of a ship or carriage, by swinging, waltzing, and the like, I believe to be the relaxation of the œsophageal sphincter by the vibration. The body being at rest, or rotated on itself, and the œsophagus hanging somewhat loose in it, the jar is strongly felt, and the involuntary plexus supplying these muscular fibres is temporarily paralyzed by it. In fact, a succession of small strokes produces the same effect in unaccustomed nerves as one single severe concussion.

In both cases use begets hardness : those who are exposed to much knocking about—wrestlers, prize-fighters, huntsmen, &c. —will get to stand blows that would once have stunned them ; and the jar and swinging of the gullet and stomach in time ceases to be followed by relaxation of the sphincter. And some persons and animals, from perhaps a peculiar structure (I will not call it a malformation) of the parts, never experience seasickness at all.

The earliest notice one has of this œsophageal palsy is faintness or giddiness, which in a healthy and normally sensitive persons always precedes sickness, whether arising from the poisoning of those nerves by an emetic, from blows on the stomach or head, or from swinging motions.

The relaxation of sphincters is always followed by the expul-

[1] See "Lectures chiefly Clinical," 2d Introductory Lecture, "Disease and Cure."

sion of the contents of hollow organs. Directly the anus is opened, the abdominal muscles act in forcing out the feces. So also with the bladder. And immediatly after the relaxation of the œsophageal fibres, the diaphragm and its colleagues energetically press upon and empty the stomach.

Even after it is emptied, they continue to be spasmodically contracted, and the unhappy landsman lies retching and roaring, with nothing to throw up except a little bile, which the squeezing has forced backwards through the pylorus. This is the worst part of the ailment, just as cramp of a stump or of a limb lying loose is more painful than when the muscles have some resistance to act upon.

Exposure to cold, either local or general, makes sea-sickness worse, by lowering the vitality of the nervous tissues—partially numbing them, in fact. Artless landsmen often aggravate their misfortunes by remaining on the wet, chilly deck in blustering weather; the more experienced sufferers have avoided a great deal by immediately going below and getting warm and comfortable before the nauseating stinks begin to be rife. Sitting with the back leant against the funnel is also of use, if you cannot get below.

I have never had any valid experience of the proposition made of putting ice down the back as a preventive, not having travelled in any weather rough enough to be a good test since it was made public. I am interested in the result, as it would a good deal affect my view of sea-sickness. Its having been set forth as a specific for cholera probably may prejudice the public against it, but should not influence the calm judgment of an experimentalist.

A thick belt or Spanish faja will sometimes keep off sea sickness, partly from the local warmth over the epigastrium, partly perhaps from the compression keeping the stomach and its neighbors mechanically steady. The benefit of lying flat on the back arises probably from a similar cause.

Temporary stimulants—ammonia, spirituous liquors, chloroform, opium—keep it off for a short time in some individuals; but oftener, I think, the reaction comes on very soon, and their last state is worse than the first. The beneficial action of stimulants lasts longer if they are combined with carbonic acid, as in effervescing drinks. Aboard our Channel packets, "Soda and B." is popular; and I have found devilled lobster and champagne a real blessing in some rough weather off the coast of Portugal: I have also tried good bottled porter *non sine gloriâ*.

The powerful effect of mental emotion in bracing us up against sea-sickness is very remarkable, and associates its pathology closely with that of other functional paralyses. This is said to be observed in a striking manner in shipwrecks, when fright renders every soul alert, though before there was any danger they had been exclaiming that they recked not what became of them. Of that I have no experience; but I remember once lying prostrate with nausea in a Peninsular steamer, when the captain, knowing I was a doctor, begged me to come and attend to an engineer who had got rolled into the machinery. Only one finger was crushed, but the binding up that and the encouragement of the frightened man quite cured me, though to an unapt surgeon the mixture of blood, grease, and coal-dust, entailed by a machinery accident, is not agreeable.

Almost always, the inconveniences of sea-sickness in a previously healthy person cease with the cause: landing or smooth water sets all to rights. But sometimes, as in the following case, there is illness afterwards :—

Case CCXXIII.—S. S—, a middle-aged gentleman in fair health, accustomed to suffer in rough weather, went for a day's trip from Sorrento to Capri in an open boat. There was a good deal of wind both going and coming; but, contrary to his custom, he was not sick. He remained well that night; but on the morrow he was attacked with spasmodic pain in the right side of the epigastrium, so that I almost thought he must be passing a gall-stone; but the bowels were opened naturally with formed feces of normal coloring. The pulse also was unaffected. He made himself vomit with warm water and putting his finger down the throat, and brought up unaltered food which he had eaten the day before. This vomiting was of no immediate use in giving relief, but it established the diagnosis of stomach-ache *vice* gall-stones.

The pain spread over the epigastrium, and as it spread became less, and gradually ceased the next night, helped probably by hot fomentations. But a certain soreness remained for a couple days more.

I fancy in this case the paralysis had affected the muscular fibres of the stomach more than those of the œsophagus, and so a morbid condition was engendered, described in a former chapter as spasmodic pain of the epigastrium, or stomach-ache.[1] The gastric glands were also paralyzed; so the food was undigested, and not being of a fermenting nature, was unaltered.

Sometimes sea-sickness passes into a condition of chronic vomiting.

CASE CCXXIV.—Eliza W—, a young single woman, was quite well till the beginning of October, when she came up by a Hull steamer to London during the equinoctial gales. She was violently sick on the voyage, and fancied she twisted something inside her. On landing the sickness did not cease, but continued till her admission to St. Mary's, November 14th, 1863. She had also got very hysterical, and said that one day she was quite paralyzed. Her tongue was very foul, her pulse natural, her bowels constipated, the urine painful to pass. Her right eye also became painful, and she could not raise the lid.

Under Quinine and a Chloroform poultice she got much better by the 26th of November. Then on its being found that the pupil of the painful eye was dilated, and the internal *rectus oculi* paralyzed, it was thought right to leech her temples.

On the 17th of December she was attacked with vomiting again, and on the possibility that the brain was inflamed the head was shaved and blistered, and Iodide of Potassium given. Apparent relief followed, but the patient got very hysterical. Finally the vomiting was stopped with Valerian, but not till the end of February.

I have never heard of such a thing as chronic vomiting arising from sea-sickness in a man; and I suspect that the pathological interpretation is the passing of the temporary morbid condition into that which was described in a previous part of this chapter as hysterical vomiting; hysteria being much more common in the female sex than in ours.

I should presume Shower-baths and Valerian would be the best cures; but I have not the authority of experience, as the contingency is rare.

SECTION XI.

Review of treatment.

Ice or iced water swallowed is often most useful in the acute vomiting of fevers and of cholera. It stops the straining and relieves the distress. I presume it acts on the same principle

[1] Page 162.

as a shower-bath, by a revulsive shock to the nervous system. It is also a good astringent in bloody vomiting.

Rest in the horizontal posture and absence of excitement is a powerful remedy. It was adopted in all the cases quoted in previous pages.

Milk and Lime-water as a sole diet for a few days is an application of the same principle of rest. It also was a *processus integer.*

Leeches will be seen to have often stopped vomiting. Not only in gastric ulcers, but in all anatomical lesions, including congestion of the stomach, their utility is readily understood.

Brandy, plain or burnt, in teaspoonful doses, is a favorite domestic remedy. It relieves the faintness which accompanies vomiting, and perhaps may be some check in acute cases, such as sea-sickness (q. v.). But it is obviously unsuited for severe or chronic disease.

Champagne and other *effervescing* stimulants come into the same category. I fancy the effervescence diffuses the stimulating ethers more quickly than when they are taken flat.

Chloroform internally may also be classed with them. Externally on the epigastrium it has not appeared to me of certain use.

Hydrocyanic Acid, when the vomiting arises not so much from a fault of the stomach itself as from a secondary condition of the nervous system, as in pregnancy, diseased heart, abdominal tumor, and in pulmonary consumption.

But in pulmonary consumption the most powerful remedy is *Opium*. In gastric ulcer also it is invaluable, and in painful malignant tumors. The *Valerian* and *Shower-baths* are both useful to the same class of cases. There are no remedies by which I have oftener stayed chronic vomiting; simply because nervous debility is the most usual cause of it.

Carbonates of Magnesia, Soda, and *Ammonia*, and *Hyposulphite of Soda*, are especially indicated where there is acid or alcoholic fermentation of the vomited matters, whether *Sarcina ventriculi* be found there or not. They act palliatively in arresting the decomposition.

Creasote is an uncertain remedy. I confess I cannot find what cases it is suited to. It has never done any good where the other remedies have been fruitlessly tried.

The administration of food in cases of chronic vomiting is a matter of much moment. We must not let our patients sink for want of it. Even when milk and Lime-water does not check the retching, it is by far the best diet; and in teaspoonfuls it can almost always be kept down.

The risk of being starved to death from vomiting is not purely hypothetical. A young woman came under my care at St. Mary's in 1857, who had been deserted by her lover. She had had violent hysteria, and an utter inability to keep any-thing on her stomach for some days already; the pulse failed, and the tongue was dry and brown. An attempt was made to restore life by means of nutritive enemata, but in vain. At the post-mortem examination every organ was in a completely normal state, and the catamenia were still flowing from the uterus. She had died of inanition only. The nutritive enemata were proved right, however disappointing the result was.

The treatment of sea-sickness has been discussed a page or two back.

241

CHAPTER VI.

FLATULENCE.

Section 1.—General Remarks. Section 2.—Eructation. Section 3.—Intestinal flatulence. Section 4.—Colonic flatulence. Section 5.—Treatment of the several sorts.

SECTION I.

General remarks.

WHEN we speak of flatulence it must be remembered that we must not set down all the air contained in the intestinal canal as morbid: we are not like old-fashioned nurses to be always looking upon "wind" as an evil. A certain amount of oxygen is wanted to aid in the acidification which is necessary to digestion; and as this oxygen is to be derived from the atmospheric air, it implies the presence of still more nitrogen. Carbonic acid is a sedative to mucous membranes, it is the natural atmosphere of all internal parts, and they become irritated and inflamed if they are deprived of it. Growth in wounds and normal secretion in mucous membranes go on naturally only when thus defended against external influences. Again, it is an important agent, indeed it may be called a great moving agent, in the digestion and circulation through the body of aliments needful to growth.

There are several elements of nutrition, such, for instance, as the carbonate of lime and phosphate of lime, wanted for the bones and nerves, which are insoluble in water, but are soluble in water saturated with carbonic acid. This saturation is effected by the gas which remains in the bowels as a reservoir —as a reservoir, too, where a certain amount of compression is exerted, and the taking up of the carbonic acid is assisted just as in natural springs or in artificial fixed-air machines. This use of the air in the alimentary canal is really a most important one.

16

For the nitrogen I do not know how to find a use in the nutrition or modification of the tissues. Some of it is probably
taken up by the blood, and excreted by the lungs, as in the
expired air a considerable proportion of this gas is known to
be found, forming, according to the latest experiments of M.
Barral, one per cent. of the whole, and some may perhaps be
made into ammonia. But to that which remains still in the
alimentary canal an employment may be assigned, humble
indeed, yet contributing most exceedingly to our comfort and
health. The feces when they arrive at the ilio-cæcal valve are
almost fluid, and are so largely mixed with water saturated by
salts, that they are of greater specific gravity than ordinary
water, and either sink in or become mingled with it. If now
our digestive organs are not performing their duty well, or pass
the mass on too quickly, it comes into the external air in a
very similar state to that above described. It is a heavy, unformed, half-liquid pulp, diffusing itself inconveniently; but if
partially dried by the gas present, and lightened by the admixture, it is much less offensive to the senses, and easier retained
by the sphincter ani.

It is only then when in excess that I would speak of air in
the alimentary canal as "flatulence."

Eructation.

In approaching the subject of eructation it must be remarked
that gaseous contents of the hollow viscera are differently cir-
cumstanced from liquids and solids; their high degree of ex-
pansibility by heat and their low specific gravity give them an
inherent force which urges them outwards without any aid from
the muscular system. Other contents of the stomach require
the action of the expiratory muscles to expel them, whereas
gas warmed by the body tends to rise through the œsophagus
directly that tube is relaxed.

The essential condition is the relaxed and open state of the
cardiac end of the gullet. The air, instead of being retained by
the contraction of this powerful sphincter, finds its way upwards
in greater or less quantity. The passage of the bubble towards

the mouth, except in completely paralytic patients, causes a reaction, and by the time it gets to the fauces it is compressed by the stimulated muscles, and is suddenly expelled. Hence the noise is greater than is caused by the mere bubbling of air up the throat, such as you produce in moving a dead body, or an apoplectic patient. There is a combination of relaxation with spasm, the former taking the initiative.

The relaxation is by no means so complete as in vomiting. The bubble of air is allowed to pass, and then the œsophagus contracts again immediately.

The following table exhibits a comparison of several analyses of the air found in different parts of the healthy human intestinal canal :—

	In Stomach. Volume per cent.		In Ilia. Volume per cent.	In Colon. Volume per cent.	
Carbonic acid	14	(Chevreul)	24.39 (Chevreul)	43.5 –70	(Chevreul)
	25.2–27 8	(Chevillot)		23.11–93	(Chevillot)
Oxygen.........	11	(Chevreul)	0 (Chevreul)	0	(Chevreul)
	8.2–13.0	(Chevillot)	2–3 (Chevillot)	2–3	(Chevillot)
Nitrogen.......	71.45	(Chevreul)	20.08 (Chevreul)	18.40–51.03 (Chevreul)	
	66.8–59.2	(Chevillot)		95.2 –90.0 (Chevillot)	
Hydrogen......	3.55	(Chevreul)	55.53 (Chevreul)		
	a trace	(Chevillot)			
Sulphuretted hydrogen ...			•	1.0	(Marchand)
Carburetted hydrogen ...				5.47–11.6	(Chevreul)
				28.0	(Chevillot)

With regard to the gaseous contents of the stomach, as exhibited above, it may be observed readily that more than four-fifths is atmospheric air, and the rest is carbonic acid in much less proportion than in the breath which is passing out of the trachea by expiration, and which constitutes the air of the mouth and saliva. This fact gives us a strong hint of its source. It is evidently in a healthy person swallowed with the food and frothy saliva in such quantities as to fill the organ up to the points of normal distension.

I think too in the majority of cases also, where the collection and evacuation of air from the stomach is so abundant and inconvenient as to be considered a disease, that we may trace out the same source of it. Observe paroxysms of sobbing, globus hystericus, epilepsy, or chorea, and you will see great quantities gulped down. Watch those who are suffering from heartburn, and you will see them swallow air or frothy saliva, as if to relieve their discomfort.

Other persons have a careless, vulgar way of eating with the mouth open, which makes them swallow a quantity of air. Others have a trick of half-unconsciously gulping it down; and a very silly aspect it gives them, something like that of a gobbling turkey-cock: you may notice them bridling up and tucking in their chins. I fancy the feeling leading them to do it must be something like that which makes horses crib-bite—a sort of modified heartburn; but it is more trick than anything else.

In health all the air swallowed is readily absorbed. There are many individuals who never pass it away, upwards or downwards, for months together; indeed, so long as the perfect type of health is preserved, it may be said to be never excreted. After a meal their abdomen is as usual distended with air, but it is all removed by absorption before the next.

In many morbid conditions this is not done. When the vitality is lowered, probably the function most generally interfered with is absorption. The air collects, is swelled by heat, and expelled, although in no excessive quantity. Should the œsophagus be easily relaxed, there is eructation; should it be contracted, there is intestinal flatulence.

So far, the bulk of air swallowed has been supposed to be increased only by heat and expansion. But in some cases it is further augmented by gases disengaged from decomposed food. The occurrence of alcoholic fermentation in the digestive canal is proved by instances of vomiting, in which the matters ejected are visibly undergoing this chemical change. They are frothy with carbonic acid like yeasty beer, and they continue frothing up even when left to stand after ejection. (See CASES CXCII, CXCIII, CXCIV, CXCV.) We can easily imagine what a disturbance in the stomach this must make, and are not surprised at the ejection of such a turbulent guest.

Fortunately this spread of alcoholic fermentation through the saccharine contents of the stomach is rare. Its features are so marked, and the discomfort it causes so great, that we should be sure to hear more about it were it common. The fact is, that even where it begins and gives rise to the disengagement of some carbonic acid, it is rapidly stopped by the conversion of the sugar into lactic acid, a kind of fermentation more congenial to the temperature of the body. So that the "acidity," which in a former chapter has been spoken about as an evil, is a defence against one much more serious.

It will be seen from what has gone before that I class the cases of eructation which come before us into three groups:—

1. Those where there is simply a relaxed œsophagus, and the air, though only in natural quantity, breaks upwards.

2. Where there is an excess of atmospheric air swallowed from habit, or in the attempt to relieve an uncomfortable feeling arising either from the stomach itself, or some of the neighboring viscera, as the heart for instance.

3. Where carbonic acid is formed by alcoholic fermentation, unchecked by acetification.

This grouping has a bearing mainly on treatment, and a reference to the cases recorded will show readily to which class they belong.

In the majority of patients it is the escape of air from the stomach that is complained of; but sometimes a retention of it takes place with considerable inconvenience. It may (especially in elderly persons and cardiac patients) collect in such quantities as to cause a paralysis of the muscular coat of the stomach, and put them in considerable danger by impeding the action of the heart and diaphragm, and causing deadly faintness. I use the epithet "deadly" from recollection of the following case.

CASE CCXXV.—When I was physician to the Chelsea Dispensary, I saw occasionally for shortness of breath a fat single woman of from forty-five to fifty years of age. She complained of eructations, and of the upper part of the belly being swollen with wind, which I attributed to over-fat diet and sluggish habits. One day they sent for me suddenly to say she was dying, and when I got to the house she was dead. On a post-mortem examination, the heart was indeed slightly dilated and perhaps pale, but appeared more equal to work than the majority of hearts we see. But it was really quite difficult to get it out of the chest, so pressed-up was the diaphragm by the stomach enormously distended by its gaseous contents, devoid of smell, and certainly not, therefore, the product of post-mortal decomposition. This distension seemed to have been the cause of death.

More generally matters do not go quite so far as that.

CASE CCXXVI.—Mr. James L—, a hale-looking man of 63, came for my opinion at the beginning of March, 1867, about excessive dyspnœa by day and orthopnœa by night, with which he suffered. Sponges and other instruments had been put down the throat (? into the larynx) without any benefit. On examination of the pulse, this want of success was readily explained by an excessive irregularity and intermittence. The heart was also irregular, but not so intermittent as the pulse. The sounds were normal, but diffused.

As the urine was also albuminous, I gave an unfavorable prognosis.

But, to my surprise, a few days' dosing with Strychnine, Digitalis, and Iron reduced the action of the heart to regularity, and so far relieved the dyspnœa, that the patient was able to take exercise with ease. He said he had broken off wind from the stomach in several great bursts, and that new relief had followed each explosion. It was necessary to modify my prognosis. In three weeks' time he was really as hearty and well as any one can be at sixty-three.

This sudden relief of the stomach by paroxysmal explosions is exemplified in a former case, also an elderly man, where the cure of a long-continued waterbrash followed thereon (CASE CXXV).

The pathology of such cases seems to be that there is a paralysis of the gastric walls; air is drawn into the dilating organ, swells from heat, further dilates it and increases the paralysis, till such time as the sudden contraction of the circular fibres is brought about by nature or art, and then all is expelled at once, sounding the triumphal note of cure rather than being its cause.

It is, however, in the flatulence of dyspeptic hysteria that the explosions are most remarkable; they occur again and again, and are graphically described in CASE XCVI as "roaring eructations." When they are repeated so quickly as this, the air can of course be collected only by being swallowed.

The pathological etiology of eructating flatulence is to be sought in causes which lower the local vital force of the gastric involuntary nerves, and so make the muscular action intermittent and irregular. Such, for example, is starvation in CASES I, II, CX, &c., the pressure of the last in the shoemaker CASE LXXII, excess of tea-drinking in CASE XCVI, extraordinary fatigue in CASES IX, XIII, &c.

SECTION III.

Intestinal flatulence.

A reference to the short table given in the second section will show that in the gaseous contents of the ilia there is an increase in the quantity of carbonic acid relatively to the oxygen, or, if we like so to regard it, a decrease of the latter. At the same time hydrogen, scarcely present in the stomach, forms a good half of their bulk. This hydrogen cannot be swallowed air, and is not likely to be excreted from the blood; for we do not know of any gas besides carbonic acid owing its origin to any important amount from the circulating fluid. It must, I conceive, arise from the chemical changes going on in the remains of the food. I do not think any large quantity of air passes the pylorus, but that the bulk of the gas in the ilia comes from decomposition of their contents.

In a state of health this gas is reabsorbed nearly as soon as formed, so that only for a short time after meals is the abdomen puffy; but, as I explained before, lowered vitality promotes the collection of air by arresting absorption.

Lowered vitality also increases the extent of decomposition, by diminishing the flow of bile. The action of this secretion on food is exhibited in the experiments made by MM. Bidder and Schmidt upon dogs.[1] They found that when the flow of bile into the intestine was cut off by tying the ductus communis choledochus, rapid chemical changes took place in all sorts of food. When the animals were fed on flesh, the feces smelt like carrion; there was a continual rumbling of the abdomen and an evacuation of fetid air. When they were fed on bread only, odorless gases and sour feces were passed. No further injury beyond emaciation and weakness followed during the eight weeks of the experiments. From them we have a right to infer that one of the chief functions of the bile is to act on albuminous matters as an antiseptic, preventing their putrid decomposition, and preserving them safely to be exposed as much as possible to the absorbents of the alimentary canal; and that at the same time the excessive formation of acid from vegetables is checked,

---

[1] Bidder and Schmidt, " Die Verdauungssäfte," p. 230.

so that it may proceed gradually and as required by the digestive process. In fact, the condition produced in dogs by mechanically stopping the functioning of the liver answers exactly to the intestinal flatulence of dyspeptics in our species.

It may be observed that it is a long time after a meal, in fact just before the next meal, that the bile is normally poured in greatest quantities into the duodenum: in dogs in twelve or fourteen hours, in men about four or five. Now this is just the period when it is most wanted to prevent decay, and just the period when intestinal flatulence from its deficiency most usually occurs.

Excess of gas in the small intestines is the most troublesome sort of wind. Should it escape upwards through the pylorus into the stomach, it is apt to cause vomiting; or sometimes it constitutes a most nauseous eructation of sulphuretted hydrogen. Luckily this is rare. There seems too to be some difficulty about the passage of air downwards through the ilio-cæcal valve. Hence intestinal flatus often rolls about in the abdomen from the changes in position which the motion outwards of the alimentary masses involve, and causes the well-known and distressing grumblings of the belly or "borborygmi," aptly called in English a "glug-glug." The abdomen will be distended for several days with it, without its being expelled or absorbed

Its escape into the colon, even without making its way out of the body, gives immediate relief (see CASE XX, page 46).

There is very often considerable pain in one side or the other, most generally the right hypochondrium. The patients, especially if they are old Indians, will say they have got "liver." Where there is most pain in the side there is least grumbling of the belly: so I suppose it must arise from the long-continued immovable distension of one part of the gut.

Flatus in the intestines is troublesome during the day, from the tumidity of the abdomen, and noise on motion, and pain in the sides; but when it comes at night it causes still more inconvenience by preventing sleep. It is hard to explain why this should be; in many cases there is not enough pain or discomfort to account for it, yet a complete wakefulness and apparent want of wish for sleep prevail. It is to be remarked, also, that this insomnia is in many instances made worse by Opium.

Sometimes the patient will go to sleep easily and naturally on first lying down, and will then wake up in an hour or two, finding the abdomen tumid and uncomfortable, and will remain entirely without rest for the remainder of the night; or if there be a lapse into unconsciousness for a few minutes, the uneasy sleep seems rather to aggravate than· to relieve the feverish restlessness, and to cause headache.

During this unnatural repose men are often annoyed with disgusting erotic dreams' and abnormally frequent seminal emissions. I have never ascertained whether any analogous effect is produced in the female sex. The line of causation cannot at present be traced, the bowels and the generative organs appearing to have so little to do with one another.

The persons most liable to this troublesome affection are women, especially those of weak muscular fibre, anæmic diathesis, and a tendency to form fat. We may attribute it, under these circumstances, to a naturally sluggish portal circulation, which does not so quickly absorb the contained air as a freer current through the bloodvessels would do.

CASE CCXXVII.—Mrs. R—, aged about 40, has been from time to time a patient of mine during the last six years, on account of increasing corpulence, inability to walk without violent perspirations or exhaustion, palpitation of heart and anæmic irregularity of catamenia. Last week (March, 1867) she came to town in considerable pain all over the abdomen, starting from under the right ribs. The abdomen was very large, and I thought at first she had been getting fat again: but she denied having increased in corpulence as to the other parts of her body lately. It was very tympanitic on percussion.

This swelling of the abdomen made her very short of breath, and at night she could hardly lie down, and was frequently woke up by discomfort.

The bowels were fidgety in their action; sometimes confined, but more usually open in small quantities several times daily. The feces were semi-fluid, light-colored, and very offensive. Flatus was not now passed *per anum* with them, though that escape was habitual in ordinary health.

The appetite was deficient, the tongue coated. The catamenia had been the last few periods rather more profuse than ordinary.

I am giving her Valerian and Strychnine to strengthen the peristaltic action, and a small dose nightly of Aloës to augment temporarily the excretion of bile.

Remark here that when she was in health the gas escaped by the usual passage, but that in illness it was retained; and was

retained immovably, as shown by the pain in the right side.
The feces are just those of the dogs experimented upon by MM.
Bidder and Schmidt. The sulphuretted hydrogen and hydro-
sulphate of ammonia formed by the decomposition of animal
viands unguarded by bile seem to be purgative poisons, and
where they are found the stools are semi-liquid.

In looking over the foregoing pages for chronic conditions
in which intestinal flatus was conspicuous, I find Bright's disease, CASE IV; Old age, CASE XII; Rachitis, CASE XLIV.
Immediate causes mentioned are—Anxiety and application of
mind, CASES XIV, XV, LXXVIII; Loss of blood, CASE XLIX;
Over-eating, CASES LX, LXII, LXIV; Irregularity and other
bad habits of meals, CASES LXXVI, LXXIX.

In many other similar cases recorded in Chapters II and III,
where flatulence is spoken of, the subject of the present section
is intended.

Disease of the stomach seems as a rule to exclude intestinal
flatus. Thus it may be observed to be conspicuous by its absence in the Chapters on "Abdominal pains" and "Vomiting."

SECTION IV.

Colonic flatulence.

Flatulence in the colon may be distinguished from that in
the small intestines by its position ascertained by percussion,
by the absence of rumbling (except a little bubbling through
the ilio-cæcal valve just before it escapes), and by its passing
out freely per anum.

A reference to the table in page 243 will show in the normal
gaseous contents of this part of the bowels the presence of sulphuretted and carburetted hydrogen. But in health the quantity of the former is not enough to overcome the prevailing
odor of feces. In some cases of disease the sulphuretted hydrogen, arising from the decomposition of albuminous matter
unchecked by the normal flow of bile, is in excess, and may
then be readily detected. In other cases scentless carburetted
hydrogen and carbonic acid seem the prevailing gases. In the
former there is albuminous, in the other starchy dyspepsia.

When much sulphuretted hydrogen is present, there may be a diarrhœa of feculent matter; but colonic flatulence is more commonly accompanied by costiveness and constipation. The colon does not appear to be so sensitive to poisons as the ilia.

It is intestinal wind that is generally complained of by patients who are bad enough to be driven to a doctor, and is that which is generally alluded to as "flatulence" in the cases in an earlier part of the volume.

Colonic flatulence is not nearly so distressing as intestinal, and does not cause so much wakefulness or other nervous disorders.

SECTION V.

· Treatment of the several sorts of flatulence.

Eructation may in some cases be stayed by solely a direct restorative treatment of the cause. The defective digestion may be replaced by artificial gastric juice. For example—

CASE CCXXVIII.—James B—, a laborer of 50, was taken in at St. Mary's April 20th, 1856, for a catarrhal cough of ten weeks' standing, with some congestion of the lower part of the lungs. The object of admitting him was to give his cough the benefit of the regulated temperature of the ward, with rest. No cough medicines were considered needful. But he complained that after meals he had throughout his illness been troubled with wind breaking up from the stomach. It was tasteless and inodorous.

Fifteen grains of Boudault's Pepsine Powder was administered daily with his dinner. On the 25th the flatulence was relieved, and he went out well on the 1st of May.

Where excess of air is swallowed from abnormal sensibility and breaks up in eructations, Valerian and Ammonia are useful, but above all shower-baths. I do not know any disease in which their value is more marked. The more the patients can bear of them the better, and the sooner they can get educated to take them in full quantity and cold the better. First let them be administered tepid, then with the shower cold and the foot-pan warmed with hot water, then make them all cold, and each day let the quantity of water be gradually increased till the full extent that the bath will hold be arrived at.

Examples of this principle of treatment are given in Cases CXCVII, CC, CCIII, CCV.

When the eructations depend on the formation of carbonic acid by alcoholic fermentation, the Hyposulphite of Soda is indicated. See Case CXCV.

Eructation from this cause is rare without vomiting.

The aim of the treatment mentioned is to prevent decomposition of organic matters. In the laboratory we find that nothing is so powerful in this respect as Sulphurous Acid; and accordingly it is used in various processes of the arts for the purpose. Sulphur is burnt by wine-growers in casks used a second time to arrest the fermentation which is apt to be going on in the liquids soaked up by the cracks or porous parts of the staves, and the acid vapors effectually do their duty. The agents of the Board of Health find no disinfectant for sewers so quick and certain in its action as Macdougal's, the chief ingredient in which is Sulphite of Lime. Muscular tissue may be prepared on the same principle, and keeps as well as when salted or dried; and we may test even on such a delicate substance as yelk of egg how fresh it keeps with sulphite of salt. The same effect is produced by taking as a medicine Hyposulphite of Soda; the fermentation of the contents of the stomach is arrested, and the evil effects of that fermentation prevented.

But it must be remembered that the digestion of the meat is also checked. Dried, salted, or otherwise chemically prepared victuals are not so soluble as fresh; indeed, if completely dry they are not soluble at all; and to continue the Hyposulphite of Soda long would put the patient into the condition of a sailor reduced to salt junk.

A safer, but equally powerful arrester of chemical changes is Charcoal. When soup has begun to turn in hot weather, economical cooks heat it up with a little bag of Charcoal in it, and it becomes quite sweet. This shows that the carbon does something more than merely condense the gases formed. The same agent will accomplish the same result in the alimentary canal. I myself have used it truly only in cases where decomposition producing flatulence occurs in the intestines; but I should not hesitate to give it in gastric fermentation also, if Hyposulphite of Soda chanced to disagree or had failed in its effect.

I

The general treatment of indigestion by Quinine and Strychnine, as several times here advocated, is specially indicated in
eructation, inasmuch as the most failing function is that of the
contractile fibres of the œsophagus and stomach, and to these
our remedy first arrives. An example of their rapid action
may be read in CASE CXXV.

As stated in the last leaf but one, *intestinal flatulence* exhibits
best the power of charcoal, because the air has more difficulty
in being got rid of without some such help, and the air formed
by decomposition is peculiarly copious and troublesome.

In ordinary cases there are usually joined several other remedies to the Charcoal, such as Quinine and Strychnine (for the
reason given a few sentences back), and Soda and Valerian, or
Galbanum, or Assafœtida, so that the action of the Carbon is
complicated. I have therefore selected for the nonce, to exhibit
its special and independent action, a rather out-of-the-way instance, in which the collection of air took place in consequence
of a mechanical lesion entirely preventing its passage (for the
small exit must have been always blocked up by feces), and in
which the only antiseptic and absorbent used was Charcoal.

CASE CCXXIX.—Elizabeth C—, aged 63, was admitted to St Mary's under
my care January 15th, 1857. Though thin and not muscular, she had always
been a hard-working active woman, and had borne thirteen children, of whom
but two were dead, one of phthisis and one of scarlatina. She herself could
recollect no illness except scarlatina and child-bearing, till five months agone,
when she noticed that the left side of the abdomen was often swelled, and
that the swelling was relieved by a copious explosion of wind by the anus.
A like swelling she also perceived some time afterwards on the right side,
since which she had not so often been relieved by the passage of wind. She
had also frequently a feeling of numbness and involuntary twitches in the
legs.

She lay on her back, when I visited her, with the abdomen raised up by a
great collection of air. It measured thirty-eight inches in circumference. No
solid tumor could be felt. The bowels were very constipated, and under the
influence of purgatives only a little fluid feculent matter, but no air, was
passed.

An attempt was made to relieve her by passing a tube up the rectum, but
no air was let off even thus. A many-tailed bandage was bound tightly round
the belly, but no diminution in size followed. Turpentine, too, was administered, but it was fruitless.

On the 22d I ordered a drachm of Charcoal to be given every other hour, the bandage being still kept on. On the 25th she was much better, the distension being much less. On the 27th she had increased somewhat, so I added 1-12th of a grain of Strychnia to the powders on those occasions daily. From that time we continued to find the abdomen softer, the patient lost her pain and gained strength, but with occasional relapses of distension.

On the 20th of February she was able to get up. Her bowels were regularly opened by a simple enema, with sometimes a few drops of Cajeput Oil. Her tongue was clean, and the general health was good, and in the beginning of March she was actually assisting in the work of the ward. She herself pronounced that she was well enough to return home, and arranged to do so on the 23d. However, early in the morning of that very day she suddenly died, the only warning of her being worse having been a certain relapse of distension on the 21st.

On post-mortem examination there was found in the lower part of the ileum, on the right side, an occlusion, as of a hard contracted scar, without any peritoneal adhesions. The occlusion, at first view, seemed quite complete ; but on further manipulation a dissecting probe was passed through it by a winding passage. Above this the intestines were greatly distended with air and semi-fluid black feces. But what surprised us was the *entire absence of fetid odor* in all this matter so long retained. It was not nearly so unpleasant as that found in a corpse accidentally killed in full health.

The last observation is my reason for my citing here this somewhat long and painful case. If the Charcoal can so act where a mechanical impediment confines the gases to the intestines as by a ligature, and half kills them by strangulation, it must be still more powerful when it is aided by the vital force still remaining only slightly arrested, as in ordinary cases. For, in truth, people may be very flatulent without being very ill.

Charcoal being tasteless is not disagreeable to take when you have got over the grittiness in the mouth. The only other objection I have had raised is its color. A wit of the Midland Circuit told me I was turning his "colon" into a "coalhole."

It is scarcely needful to say that easily fermentable articles of diet must be for some time shunned, if the patient would avoid a recurrence of the complaint.

Great advantage arises in intestinal flatulence from the use of such expedients as restore the flow of bile in full quantity ; a chief business of that secretion being the prevention of chemical decomposition in organic matters.

Temporary use may be made of drugs for this purpose.

The Salts of Mercury (viz., the gray oxide and calomel) were found in some experiments made by Dr. Handfield Jones on animals to increase the production of yellow matter in the hepatic cells. But when this metal was given there was also great sanguineous congestion of the liver, which, on the contrary, was pale after an administration of drugs which had not augmented the yellow matter.

Muriate of Manganese and *Colchicum* had also the like effect.

Nitro-muriatic acid during life caused a flow of bile *per anum* in a cat; but there was no excess of yellow matter in the hepatic cells *post-mortem*.

Aloes, Oil of Turpentine, and *Rhubarb*, acted much as Nitro-muriatic Acid.

Antimony promoted in the liver, as in all the mucous membranes, a copious flow of water and mucus.[1]

We have thus in our Pharmacopœias most powerful agencies for modifying the quantity and quality of the bile. And it cannot be doubted that further inquiry may extend widely our knowledge of the nature of our already existing numerous tools, so as to confer incalculable benefit on rational medicine. Chemistry cannot render a reason of their mode of action; it has in it something essentially vital, or if you like the term better, essentially physiological. Still we must bear in mind that, as far as we can see, it is temporary; and, since no one would wish to continue their use for life, we must mainly depend in the end on more direct restoratives of life for final cure, such as are pointed out in the experiments on food.

Mercury certainly is a powerful temporary relief, but the sanguineous congestion of the liver seen in Dr. Jones' experiments should warn us against trusting to it. Again and again the rough clearance has to be resorted to, till increasing necessity for it alarms the patient, and points to the anæmia and weakness which are the inevitable consequences of an habitual employment of that drug.

I understand there is the same objection to Colchicum. Manganese I know nothing about.

[1] "Medico-Chirurgical Transactions," vol. xxxv. p. 249, and "Medical Times and Gazette" for March 19, 1852.

Nitro-muriatic Acid and Aloës I am pleased with as bile promoters. The longer they are used the less they are required. Turpentine and Rhubarb are too nasty for continuous use.*

The chronic action of Antimony on the digestive viscera I have not experience of.

But, as I said before, it is to a renewal of life by nutriment that we look for cure. To which undertaking science contributes the following observations. It was found by Drs. Bidder and Schmidt that a *full diet* augmented not only the quantity of the bile, but also the amount of material therein. Thus whilst a cat on ordinary diet secreted 0.807 of a gramme per kilogramme of weight hourly, and of solid material 0.045 of a gramme, on very full flesh diet the secretion was in one cat 1.185 gramme of fluid, and 0.062 of a gramme of solid, in another 1.003 of fluid, containing 0.063 of solid. The same fact was fully confirmed by observations also upon dogs and geese, the details of which correspond to the above.[1]

Flesh diet causes the secretion of more bile than vegetable food. For example, in an experiment made by Dr. Nasse on a dog,[2] a diet of bread and potatoes caused a daily secretion of 171.8 grammes, in which was 6.252 of solid matter; whilst meat made it amount to 208.5 of fluid, or 7.06 of solid residue.

Water increases the quantity of the bile within an hour after it is drunk, and not only the quantity of fluid, but also of the solid contents, though in a less proportion. Thus a dog weighing about 5 kilogrammes, which after a meal of 185 grammes of *beef alone* secreted in an hour 2.283 grammes of bile with 0.135 of solid matter in it; after a meal of 25 grammes of *beef and 158 of water* secreted 4.030 of fluid, and 0.117 of solid bile. And the same dog, after 185 grammes of *water alone*, made no less than 5.165 of bile, or 0.143 of solid matter. And the same thing was observed in the three other similar experiments.

To this Dr. Nasse adds, that though water increases the fluid

[1] "Die Verdanungssäfte."
[2] Nasse, "Commentatio de Bilis quotidie à cane secretâ copiâ," &c. Marburg, 1851.

bile and also the organic solid constituents, it does not have the like effect on the amount of salts.

On the other hand, it was found by Bidder and Schmidt that *fatty food* instead of increasing, as might have been expected on chemical grounds, the quantity of bile, extraordinarily diminishes it. Thus in a mean of three experiments on cats, the hourly discharge after a diet of pure fat was of bile 0.327, of solid matter in it 0.036 of a gramme. We might have supposed that the formation of a substance which is the most hydro-carbonaceous in the body would have been promoted by a peculiarly pure hydro-carbon aliment: but such is not the case. Nay, so far from it, that the numbers given above correspond most closely to what would probably have been the quantity of bile secreted by the animals in a state of complete deprivation of food. Fat appears to be eminently " bilious," as the vulgar tongue expresses it, that is to say, it diminishes the vitality of the liver.

Alcohol also by arresting metamorphosis[1] must be hurtful, as tending to diminish the normal formation of bile.

The rules then of diet in intestinal flatulence should be—(1) to Use a full allowance of lean meat and water ; (2) to Avoid fat, butter, and rich sauces ; and (3) to Diminish the allowanc e of alcohol.

None of these special items of treatment however diminish the importance of *general treatment*. Iron when there is anæmia, and the nerve-tonics Quinine and Strychnine when there is not, are what I use myself, and it is very rare that I wander from these old friends into the fields of experiment without regretting it. Indeed the general treatment will oft-times render needless the special, as it did in the following recent instance.

CASE CCXXX.—A maiden lady past forty, of spare wiry build and sharp decided manners, came in the first week of 1867, on account of the inconvenience, nay possibly scandal, which she suffered from a continually increasing swelling of the abdomen. It varied from time to time, yet had steadily augmented for several months. Latterly, when it was largest, she had felt palpitations of the heart and in the epigastrium, especially when sitting still for longer than usual. The ankles also frequently were puffy, taking the mark

---

[1] See " Lectures chiefly Clinical," Lect. L, "On Alcohol."

of the boot in an unaccustomed way. The catamenia were quite regular and sufficient, the tongue was clean, the bowels were natural in action; and the only abnormal secretion was the urine, which was sometimes thick, and caused pain apparently from excess of acidity. There was no hysteria.

The abdomen was tight and round, and large for a person of her stature. It was very resonant on percussion throughout. There was no glugging on pressure, and the patient stated that air very rarely escaped *per anum* even at stool.

I desired her to wear an Elstob's belt, and to take ⅟ of a grain of Strychnia with three grains of Quiniæ et Ferri Citras twice a day; also at night another similar dose of Strychnia in a pill with two grains of Aloës and Myrrh mass, and the same quantity of Extract of Hyoscyamus.

I saw her twice more at intervals of about ten days, and as the belly became softer was able to make sure of the absence of any solid substructure to the tumefaction. She went home quite slim and comfortable, but still denied having expelled any explanatory amounts of air by the external vents.

As to the cause of her disease, I found that she kept house for her father, an old and infirm medical man in a remote province. Whenever she consulted him about any little ailments, he always gave her purgatives, and the more abnormal the alvine secretions became, the more purgatives he gave her.

I suppose this treatment had produced anæmia and reaction of the ilia, which as a chronic condition, continually tending to increase itself, had gone on from worse to worse till the parietal muscles had also become relaxed. It is true that general anæmia, indicated by a pallid condition of the lips, tongue, and mucous membranes, had not as yet arisen; but yet it might have existed in the local organs affected. I have noticed that persons of middle age become generally anæmic in consequence of imperfect digestion much more slowly than the young. For the life of the latter rapid renewal of the tissues is such an essential feature, that a check to the rapidity tells much more severely on their health than it does in the case of adults, and still more of old people. Probably this is the reason why consumption is so "galloping" in youth, so sluggish in old age; and why starvation makes such havoc among young paupers, while on the same dietary their grandparents may struggle on with a half life to the extreme term.

The intention of the belt prescribed in the above case was to strengthen the contractile force of the abdominal muscles, just as a bandage is put round a woman after childbed to restore her pristine figure. It affords the fibres a firm basis of contraction, so that they recover from the half-palsied state, and do

not again fall into it. Its pressure therefore should be even,
and need not be very great. I mention this because patients
are apt to mistake the object of the appliance, to suppose that
it is designed to expel the retained air by compression, to be
disappointed when it does not do so, and to tighten it too much.
Now too great constriction acts unevenly, and produces the bad
local effect of ill-fitting stays.

Flatus in the colon requires the same medicinal and dietetic
treatment as that of the intestines. When there is a tendency
to congestion of the rectum and to piles, as not unfrequently
happens, cold water enemata are useful, and in elderly persons
a carminative, such as Extract of Rue, or a few drops of Ether,
may be added to the enemata.

When the abdomen is not much more dilated than natural
by flatulence, efforts should always be made to retain the air
inside the bowel till the period of fecal evacuation. For not
uncommonly the parting with it induces a condition of constipation. If retained, it may roll about uncomfortably for a time,
but will soon either become absorbed or mixed up with the
feces, and so assist their normal evacuation, as described in a
previous page.[1] The proof of that is that it is not afterwards
passed.

[1] See page 242.

CHAPTER VII.

DIARRHŒA.

Difference of diarrhœa from mere frequency of evacuation.—Subdivision of forms.—Their causes and indications.—Supplementary and reflex diarrhœa. —Infantile.—Diarrhœa in typh-fever.—Ulceration of bowels.—Mucous flux. —Copious solid diarrhœa.—Acid diarrhœa.—Use of opium.—Riding in chronic cases.—Cautions to travellers.

WHEN the absorbing power of the intestines is defective, the consequence is an excess in the *quantity* of matters which pass through them; that which ought to be taken up is carried along out into the normal draught, and so constitutes a true diarrhœa. It is of great practical importance to distinguish this from the mere frequency of evacuation, which is quite consistent with a natural or even with a deficient amount of feces. The number of motions, or the number of times an inclination is felt to void them, is often increased, while even less than the average quantity may be passed in the twenty-four hours. This affection is of the nature of tenesmus, and arises from some tissue lesion of colon or rectum; whereas true diarrhœa, as aforesaid, depends upon defective function of the ilia.

The arrest of function, as declared by the prevailing contents, of the stools, constitutes the best principle of division which has been moreover adopted in the chapter on Vomiting; and according to it we may speak without much danger of being misunderstood of *Crapulous, Bilious, Serous, Dysenteric,* and *Choleraic diarrhœa.*

Crapulous diarrhœa is simply an excessive quantity of food taken, or the natural quantity arrested in its solution by suspension in the gastric function. I call it "crapulous," because it is most usual after a debauch; but in weakly persons it is not necessary that the intemperance should be absolute; that which is moderation for others may be an excess in them. An examination of the feces exhibits a quantity of undigested food as the prominent feature, sometimes fetid and fermenting, and deficient in the bile which should prevent decomposition.

Bilious diarrhœa is the next simplest form of the disorder. Bile, normally poured out by the liver to the extent of from three to four pints a day, if not concentrated by the intestinal absorption, adds largely to the excrements, where its presence is declared by its well-known smell, and by a color exhibiting various shades of yellow, brown, and olive-green, according to its absorption of oxygen and mixture with feces.

The arrest of the absorbing powers of the intestines and the consequent rejection of bile, mixed at first with feces, and augmented by the exudation of water from the intestinal parietes, is what so often happens in comparative health from the impression of cold, from irritation of the alimentary canal by unwholesome food, and from mental emotion. It is possible also that the qualities of the bile itself may be altered in some cases, or its quantity may be increased. It may be poisoned by drugs, as by Calomel or by Senna, and so rendered incapable of absorption, and be poured through the ilia without their being in fault. Again, congestion of the liver and of the portal veins, such as is especially frequent in Europeans resident in warm climates, causes the bile to be at one time deficient, and afterwards to be poured out in excess. Or irritation of the stomach and duodenum may cause it to be retained in the liver and gall-bladder till it is unfit for absorption. In such cases bile is rejected *per anum* and constitutes the matter excreted in true bilious diarrhœa.

We should distinguish this symptom from a different one sometimes confounded with it—viz., the presence of a light grass-green matter in the stools. This is not bile at all, but altered blood, and denotes inflammation of the mucous membrane, a state requiring very opposite treatment from that proper for bilious diarrhœa. Our best aids to diagnosis are—first, the smell: in real bilious stools the odor of the hepatic secretion can always be perceived, in spite of the feces mixed with it; while in the grass-green stools the smell is not of bile, but more or less putrid. Secondly, the microscope exhibits in the mucus the usual globules mixed with small shreds of fibrin and blood-globules.

In *Serous* or *Watery* diarrhœa it is probable that there is an increased exhalation of aqueous fluid from the bloodvessels of the intestines, as well as an arrest of its absorption. In this form, when pure, if the feces are retained by a voluntary effort, they may be concentrated nearly to their normal condition by the removal of the water, and thus a test afforded that their state depends mainly on the addition of this constituent. For that which can be so readily taken up again into the blood cannot be of a nature very foreign to it. For example, in the diarrhœa produced by a saline purgative you may feel several pints of fluid rolling about in the bowels; but if you resist the inclination to stool, it goes off at last, and you void afterwards little more than the ordinary amount of semi-solid feces. It is not so in bilious or inflammatory diarrhœas: you cannot cause the absorption of the fluid-by forcible retention.

Watery diarrhœa, when not arising from the action of drugs, indicates a congested state of the venous plexus of the alimentary canal, and a consequent morbid proneness to deficiency in absorption. The vitality of the mucous membrane is deficient; and if it be not restored, local death, exhibited in the form of ulcers and sloughs, must result. The exhalation, however, tends to become habitual, and so continues beyond the period of congestion, so that the whole mass of blood is relieved of its water, and in this way sometimes dropsical swellings may be reabsorbed and pass off through the bowels.

In *Dysenteric* or *Muco-purulent* diarrhœa, water is also in excess, but the characteristic feature is the presence of mucus or pus mixed with it; in which also there are shreds of fibrin, blood-globules, and flakes of the epithelium of the bowels.

Should any of these products of inflammation be unmixed with feces, then it is probable they come from the colon or rectum; but if they are mixed up with a large quantity of watery fluid, and still more if that watery fluid shows itself to be the serum of the blood by coagulating with heat, then there is little doubt of their source being the mucous membrane of the ilia. The fluid in muco-purulent diarrhœa is always highly alkaline, and if it be examined with the microscope, phosphatic crystals are found scattered through it. If allowed to stand, it separates into two parts: the one *serous*, varying from transparent whiteness through all the shades of yellow to deep brown; or, where blood be present, to red and black, in which are the flakes of fibrin, the ammoniacal crystals, and floating globules; the other *sedimentary*, consisting principally of gray, granular matter, the débris of food mixed with more or less of the coloring matter of the bile and half-digested blood.

The degree of serosity and the proportion of the products of inflammation in the first, show the extent to which inflammation has gone in the mucous membrane; whiteness, bloodiness, putridity, alkalinity, being bad signs; yellowness, opacity, the smell of bile, and the absence of putridity, being good.

The second, or sedimentary portion proves the condition of the general system rather than of the ilia in particular. If it be copious in proportion to the fluid, then the normal function of destructive assimilation is shown to be little interfered with; if scanty, then we know this process to be arrested, the effete tissues are not removed from the body, and we have to do with a more grave state of affairs. The quantity of solid matter is the best test of an advance towards health, or departure therefrom, in all cases where there is this state of the bowels.

The most common examples of muco-purulent diarrhœa are found amongst acute diseases, in low fever, enteritis, and dysentery, especially in the teething dysentery of children. Amongst chronic diseases, in ulceration of the bowels, whether a consequence of phthisis or low fever, it is the most usual course for the symptoms to take.

Bloody diarrhœa, where the blood is in small streaks in the mucus, or slightly mixed with the serum, or mixed with the grass-green mucus above described, shows recent inflammation. When it is in clots, either black or fibrinous, with the globules partially washed away, that a bloodvessel of notable size has been opened, probably by ulceration. Should pus be mixed with it, the diagnosis of ulceration is confirmed. Black semi-digested blood, precipitated by standing with the sediment of fluid stools, comes from high in the alimentary canal, not seldom from the stomach itself.

Putridity of the stools in diarrhœa always shows that there is an imperfect quantity of bile, one of the most clearly ascertained functions of the hepatic secretion being to prevent decay of albumen. Putridity may arise from two sources—namely, the food taken, or the secretions into the canal. A close examination of the stools will generally distinguish them; for if it is. non-digested food which is decaying, then the solid constituents of the feces are bulky, pale, containing large lumps of still paler substance, which under the microscope will be found to consist of muscular fibre, fat, and other parts of victuals, often swarming with infusoria. Whereas, if the fetor arise from the decomposed albumen of the scrum, it will be observed to exhale from the more fluid part of the motions, which smell like the washings of macerated flesh, while the solid part is scanty and comparatively unaffected. This shows a much more serious state of the vital powers, and in severe complaints is often the harbinger of death, especially if joined to a peculiar mouse-like smell in the sweat.

In *Choleraic* diarrhœa the whole of the blood is so altered in its physical qualities that little of it remains capable of support-ing life, or of absorbing the wherewithal to support life. The functions of the liver and kidneys are suspended for want of live blood, no bile appears in the stools or vomit, no urine in the bladder.

For the purpose of comparing the degree in which life is de-ficient in the different forms of diarrhœa, I subjoin a table in which the first column is occupied by the several functions, the loss of one or more of which characterizes those different forms. It will be seen that the sign of minus can be placed against one after the other till the normal condition of all is finally lost, as an essential, not accidental, part of the disease.

Full life.	Crapulous diarrhœa.	Bilious diarrhœa.	Serous diarrhœa.	Dysenteric diarrhœa.	Choleraic diarrhœa.
1. Blood healthily sufficient for the support of the whole system.	1. Normal.	1. Normal.	1. Normal.	1. Normal.	1. — (Mass of blood poisoned and half dead primarily affects the whole system.)
2. Coats of capillaries elastic and retentive of the contained blood.	2. Normal.	2. Normal.	2. Normal.	2. — (Coats of capillaries wanting in elasticity, becoming congested and ruptured, let out the blood. The other stages of inflammation may or may not follow.)	
3. Exosmosis and endosmosis of serum through the mucous membranes equal to one another.	3. Normal.	3. Normal.	3. — (Exosmosed serum passed away by stool, instead of being reabsorbed.)		
4. Bile made in full quantity; its fluids reabsorbed by intestines, and its colored solids only rejected per anum.	4. Normal.	4. — (Constituents of bile not absorbed, but passed away, forming the bulk of the dejections, and shown by the smell to be unaltered.)			
5. Food fully digested.	5. — (Undigested materials of diet passed away with fæces, and hurrying downwards the bile and other contents of the intestines.)				

In some rare instances of mucous flux the stools are acid from time to time. There is nothing special in the pathology of this. It arises simply from so much acid being formed from the decomposition of undigested food, that it cannot be neutralized by the alkaline juices. The acidification takes place apparently in the cæcum, during the delay of the decomposing aliments there; for considerable pain is often experienced in the right iliac region, and in the course of the colon, just before the evacuations.

In all forms of diarrhœa from affections of the small intestines, the evil is twofold: first the aliment, which ought to contribute to the support of the system, is hurried through the abdomen, and so the supplies are cut off; and secondly, destruction is carried on at an increased rate by exhalation from the mucous membrane of the bowels. The stick is being cut away at both ends, and hence there is nothing which produces such rapid emaciation. Where so-called "diarrhœa" is reported to you as lasting for any length of time without emaciation, always let your suspicions of the correctness of the nomenclature be roused, and observe carefully whether the quantity of excrement be really in excess, or whether the ailment have not rather the nature of tenesmus, and arise from the colon or rectum. You will generally find such to be the fact, and must vary your treatment accordingly.

Sometimes diarrhœa seems to be the transference of a tendency to exudation of serum from another tissue to the alimentary canal. Such is that which sometimes comes on of its own accord or may be artificially induced in ascites, and which certainly sometimes diminishes the abdominal collection. Such is the diarrhœa of uræmia, which, however, does not usually relieve anasarca, but rather increases it from the weakening of the blood which follows. Hence it is a very bad, almost a fatal symptom, in Bright's disease.

The most important part of treatment is the diet. It must be such as does not need a perfect state of the digestive organs for its absorption, while at the same time it is nutritive to the patient. The best of all is milk and Lime-water. In feverish cases it may be iced, and Soda-water may be occasionally substituted for the Lime. Keeping a person solely on this diet is often alone sufficient to cure all sorts of diarrhœa not dependent on a permanent chronic cause; and even where there is such a cause for it, much temporary benefit is derived, and a sounder starting-point for medical treatment than the previous state is gained.

In a temporary diarrhœa without other disease, the loss of the normal supply to the body is not of much consequence; a short starvation perhaps does good to a person otherwise healthy. But in severe acute disease, or in long-continued chronic diarrhœa, this is an important consideration, and care must be taken to allow for it. Since food in the usual quantities at once cannot be borne, and is rejected undigested, give it very frequently in small portions. The alkaline milk diet I have just recommended allows this to be done most conveniently. A jug of the liquid must be kept close at hand and sipped from time to time, so that as much nutriment may be taken in the twenty-four hours as would be done by a healthy person.

It is a good rule when there are lumps of feculent matter in the stools and a smell like that of normal excrement, to give purgatives, and when there is no normal smell present, to abstain. For it is only the remains of previous constipation that require to be got rid of, and when they are not present, harm is done by purgation. I have known cases of chronic diarrhœa much injured by the routine practice of so beginning treatment.

Where the products of acute inflammation are found mixed in the stools, such as white and opaque mucus, flakes of fibrin, epithelium, blood-streaked mucus, bright green matter, &c., as above described, then Leeches, Fomentations, warm Hip-baths, and poultices to the abdomen are appropriate and should not be delayed. In children, the whole abdomen and loins may be fastened up in a large circumambient Poultice, which they cannot wriggle away from, a Leech put on near the navel, and the bite allowed to bleed for a little time. The articles of materia medica I have most trust in are Opium, Ipecacuanha, and Carbonate of Soda. A syrup may be made of ten drops of Laudanum, two grains of Ipecacuanha powder, and a scruple of Soda in an ounce of half treacle and half water, and doses of not over a teaspoonful given at hour intervals. I have found this

answer better than the old plan of administering Calomel. In
teething infants this treatment is of the most marked utility.
I suppose the anodyne soothes their neuralgia. In their case,
too, lancing the gums will sometimes stop a most violent
diarrhœa where the stools show evident proofs of the inflammatory condition of the ilia. The action of the lancing is probably
much the same as that of Leeches, viz., a relief to the congestion of the mucous membrane. Upon the protrusion of the
teeth it can hardly be supposed to have any influence, but that
it alleviates toothache any adult may experience for himself,
though it is impossible to get from his little patients an account
of this remedial effect.

But there is no doubt that the most active cure in infantile
diarrhœa is change of diet. Bringing up by hand or unwholesome states of the breast-milk are generally at the bottom of
the ailment. No remedy is equal to a healthy wet-nurse, or
where prejudice forbids that, as near an imitation as can be
made of human milk by that of animals, such as the donkey's,
or the cow's diluted and slightly sweetened.

In low fever the presence of diarrhœa still suggests to many
practitioners, and used to suggest to many more, the employment of Mercury. The effect of this is the increase of solid
sedimentary matter in the stools; in other words, a restoration
of the destructive assimilation going on in the body. The
motions are diminished in number and in fluidity, but not in
actual quantity. In fact more solid effete matter is excreted,
and thus the tissues devitalized by the typh poison are removed,
and room is made for new nutriment. This increase of solid
matter is taken as an evidence and test of benefit accruing from
the use of Mercury, and as a prognosis of good. But I must
say, without reserve (and am glad of the opportunity of so
doing), that I think this an unwise hurrying of nature; for only
the destructive assimilation is augmented, not the constructive,
and thus the powers of the body and its resistance are lowered.
Now the use of Hydrochloric Acid both stops diarrhœa and
increases at the same time absorption in the intestinal canal.
For some years therefore I have employed no other remedy in
low fever, and with decided success, as I have more largely set
forth in my published " Clinical Lectures."

Where, in the absence of fever, blood is passed by the bowels, the two most powerful means of checking it I have found to be Turpentine and Acetate of Lead, especially the latter. Its direct influence as a poison on the bowels would have led to an expectation of this. If the hemorrhage has gone on for some time, I am inclined to think it must be sometimes due to a clot distending the bowel, and preventing it contracting upon the bleeding spot, for certainly a dose of Castor oil, in the results of whose action a quantity of pale clots were exhibited, has several times in my experience stopped bleeding.

Diarrhœa from ulceration of the ilia tends to prolong itself; for the weaker the system is, the more irritable are the sore places, and the less can the morbid actions they set up be resisted. It is wrong, therefore, to let it go on an hour longer than we can help. The readiest means for arresting it are such as blunt the sensibility of the ulcerated spots. Milk-and-lime-water diet should be used first, then Chalk and Opium, which appear to act on the sore mucous membrane just as they do on a raw blistered surface of skin. If these fail, Sulphate of Copper should be used in doses increased from a quarter of a grain up to two grains.

Where there is a simple flux of transparent mucus without fever or pain on pressure, and no fibrin or blood in the motions, the vegetable astringents, such as Logwood, Bark, Kino, and Tannin, are often of great use. In such cases, too, I have prescribed Iron, in the form of the Tincture of the Sesquichloride, with seeming benefit. I must, however, say, that I feel doubtful in the greater number of cases whether this form of flux be not due rather to the colon than to the ilia.

Where the solid matter is pale, fetid, and consists mainly of undigested food, inspissated bile may be given with benefit; the stools become less fetid and less frequent under its employment. This is particularly the case in children whose mesenteric glands are diseased. Pepsine also diminishes the fetor of the motions in the best way—namely, by promoting the normal solution of the food, and acting as a direct restorative.

Acid diarrhœa indicates the free employment of Chalk.

The use of Opium in diarrhœa must never be made a matter of routine. As a general rule, I have found it beneficial without consequent harm in cases where there was tenesmus and frequent stools; but where the feces are bulky and copious it appears to impede the natural secretion. Where the stools also are putrid, caution is required in its use. In the diarrhœa which so often accompanies and proves fatal in uræmia, it checks the debilitating flux, but is apt to bring on coma.

In some cases of diarrhœa from chronic mucous flux of the intestines, without ulceration or acute inflammation, I have known horse exercise to be serviceable. I suppose it is the gentle agitation of the abdomen, combined with the air and amusement, that proves of use.

In recommending the recreation of travelling to invalids subject to diarrhœa, you must be very careful of the route you select. The epidemic influence of cholera which has overspread Europe during the present generation, visiting almost every square mile of its surface several times during the last few years, has in many places left behind it a chronic endemic poison. The natives are insensible to it, but few strangers escape becoming affected more or less, according to their idiosyncrasies. Strong persons find it only an inconvenience, but an invalid is put in some danger, and certainly loses all the advantage of the tour. This is especially the case in the mountainous districts of the south of France, the Pyrenees, and Dauphiny, and in the volcanic regions bordering the Rhine, the Eifel and Moselle country, as well as those in the centre of France, the ancient province of Auvergne. All these places are attractive from their picturesque beauties, and therefore it is necessary that travellers should be warned of this evil attendant upon choosing them as the scene of a tour. It must not be supposed that this diarrhœa is solely the result of the foreign modes of cooking. I have known English biscuits and porter, and boiled eggs, adopted as a diet without relief, though of course nothing foreign could have got into them. I believe the cause to be as I have represented it—namely, a poison left endemic since the passage of cholera through the country, but to which the natives have become acclimatized. That it is of late years only that this diarrhœa has been prevalent is shown both by local report and the omission of all mention of it from the well known work on "Climate," by Sir James Clark.

One source from which strangers contract this diarrhœa is an evil capable of, and loudly calls for amendment : I refer to the filthy privies in continental inns. A gentleman, lately eminent in our profession and of good judgment, told me that, during a Pyrenean tour some years ago, he entirely escaped the diarrhœa which everybody else without exception suffered from, by adhering to a strict rule of never entering one of these digusting holes, but worshipping Cloacina under the pure light of the stars. Invalids and ladies cannot so well manage this, unless they are rich enough to travel with carriages and servants and locomotive water-closets.

Those who have already suffered are by no means exempt; indeed they would seem by the occurrence of such cases as are illustrated in CASES XVIII, XIX, and XX, to be peculiarly exposed to relapses from both internal and cosmical influences.

In Italy I have found that the best remedy for the diarrhœa which so often attacks travellers from over-fatigue in summer and autumn, is lemon-juice and the horizontal posture. Lying down for a couple of hours on the back, and drinking two or three glasses of strong lemonade, with very little sugar, generally stops it. If that is not successful, Opium must be had recourse to; but it is seldom required in that land of lemons.

CHAPTER VIII.

CONSTIPATION AND COSTIVENESS.

Definitions.—SECTION 1.—CONSTIPATION.—From mechanical obstruction—Nervous exaggeration of the sphincters—Catarrh—Atony of colon—Insoluble articles of diet—Remedies. SECTION 2.—COSTIVENESS.—From deficient excretive life—Quality of stools—Occasionally interchanged with Diarrhœa—What diseases it accompanies—Effect on nervous system—Indications of treatment—Inconveniences of purgatives—What sort of purgatives are to be adopted—Dietary—Water—Watering places—Cautions respecting the use of them—Hydropathy.

THE words which head this chapter are sometimes employed as synonymous; but I do not wish them so to be understood here. By the former I would imply injury to the health from the quantity of feces retained in the alimentary canal; by the latter a deficiency in the quantity expelled by reason of a deficiency in the quantity found.

SECTION I.

Constipation.

The expulsive power is relatively or absolutely in default—the feces, normal or abnormal in quality, collect in some part of the bowels, and give proof of that collection by being occasionally passed in considerable quantities at a time. In the stools there are portions drier than the general mass—scybala of various sizes, dark brown or black, and usually with less smell than ordinary feces.

The most complete type of constipation is that which arises from mechanical obstruction, to discuss which however would lead us too far from the plan of this volume. It has not much connection with ordinary causes and effects of indigestion. It is the case I alluded to in describing the expulsive power as "relatively" in default.

It is also relatively in default in cases of hysteria and nervousness which spasmodically contract the sphincter ani and rectum, so that the fecal mass is kept back, and for its due expulsion there would be required a more than ordinary force, which in point of fact is not likely to be forthcoming in such cases.

And not uncommonly a catarrhal state of the upper parts, say of the stomach, will originate a relative deficiency of expulsive power, by enveloping the alimentary mass in a slimy coat, so that to push it on extraordinary peristaltic force is needed.

But the most common case is an absolute deficiency of power presented by a weak state or atony of the colon. This is a state frequent among those who lead a sedentary life, the anæmic, those debilitated by long acute illness or confinement to bed, and may be suspected wherever we observe a pale greasy skin and weak limbs. Old people very frequently suffer from it; so frequently indeed, that a diminished propulsive force in the large intestines may be considered as a normal consequence of advanced age.

No class of persons oftener suffer from constipation than old Indians. Their sedentary life and high feeding are partly chargeable with their liability. But in addition to this, the en-demic diseases of the country are often the exciting cause, and I have distinctly traced the commencement of a constipated habit of bowels to attacks of dysenteric fever brought on by malaria. The inflammation of the colon seems to leave behind it a local paralysis of the part: it acts in fact like the habit of taking artificial purgatives. So that the Anglo-Indian who suffers in this way must not be always accused of previous excesses or laziness.

Neglect of the natural call to evacuate the bowels also pro-duces this sort of torpidity by too long-continued dilatation even in young and strong persons. This neglect of the natural call may be from laziness, may be from a painful condition of the evacuating organs. (See CASE CLXIII.)

Where there is an individual tendency to atony of the colon, the tendency is aggravated, and sometimes first made apparent, by certain articles of diet, especially those which contain much insoluble matter. It is a mistake to suppose that these "irri-tate" the bowels, or pass quickly through them. The reverse is true; and, as a general rule, the regular transmission of the

18

mass is in proportion to the completeness of its digestion. No sort of food is so apt to be followed by constipation in atonic persons as that which contains a large amount of matter incapable of being acted upon by the digestive juices, such as skin and gristle, the husks and stones of fruit, and half-cooked vegetables, in which, besides cellulose, there is the equally impracticable substance, unbroken starch. All substances capable of being squeezed into a tough mass, such as puff pastry and new bread, come under the same class of insolubles; and gum and gelatine are liable to the same imputation according to some observers.

The most successful practice in simple constipation is the free use of cold-water enemata, and a long-continued course of Quinine and Strychnine. When there are no piles, this may be advantageously combined with the use of Aloës. The treatment does not forbid the administration of whatever else may be needful to relieve the disease in which constipation occurs; which disease of course requires to be removed before the local symptom will be free from risk of relapse. It is scarcely necessary to say that nothing will avail if the bad habits which have induced the constipation are persisted in.

If there be piles, the introduction into the rectum of a greased rectum-plug for a few moments, so as to empty. them, washing with cold water, and lying down on the back for ten minutes after evacuation, give great relief.

Constipation may often be much alleviated by oleaginous articles of diet, such as butter, bacon, &c., being taken with the usual food. This is especially the case with old people, who are apt to be too abstemious in this respect. We should not fail to impress upon them the physiological fact of costiveness being a normal condition of advancing years, and lead our patients to adjust their expectations to their age. They must not demand from sexagenarian bowels the same sensitiveness that is due at two-and-twenty. Daily evacuation, which should be the rule in youth, is an excess in an old man, and still more in an old woman. Thrice a week is enough for even robust persons.

If the constipation arise from impediments to the movements of the bowels upon one another, such as adhesions, scars of old ulcers, compression of the area of the gut, tumors, retroversion of the uterus, and the like, a more soothing treatment should be adopted. Then the enemata should be warmed, and have an ounce of olive oil added to them. If there be local pain, a little Opium may be dissolved in the oil, and some Leeches applied to the spot corresponding to the seat of pain. Hot fomentations and poultices of fresh laurel leaves also give great relief.

The depending position of the cæcum makes it the commonest seat of fecal collections; and if it is found difficult to fix on any other spot, it is wise to take it for granted that this is the failing one, and direct our local applications accordingly. We should not be satisfied with the one or two very copious stools which will follow these efforts; the treatment must be persevered in until the bowel has recovered its tone, or there will be great risk of relapse.

When there is much flatulence with the constipation, Turpentine and Rue may with advantage be added to the enemata.

SECTION II.

Costiveness.

In costiveness the absolute quantity of feces is always too small. It is in fact a deficient excretion into the alimentary canal.

That the greatest part of the matters which ought to be thus excreted come from the liver we have not the means of knowing, but the main point, that they are derived from portal blood, we are justified in asserting; so that the solution of the former question is of the less importance. And, at least, that a great deal of the color of feces is due to bile we may know from the phenomena attendant on obstructed gall-ducts.

But even when there is complete occlusion of the communication between the liver and intestines, the feces by no means consists entirely of undigested food; there is in them a great proportion of yellowish-gray granular matter which appears also in the healthy state, and still makes up the bulk of the solid excreta.

In deficiency, therefore, of the excretive powers of the intestines generally (*vulgo* "costiveness" or "biliousness"), there is a different substance retained than is the case when local lesion of the liver or gall-bladder obstructs the passage of bile. There is a partial retention of the whole matters destined for depuration from these quarters, instead of a complete retention of one constituent.

Hence there is not, as happens in mechanical retention of the bile, the well-known stain of jaundice communicated to the blood and skin, nor are the stools clay-colored. But there is a dinginess of complexion, and the stools are scanty. The skin is greasy and opaque, the countenance sometimes puffy and bloated, sometimes thin and pale, the lower eyelid especially sallow and discolored. The sebaceous follicles on the alæ nasi are stopped up with black matter.

There is seldom any decided emaciation, nor is there always even loss of muscular power; but still there is great sluggishness of body and apathy of mind, and often a miserable want of decision and energy. Digestion is accompanied by a good deal of discomfort and flatulence, but rarely by actual pain, and the distress does not begin till several hours after eating, so as to be with difficulty referred to any particular meal.

In the least complicated cases of checked intestinal secretion the stools are dark, hard, and dry; but their appearance may be varied by several circumstances.

Sometimes there is an augmented secretion of mucus, and then they are intimately mixed up with it, forming a black, slimy, almost gelatinous mass.

Sometimes, from the appetite not suffering, the patient will eat largely, and then there appears irregularly from time to time a quantity of fetid, semi-digested food, constituting a sort of diarrhœa accompanied with pain and colic. And this diarrhœa will often be the occasion of your patient's first coming to you, so that you might be deceived into a false impression of the case.

The congestion of the portal vessels in the upper part of the alimentary canal is often followed by the same state in the lower, and thus piles are formed, which add much to the general distress.

Costiveness is a common accompaniment of anæmia, chlorosis, and debility in both males and females, of diseased hearts, especially where the muscle is dilated rather than hypertrophied, of contracted liver, and, in short, of anything which makes the abdominal circulation sluggish. Sometimes it is found in cases of pulmonary tuberculosis, but hardly ever in consumptives under middle age. In their old age it may, like constipation, be considered the normal state of the abdominal viscera. All those pulmonary cases in which I have seen it last long enough to be a marked feature have been examples of senile phthisis. It is often the first and most characteristic phenomenon of that change of system which takes place in females after the cessation of the catamenia. The stools get gradually more and more scanty as the uterine secretion diminishes, as the pulse grows feebler, as the feet and hands are more liable to get cold. There is evidently lessened vitality throughout the whole body.

Habitual constipation, especially if it be habitually attempted to be relieved by purgatives, often induces costiveness. The "bilious" aspect of an old Indian is almost proverbial. You will usually find these persons have been blue-pillers from early life.

One end of this state of things, if left unchecked, is gradual progress from bad to worse. The decrease of destructive assimilation loads the tissues with effete matter, useless for the purposes of life, and a constant source of general discomfort. This impedes the constructive assimilation of food as well— growth is arrested, the blood is not renewed, and hence progressive anæmia, weakness, want of nervous and muscular power, and possibly in the end the degeneration of one or more of the viscera, and death from that cause.

A very striking attendant on the loss of destructive assimilation, is the depression of spirits; melancholy is so named from the dark, scanty stools, which were observed by the Greeks to be associated with it. It appears to me to be an almost universal rule in disease that the general discomfort is proportioned to the arrest of this vital destruction, and I am inclined to attribute it to the influence on the nerves of general sensation of effete matter which is retained. In all maladies, both acute and chronic, may be observed the truth of this law. Mark the ushering in of a fever: the malaise is excessive, there are pains in the back, in the head and the limbs, or a sense of what the

patients graphically call "all overishness;" but when they get
worse, and destruction begins, when the effete matter passes off
as urea and increases the specific gravity of the urine, then no
aggravation of local symptoms, however much it may alarm
their physician and make his prognosis graver, prevents the
general feeling of relief. Or watch a case of consumption :
the deposit of tubercle may be insignificant, and is at all events
in its first stage, yet the patient is despairing of recovery.
Why ? Because the skin is sluggish, the bowels costive, the
urine of low specific gravity ; because, in short, there is retention
of effete matter in the system. But let this tubercle soften ; let
there be night sweats, copious expectoration, diarrhœa, everything that prophesies ill, and who so full of hope as the sinking
sufferer himself ? Morbid states where destruction is in excess
are the most fatal, but those where retention preponderates are
invariably the most distressing.

Costiveness must be regarded as a disorder of the whole system, and not of the intestinal canal alone. The only effectual
remedies are those that are advised under that conviction.

The objects of treatment must be: first, to relieve the body
of the immediate presence of effete matter; and, secondly, to
prevent artificially its reaccumulation till such time as a complete renewal of the tissues has taken place. Then the body
ought to be able to take care of itself, and then, and not till
then, a cure may be said to have been performed. The attention to local disorders, arising from the successful study of
morbid anatomy, has too much made us forget this main object
of all medical work—the replacement of morbid tissue by
healthy. "Renew my age," was the chief earthly blessing
prayed for by the inspired prophet; and physiology teaches us
it should be the motto of the rational physician; for if he omit
to rebuild the healthy, his care for the destruction of the unhealthy is all thrown away.

Purgatives may very fairly begin the treatment; for the immediate relief they give to the feelings of discomfort is great.
But they must not end it. And let not the relief be set down
to the mere "clearing out of the bowels;" it is the cleansing of
the blood which is the real object of the remedy, and the real
cause of the relief. An inspection of what comes away shows

it has been newly formed; it is fresh bile and other natural constituents of recent feces, not of those which have rested long in the canal.

Nothing is easier than thus with a vigorous Blue-pill and Senna draught to drive away, as with a charm, the patient's discomforts; and he is ready enough to cry out that no more medicine is wanted. But what is the consequence of leaving off treatment? The renewal of the blood and tissues not having had time to regain its original activity—there not being enough new-made blood to carry on vigorous life—the effete materials again collect, and the disease takes a fresh starting-point. Again and again the coarse expedient is called for, and at last fails to effect its object of giving relief.

To avoid this evil consequence it is best to give no quickly-acting complete purgatives which directly deplete the abdominal plethora by serous exudation; but rather such as cause a gradual increase in the solid matter of the stools. Aloës and Rhubarb are the best of these; and I find it also beneficial to combine with them resins which act as a tonic to the surface of the mucous membrane, and prevent the exudation of serum and mucus. Four grains of Aloës-and-myrrh pill, every night, will in a week produce all the good effect of strong purgation; and it will produce the good permanently instead of merely for a time.

All accessory food that has the property of arresting destruction must be left off. Wine, beer, tea, and coffee, must, on this account, be excluded from the dietary; and milk, cocoa, whey, soda-water, Seltzer-water, &c., substituted for them.

Perhaps it is on account of their temporary arrest of destructive assimilation, that general tonics, such as Cinchona and Quinine, rarely agree well in those cases. I find it better to give pure bitters, such as Oak-bark, Quassia, and Gentian, which seem to act chiefly on the mucous membrane. Their use is to increase the appetite; and, when that object is attained, I leave them off; or, if it is attained without them, I do not begin.

Water is a very accessible remedy, and certainly a very rational one, when the destructive assimilation is deficient. The conclusive experiments of Dr. Böcker and of Dr. Falck,[1] show the increase of all interstitial metamorphosis by this agent to be in close proportion to the quantity taken, within certain bounds; and all who have heard or read of the agreeable sensations experience by patients during the water cure, cannot doubt its power of removing morbid accumulations of effete matter in the tissues. In this lies its strength; for, as Dr. Böcker observed, "the demand for new tissue, as expressed in the sensation of hunger, keeps pace exactly with the extent of the metamorphosis." And if this demand is rightly supplied, the result must be a complete renewal of the body.

The testimony of experience to the use of water as a remedial agent, is shown in the patronage bestowed from the earliest times upon numerous springs whose saline constituents are even less abundant than those of ordinary drinking water. Pfeffers, historically famous for freeing Martin Luther of his demon-haunted hypochondriasis, is still the resort of the invalid. It is situated in a most gloomy hole; and the copious hot stream that boils out of the rock is almost chemically pure. So that really the pure warm Vesta of the fountain, innocent of salt, should have the whole credit. The same may be said of the well-known Gastein and Wildbad, the crowded Baden, imperial Plombières, of the French Aix, and our own long-frequented Buxton, for, practically speaking, the influence of the saline particles they contain must be reckoned for nothing. It is certainly nothing as compared with the effects of moderate doses of water in Dr. Böcker's experiments.

As physiologists we cannot be surprised at the benefit derived from the simple expedient of drinking water beyond the demands of thirst, in all diseases of arrested metamorphosis. Taken several times a day between meals it is a most efficient remedy. Warm hip-baths are also of great use, and can be borne even from the first by those reduced to extreme anæmia and lifelessness. Afterwards, the cold sponge-bath, preceded and followed by friction to the skin, is a most active promoter of life in the skin and capillaries. The raising the specific gravity of the water by the addition of salt prevents the chill which fresh·

[1] See " Zeitschrift der K. K. Gesellschaft der Aertze zu Wien," April, 1854; and Vierordt's "Archiv," i. p. 150, 1853.

water is apt to impart. So that even persons with cold hands and feet, and very sluggish circulation, indicated by weak heart and pulses, can bear to be sponged with sea-water or brine.

Alkalies and neutral salts have the same action on the moulting of effete tissues that water has. Hence the repute of many really strong mineral wells. But care is needed lest the same result should follow their use which is threatened by the unguarded use of purgatives. In cases where there is arrest of metamorphosis without organic change in any of the viscera, I find that the weaker the spring the better for the patient.

While pulling down an old house, we must remember to be building up the new. Let full supplies of albuminous material be continuously kept up in such forms as the absorbents love. Let milk, mutton, and bread be the staple diet, with the smallest quantity of anything else that human weakness will submit to. If the patient be one of strong mind, the best and bravest thing is for him to carry out advice himself. He will then have gained a victory, not only over the flesh, but over the spirit. But if he is no Stoic, and cannot attain to the dignity of being his own gaoler, we need not be afraid of sending him to a hydropathic hotel. A little pressure will induce the owners of these houses to carry out rational directions, and the situations of most of them are well chosen for the advantages of air and amusement.

Medical men sometimes fear that in sending patients to water-cure establishments they may be abetting quackery. In my opinion scientific hydropathy, the renewal of the body by water and food, the increase of growth secondary to the increase of moulting, is very far from quackery. It is not an underhand mode of doing nothing, but a *bonâ fide* use of a powerful tool. And therefore a contrary effect than what has been feared would follow; for the very fact of medical men using the treatment as remedial, would show that science ranked it as a genuine physical power; and that, consequently, it is capable of doing as much harm as it does good; in fact, that, like all medical treatment, it needs as much prudence to prescribe it rightly as the most powerful agent in the pharmacopœia. Its being thus adopted by regular practitioners would soon remove it out of the hands of advertisers, who discredit their really valuable wares by attributing to them impossible powers.

CHAPTER IX.

NERVE DISORDERS CONNECTED WITH INDIGESTION.

Headache and Hemicrania—Vertigo—Loss of control over the thoughts—Epilepsy—Chorea—Stomach cough—Anæsthesia and Paralysis—Atrophy of muscles—Flushing of face and Nettle-rash.

THE most common morbid affection of the nervous system
arising from imperfect digestion is HEADACHE. The consequences of a debauch in a person unused to it are quite as often
splitting and throbbing of the temples as "hot coppers" and
nausea. It is not usual to consult a physician for such an occurrence, and therefore I have no illustration to quote. It has been
suggested that the state of the stomach is dependent on the state
of the brain, which is poisoned by the presence of absorbed
alcohol, and secondarily causes the vomiting just as cerebral
tumors and inflammations do. I cannot agree with that view
of the matter, because certainly we find headache accompanying
derangements of the stomach which are not the result of alcohol.
Instances are given in CASES XIV and XXXI where the comparative overlading the stomach, that is lading it with more
than in its weakness it was able to bear of vegetable food, brought
on, each time separately, severe headache. In CASE XXXVIII
indigestion of meat, and in XLV indigestion of fat, brought on
headache on each occasion. In these last two it was followed
by constipation, but that is not invariably the case. The
"bilious attacks" also, as depicted in CASES CLXVIII, CLXIX,
headache is a familiar accompaniment of the acute gastric
catarrh. This may be viewed as the acute form of the complication.

A more chronic form is exhibited in the following:—

CASE CCXXXI.—Rev. T. S— has been an occasional patient of mine since
1860, when he was 45 years old, a confirmed bachelor contented with his lot

and quite disposed to a rational enjoyment of life. He had had gout in early manhood, and lived temperately by rule to avoid a recurrence. But he had an anxious, easily worried mind, and the occasion of his coming to consult me was the occurrence as often as twice a month of intense headaches, lasting several days. They occupied the whole head, obscured the sight, and rendered him unfit for his clerical duties during the paroxysms. I found that each attack was preceded by gastric symptoms, complete anorexia, and sometimes by vomiting. A holiday trip to the seaside, when the cares of the parish were forgotten in boating, sketching, riding, and society, entirely relieved them and kept them off for many weeks afterwards. After each attack he was used to have pain in the anus and urethra, and pain on passing urine, which was acid, and deposited copious clouds of lithates on standing.

A long-continued course of non-purgative doses of Taraxacum, the habitual use of Potash-water as a drink at dinner, and some occasional short courses of Quinine, have made Mr. S— a much stronger and heartier man, and relieved him from the dominion of his headaches. It is possible too that he takes the world easier as he gets older, and being convinced of the evil consequences of worry, avoids it more.

Sorry should I be to advocate selfishness, yet truly it has a reward in this life, by preventing the stomach being disturbed by the business of others. In such cases as the above one cannot avoid seeing that the path of events is first the arrest of the gastric digestion by the accumulating influence of over-thought on the stomach, and by this latter organ retaliating on the brain so as to disable its functions. It seems a fair application of the *lex talionis.*

The gouty constitution of the patient shows him to have a weak sensitive stomach, easily put off work, and unable at such times to bear its necessary load. The alkaline drink probably promoted the secretion of gastric juice, and the Taraxacum the secretion of bile, but I question if such remedies can be trusted to alone, as I have remarked in the preceding chapter. Nerve tonics are needful to complete a cure.

The mental causes acting on the stomach need to be very slight in weakly excitable temperaments, such as women have.

CASE CCXXXII.—Mrs. James R—, aged 39, came to me in May, 1861. She was married, but childless, though the catamenia were copious and regular. About every fortnight the slightest annoyance or bodily fatigue brought on nausea, loss of appetite, and a throbbing in the temples. This was generally in the evening, and the next morning after a restless night she awoke with an intense headache, so that she could not raise her neck from the pillow.

This lasted till next night, and then went away, almost always suddenly, and she found herself quite well without any abnormal evacuation. I gave her Steel wine after food, and the intervals of the headaches seemed to grow longer; but I only saw her twice afterwards, and do not know if she were entirely cured.

CASE CCXXXIII.—Miss H—, a red-faced, dairymaidish woman of 40, had thrown upon her the charge of a large inn in a market town, where she was kept going all day among farmers and troublesome barmaids in consequence of the difficulty I found in curing her father of rheumatic gout, so as to enable him to take his share. She could get on very fairly, were it not for attacks of sickness and fluttering at the epigastrium, accompanied or followed by intense headaches at night, so severe as to awaken her up out of sleep with pain. Otherwise the bodily functions were healthily performed.

She came to me April 19th, 1861, and I gave her Iodide of Potassium and Tincture of Sesquichloride of Iron.

I saw her again on May 6th, when she said her head was much better since the last prescription, but that her legs were swelled. On inspection, this proved to be due to lumps of *Erythema nodosum*. I then gave her Citrate of Quinine and Iron. She went on with this some weeks, and was quite well as long as she took it; but on leaving it off her headaches, &c., relapsed, and she came up to London again about them. So I desired her to take Sesquioxide of Iron with her daily food as long as the untoward exertion of mind and body, to which she was exposed, lasted. This seems to have been effectual.

I have cited this last case for the sake of noticing by the way the connection of *Erythema nodosum* with gastric derangement. It is so connected mostly in cases where the nervous system suffers. Thus we shall find that hysterical women are subject to that cutaneous affection, especially when the hysteria is due to the stomach.

In these last two cases I ordered Iron, but I do not think it such an important agent as Quina, which in the second prescription of the two I have joined with it. However, the Iron did not disagree, even in the red-faced and apparently red-blooded patient. Indeed, where we see other evidences of need for tonics, our diagnosis of the anæmic condition of the blood by the color of the face must be very guarded. The tint of the inside of the lips is a safer guide.

The frequent occurrence of instances like these, where a mental responsibility which to a man would be a flea-bite, overwhelms a woman, should be a caution to those who are desirous of equalizing the brain-work of the two sexes.

It was mentioned in CASE XXXI that the patient was fearful of "apoplexy" from the giddiness which accompanies these headaches. Such a fear doubtless makes the nervous symptoms worse, and it ought to be dispersed by all possible means consistent with truth. The idea is not confined to bad observers, but was exhibited in the following instance by a medical man.

CASE CCXXXIV.—Nov. 29th, 1866.—H. W—, aged 47, a country surgeon, who has inherited and kept up a first-class, steady practice, has become lately anxious about his health, and he is fearful of apoplexy, of which his father died. So that he thought even of disposing of his practice. What causes his anxiety is that at least every month he has, generally the day after some unusual mental exertion, an attack of violent headache. It begins at daybreak, and gets rapidly worse, so that he is unable to get up, indeed can hardly raise his head from the pillow. It is accompanied with great nausea, but rarely with vomiting, and goes off rather suddenly in about forty-eight hours. His tongue is pale and flabby, the pulse beats soft and rather quick, and he notices that of late his complexion has grown paler and that he is yellower "about the gills."

Ordered to take 2 grs. of Quinine and $\frac{1}{16}$th of a grain of Strychnia twice a day. In case of being surprised by a headache to take 2 drachms of Ammoniated Tincture of Valerian every three hours till it disperses. To avoid all purgative drugs, to drink light Burgundy, and at dinner only. By all means to persist in following his profession, and to take interest and pleasure in it.

I heard from a patient of his in March, 1867, that he had become quite well and easy about his health.

The counsel to the above, to continue the practice of his profession, was given with a view of avoiding the hypochondriasis and stomach derangement often thereon consequent, which so often is produced by forsaking an active life for idleness.

The Valerian was designed as a temporary relief only, the basic treatment being Quinine and Strychnine with the hygienic measures detailed.

That such treatment will cure even very obstinate headache, deeply ingrained by time, is shown by the following narrative:—

CASE CCXXXV.—A very respectable licensed victualler, aged 44, of moderate and regular married habits, was introduced to me by his medical man, Dr. Slight, Dec. 21st, 1866. For more than eight years he had suffered from sick-headaches. They used to come on once in six weeks, but had grown gradually worse and more frequent, so that then he was never a fortnight without an attack. He almost always has warning of its approach two or three days beforehand conveyed by his friends noticing an unusual brightness and clearness of his eyes. Then the day before he feels a giddiness in the head and a coldness of the extremities. Then at night or in the early morning on comes the headache so bad that he cannot raise his head from the pillow, much less get up to business. This continues till he vomits, which brings immediate relief. He had taken at various times much *Pilula Hydrargyri*, but latterly Dr. Slight had very properly substituted Podophyllin for that " blue ruin" to chronic disease.

I continued in the same spirit the alteration in the treatment by giving him Aloës and Myrrh for a few days only, and then leaving off purgatives altogether. I prescribed also—

> ℞.—Quiniæ Sulphatis gr. xl,
> Succi limonum recentis q. s. ad illam solvendam,
> Tinct. Valerianæ flℨiiss,
> Aquæ ad Oj.
> Coch. ij max. ter die.

> ℞.—Tinct. Valerianæ Ammoniatæ,
> Coch. ij min. in aquâ soluta sumantur
> alternâ quâque horâ imminente cephalalgiâ.

On February 20th, 1867, I was informed by Dr. Slight, who came to me about other business, that our patient was much improved, and that he thought he was getting quite well.

On March 4th I saw him. He had just had one paroxysm of the bad sick-headache, but that was the only one since Dec. 21st. He had however frequently headaches of a hemicranic character of a morning, which are relieved by a cup of strong coffee.

Thinking his constitution might be becoming too habituated to the Quinine, I changed it for the Quiniæ et Ferri Citrás, of which he was to take six grains twice a day.

Since then he has been free from headache (*April 6th*).

The pain in the last case is described as HEMICRANIC when getting better, so I presume that when that peculiarity occurs that the prognosis may be considered more favorable than when it persists all over the cranium. In the ensuing instance again there was hemicrania, and it may be described as a mild curable case.

CASE CCXXXVI.—James C— has been a frequent patient of mine since early in 1861, when he was a widower of 60 years of age. Thirty years previously he had rheumatic fever and inflammation of the heart, the remains of which are discernible in an irregular pulse, a sharpish beat in the heart and a systolic murmur. He had an appointment affording him the blessing of regular occupation, but latterly he had found himself growing unequal to the mental calls upon him. Any unusual exertion brought on six-headaches, beginning with dizziness and oppression at the vertex, but usually fixing on one side or the other, and ending with vomiting.

This was invariably the case if he went too long without food. Very often a headache would begin before breakfast, but if he could manage to eat his usual meal it would go off.

Sometimes in the intervals of the headache he was much troubled with nettlerash.

I had him leeched at the back of the neck, and I afterwards gave him Citrate of Quinine and Iron; but what I found did most good was the advice to be never as much as four hours without food. He convinced himself by experience this was the best treatment.

In the autumn he took it into his head to marry, and that brought back a relapse of headaches. But a return to his former treatment took them away, and he is still able to go on with his occupation—at least he was so able when I last saw him in the spring of 1866.

Remark the nettlerash, a near relation of *Erythema nodosum*, affecting the same habits. I feel sure both these cutaneous disorders originate in the stomach, and are propagated to the surface by a sensitive nervous system.

I am sorry I had him leeched; it was a foolish concession to the opinion of another.

On the other hand, when sick-headaches are growing worse they seem usually to occupy the whole head, or the back part of the head, as described in the history given by the next.

Case CCXXXVII.—*February* 23d, 1867.—Mr. W. S—, aged 44, a clerk of sedentary habits, was always very well till eighteen months ago, never suffering from anything except occasional piles, so usual in sedentary men. At about that period he began to have sick-headaches, which now occur at least once a fortnight. They begin with what he calls, basing his nomenclature I suppose on some false medical theory, an attack of "liver;" that is to say he loses his appetite, and can eat nothing from nausea and disgust for two days. He then becomes prostrate, has a "grasping pain" at the back of the head, and in the right hypochondrium. The urine becomes dark and high-colored, and then thick.

Purgatives had been prescribed for him in a hope of warding off these attacks, and in consequence the bowels had become exceedingly costive, so as to seem to require a continued use of such drugs.

R.—Quiniae Sulphatis gr. ij,
Strychniae hydrochloratis gr. $\frac{1}{6}$,

in succo limonis et equâ solutâ bis die sumantur. Omittantur alia medicamenta.

My objection to purgatives is that not only do they make the bowels costive and become a habit difficult to leave off, but that they really increase the frequency of the disease they were prescribed for. In the following case it seems to me that a distinct history of continuously growing disease from the continuance of purgative habits is afforded.

CASE CCXXXVIII.—W. G. R—, aged 26, has never been strong, and comes of a delicate family, his mother having been a great sufferer from derangements of the digestive organs. Since boyhood he has suffered from headaches, which have been growing steadily more frequent, so that for three months there has been hardly ever a day without one. His habits are these —he goes to business after breakfast; at half-past one he dines; at three in the afternoon a sort of "hard feeling" in the stomach comes on, and nausea; in a short time this turns to a headache, with pain over both eyes and in the temples. He then takes a pill, which acts by half-past nine. He goes to bed, and the headache passes away during the night. If it lasts till the morrow, he takes a fresh purgative. After such a history it is almost needless to add that he has got much thinner during the past year, and that his urine is frequently thick, showing the imperfect vitality of the kidneys. What he brings with him is clear, of the specific gravity 1.015, acid, and not albuminous.

March 8, 1867.—Ordered him a grain and a half of Quinine and $\frac{1}{32}$d of a grain of Strychnine twice a day for three weeks, and every night for a week three and a half grains of Aloës and Myrrh pill.

April 15, 1867.—Mr. R— reports that he took the medicines for ten days, and found himself so much improved that he left them off. Then he found himself falling back gradually, and began them again, and is again progressing most favorably. His bowels are open now without the Aloës. I have ordered him Citrate of Iron-and-Quinine as a change.

It will be seen that in such-like cases I am disposed to attribute the costiveness to the disease, rather than the disease to the costiveness. I believe strong purgatives to be highly injurious, and those only permissible which increase the tone of the alimentary canal, and render it more disposed to continue its action without help than to require additional help as time goes on. Of such sort is Aloës.

The phenomenon of sick headache is closely connected with GASTRIC VERTIGO, the next nervous symptom arising from indigestion which I propose to sketch.

CASE CCXXXIX.—R. N—, aged 26, a melancholy, weather-beaten young man, first consulted me June 12th, 1866. He had been educated as a sculptor. but had lived a roving life, had made an expedition into the Central wilds of Australia, and otherwise knocked about the world a good deal. In the forests he had been subjected of course to great privations, supporting existence for some time mainly on tea and tobacco; and seemingly in consequence of that the veins of his legs and thighs had grown varicose. He wished on this account to forsake the plastic art and take to painting, as requiring less standing, and giving wider scope for inventive genius, of which he has a fair share. He had from boyhood been subject to sick-headaches, and had been used to be purged for them. The purgatives seemed to relieve immediate discomforts; but he thought the attacks were thereby aggravated, as they had become latterly more and more severe and frequent; and they were now accompanied by such giddiness that he was unable to stand or to employ his mind at all. Bright globes rolled before his eyes, and any attempt to rise brought on nausea like sea-sickness. This was occurring every ten days at least, and he was so evidently an invalid, that a marriage he was on the point of contracting was objected to by the intended-father-in-law on the score of his ill-health.

The purgatives had made his bowels very irregular and costive.

I gave him first four grains of Aloes and Myrrh pill with $_{1}^{1}$th of a grain of Hydrochlorate of Strychnia nightly, and ʒij of Ammoniated Tincture of Valerian thrice a day. After a fortnight, during which he was free from vertigines, the pill was diminished to two and a half grains of the first ingredient and $_{1}^{1}$th of a grain of the latter. The Valerian was exchanged for Quinine.

I last saw him several times in July. He had but one slight attack of vertigo, he had been able to leave off the pills, as the bowels were spontaneously opened by solid feces daily after breakfast. I recommended him now to go to a good surgeon and have his legs attended to.

In October I found he had continued well till he had his varicose veins operated upon, and then the pain and distress of the operation (which was also unsuccessful) laid him up again with his old vertigo and headache. But a week's Strychnine and Quinine set him up again. I recommended him in future to take his course of the same medicine the first ten days in each month.

Again in this case it will be seen that purgatives made the patient worse.

I think this functional disturbance of the brain is especially frequent when the indigestion takes the form of *intestinal* flatulence.

Case CCXL.—Mrs. P—, a stout lady of 52, first came to me in May, 1859. She had lost her husband three years previously, just at the period of the cessation of the catamenia, and since that time had suffered from indigestion in various forms. Latterly she had been much alarmed by the occurrence of frequent attacks of giddiness; and her son, a medical man, thought these might be due to diseased heart. I found these attacks of giddiness were always coincident with the rolling of wind about in the bowels, that they were

19

relieved when it passed away, and were also relieved by a strong purgative though they came on worse again after its action.

The administration of Valerian and Charcoal always does this patient good, but I have never seen her lately.

This very frequent degree of giddiness has been explained as a sort of drunkenness, caused by the absorption of alcohol evolved by the fermentation of sugar in the alimentary canal.

But there are several reasons against that explanation. First, in producing alcohol capable of intoxicating an adult, say five or six ounces at least, a bulk of carbonic acid would be formed enough to burst the bowels all to bits. Whereas in fact they are dilated only to the extent of a few cubic inches.

Again, when we see that fermentation has been going on in the stomach, as in certain catarrhal conditions of the organ with the tendency to parasitic growth mentioned in a former chapter, we do not find as a rule any remarkable giddiness complained of.

Again, the breath is not scented with alcohol, as it probably would be were much alcohol absorbed.

Again, the symptoms are not all like those of drunkenness.

In respect to the last observation, it is true that inexperienced persons, such as the estimable lady last quoted, may sometimes tell their physician that they feel, when giddy, as if they had been "taking too much" (alcohol); but the more habitual devotee knows the difference of the two sensations, and draws a broad line between them.

CASE CCXLI.—Herr V. J—, aged 30, a musician and teacher of music, consulted me in June, 1866, concerning a peculiar kind of giddiness, which would seize him at all sorts of inconvenient times, and quite disqualified him for the exercise of his profession. He would, in going through the streets in a hurry to keep an appointment with a pupil, suddenly become so giddy and blinded that he tumbled against passengers, and was forced to catch hold of neighboring railings for support. Vast dusky globes of mysterious gloom rolled before his eyes, he lost sight of the ground before him, so that a billowy gulf yawned under his feet, and he swayed helplessly on the brink. It was a continual renewal of the punishment of the company of Korah. "Haven't these symptoms some connection with your indulging in the gifts of Bacchus as well as singing their praises?" "No, indeed—no one knows better than I do, I am sorry to say, the effects of taking too much ; but this is quite different; it is nothing like either being screwed or *devil's trembles*." (The medical reader will identify by its initials the scientific name of the disease.) He said his belly became blown out with excessive flatulence; if he could explode, all was well; but if not, then the above-mentioned symptoms supervened. Sometimes, however, vomiting would relieve him; and if he had a succession of the attacks, a stout drench brought temporary alleviation. He said he had taken much and various purgative medicine for this purpose, and felt sure that the general effect afterwards was deleterious.

CASE CCXLII.—Another patient, A. W—, whom I saw in consultation with Mr. Hewer in April, 1867, and where the intestinal flatulence owed its origin to excessive bloodletting after some sort of fit long ago, complained to me that the ground seemed to be *rising behind him* as he walked, so that he had a constant fancy that he was going to be overwhelmed. This was a sensation not dissimilar to that noticed by the last two.

Once a shrewd fellow made a quaint play of words describing the connection of the brain and the stomach, which may serve as a reminder of the fact.

CASE CCXLIII.—Mr. H—, a railway traffic manager of 40, complained to me in September, 1862, that for seven years he had suffered from the frequent occurrence of very fetid stools, unformed and pultaceous, passed usually in the morning on first rising with a little griping. What most annoyed him was at these times he experienced great difficulty in transacting business requiring attention; the amount of fetor in the stools was an inverse measure of the mental powers; as he expressed it, " the addled eggs in the motions addled the brain."

He was used to dine in the middle of the day, and, I think, derived some benefit from my advice to him to postpone the heavy meal till evening. Ipecacuanha and Opium also seemed of use to him.

The confusion of thoughts usually comes over the subject of it just at the very times when it is most inconvenient; and it has this difference from hysteria, that the more exertion is made to exercise control, the worse it grows.

CASE CCXLIV.—A tutor of a large and rising college at Oxford, aged 26, in March, 1861, was frightened by the unaccustomed occurrence of trembling, and a tendency to lose the recollection of his whereabouts during divine service—a circumstance most particularly annoying to him from his having been recently appointed a chaplain. He had always been free from any excesses in wine, tobacco, women, or secret lust; but he read very hard for his degree in fellowship, and took a good deal of beer at dinner. He did not acknowledge to any indigestion affecting the stomach; but on inquiry it appeared that he had latterly had that peculiar looseness of bowels and fetor of stools which proceeds from imperfect solution of food, and in the evening not unfrequently was harassed with a glugging sound in the bowels. He sometimes perceived black specks floating before his eyes previous to the occurrence of the faintness.

I gave him Quinine and Strychnine, and advised his playing at rackets, instead of taking dull constitutional walks. A fortnight afterwards he came to report himself as much better.

This αποψια, or mental helplessness, would seem to bear a close relation to the vertigo last discussed. It is one of the symptoms falsely represented by the terrorist advertisements of quacks as due to former habits of masturbation, and doubtless creates on this score a good deal of secret alarm among the young and sensitive. By secret alarm it is much aggravated. As mentioned in CASE XCIII, it is sometimes caused by snuffing, sometimes by mental causes, as LXXVI and LXXVIII.

As mentioned in CASE XIX, it will sometimes alternate with diarrhœa and flatulence; and sometimes, as in the last tale related, the intestinal derangement will be by no means prominent, and only to be identified by cross-questioning.

Another form of cerebro-spinal disorder dependent on faulty digestion is EPILEPSY.

CASE CCXLV.—Late one night in June, 1854, I was summoned to see a patient of the late Mr. Tegart, Miss W—, aged 13, whom I found in a violent epileptic fit. The closeness of the sleeping nursery showing a careless, unphysiological management, I suspected corresponding neglect in the dietetic discipline as well. I accordingly administered a stout purge and the next day was shown a chamber utensil full of hard lumps of fæces, mixed with half-digested fruit and other rubbish. The patient had no more epileptic fits.

A somewhat similar occurrence to this quoted in page 70, where the indigestion of an unaccustomed amount of adipose aliment induced a single attack of epilepsy not repeated—at least not repeated during the eight months which elapsed after I wrote that narrative.

As acute morbid conditions of the alimentary canal promote acute epilepsy, so chronic morbid conditions promote chronic epilepsy, that is to say, epilepsy of a milder but more confirmed character.

CASE CCXLVI.—In November, 1863, Dr. Wallace of Parsonstown sent to consult me Mr. James F—. It appeared that he had become subject to attacks of sometimes partial, sometimes complete, loss of sensibility, preceded and accompanied by a cramp in the arms and twitching of the face. Observation of the stools elicited the fact that they frequently contained mucus. A tonic pill of Myrrh, Aloës, Turpentine, and Henbane, stayed this formation of mucus. Coincident herewith there was a marked improvement in the nervous symptoms. I then prescribed Quinine and Strychnine, but have no further note as yet.

In the next case of improvement by management of the diet there is recorded the symptom which I take to intimate the collection of mucus in the stomach.[1]

CASE CCXLVII.—Benjamin M—, aged about 40, first came under my charge October 23d, 1858. A letter he brought from his medical man described him as subject to confirmed epilepsy for two years. Several times a week he got giddy, was unable to stand, sometimes lost his senses, sometimes was convulsed, but rarely bit his tongue. After the paroxysms he always felt tired, and usually went to sleep. He had always been temperate in eating, drinking, sleeping, and matrimonial matters, and could assign no cause for his epilepsy. On examination, I found tenderness on pressure at the pit of the stomach, and the patient said that he felt as if a weight were laid on that part, especially during wet and cold weather.

To restrain the secretion of mucus, I ordered him a quarter of a grain of Nitrate of Silver night and morning, and some Bismuth and Sesquioxide of Iron twice a day. But the most important part of the prescription was as follows : "Avoid beer, pastry, fruit, sugar, tea, and coffee. In place of the latter, take milk and soda-water, with stale bread or biscuit, for breakfast. At dinner eat once-cooked plain meat, stale bread and green vegetables."

I saw him again November 18th, 1858 ; no improvement had resulted, and all I could do was to encourage him to persevere.

I did not receive another visit till July 5th, 1860, when he reported that he hardly ever had any attacks of giddiness ; indeed, never except after violent exercise. To my surprise, and at first consternation, he said he had been continuing the Nitrate of Silver, with occasional intermissions of a week or so, up to that time. No discoloration of the skin had occurred, however. I then gave him some Citrate of Quinine and Iron, which a letter from his wife reports set him in strong health in a few weeks.

He had no more fits till 1862, when over-attention to business seems to have deranged his digestion, and he had a few slight epileptic attacks while dressing in the morning. I advised a recurrence to his former dietary and to the last-prescribed medicine.

In the spring of 1866 he called to report that he had got quite well, and kept so up to that date by dint of adhering to a strict plan of diet, grounded on the one I had written out in 1858.

[1] See page 171.

The omission of a single article from the diet-table will suffice sometimes.

Case CCXLVIII.—Rev. George O—, aged about 24, married, was first my patient in August, 1856, for non-syphilitic periosteal rheumatism which completely crippled him, but vanished with remarkable rapidity under the use of Iodide of Potassium. He continued well till September, 1861, when he began to have epileptic fits. He came to me in May, 1862, when I kept him a month in London, and seeing the former success of Iodide of Potassium in one disease in his case, tried it again in another. While in London he was free from fits, but no sooner did he return home than they became worse than ever, though he increased the dose of the Iodide. So he left it off, and when I next heard he was taking, under advice, mercurial alteratives " to act on the liver" and so on, not apparently with any advantage.

I next saw him in January, 1867, when he said the fits had become so frequent and occurred at such inconvenient times, that he had been forced to give up his clerical work. I had originally advised that, and indeed it was in consequence of that advice that he had deserted me. However, that idleness did no good, till some person recommended him to give up beer, and from that date the epilepsy began to improve. The said drink had always made him flatulent, and he had found his digestion much better since he took only water. He was anxious to try some Bromide of Potassium, and I allowed him to do so, but still urged him to persist in his abstinence, as to that I attributed his improvement.

In March he called to report that he had had only one fit, and in that consciousness was not entirely lost.

Iodide and Bromide of Potassium seem to have a peculiar restorative action over the white fibrous tissues. They were first brought under the notice of the profession by the effect of Iodine in scrofulous diseases of the glands, and then by their cure of syphilitic periostitis. A notion got into the profession that they had some antagonistic, or controlling, or evacuating power over syphilitic virus; but the more recent surgical writers, such as Mr. Lee, think there is evidence against that idea, and that they benefit by curing the diseases arising out of the presence of the virus, and not by removing the virus itself. They cure the patient's tissues without specially affecting the *materies morbi*, equally in rheumatic as in venereal cases of periosteal disease. When recently affected, these patients are restored to health as readily as the subjects of secondary syphilis; and the apparent resistance in some cases to the remedy is due to the protracted nature of the ailments; just as syphilitic periostitis, when it has lasted a long time without medical aid, is very obstinate also.   On the ground that Iodide of Potassium has a special restorative power over the white fibrous tissues, I should expect most direct benefit from it in epilepsy to those cases where epilepsy is due to some lesion of the membranes of the cerebral or other masses of nerve substance, whether that lesion be temporary or permanent : and I am not surprised at its failure in the gastric cases we are now discussing.

In the next case the flatulent distension of the bowels before the fits seems to associate the epilepsy with deranged digestion. Perhaps the disease is too long ingrained for cure ; perhaps I did not pay sufficient attention to dietetics at that date; but, at all events, favorable results have not followed the use of drugs.

CASE CCXLIX.—A. G—, aged about 40, first came to me in January, 1851, after having been a patient of the late Dr. W. F. Chambers for five years for epileptic attacks, occurring about every fortnight or three weeks.   He had been taking Sulphate of Zinc (gr. iij bis die) and much purgative medicine, and I continued the prescription.

I saw him from time to time during the next two years, and found that the fits were invariably preceded by flatulence and distension of the bowels, but immediately announced by perspirations and pale urine.   By the end of 1852 the fits had become less violent, and assumed a regular periodicity, coming on every eighth night between ten minutes before and ten minutes after twelve. I do not think he made any more attempts to get well.

I am afraid one cannot in such cases find much fault with a despairing patient; so few are the instances in which an epilepsy which has assumed a regular periodicity, at the same time that it grows milder, ever is cured.   It is possible there may have been structural brain disease, and that the indigestion was only the motive cause of the epilepsy.

There is a sort of modified chronic chorea which seems dependent on chronic causes in the alimentary canal, just as acute chorea is often dependent on acute causes.   I call worms (for example) acute, because they are movable and temporary, and directly they are gone the derangements tend to get well of their own accord.[1]

[1] See the distinctions drawn between "Acute" and "Chronic" in the Introduction, page 22 and note. See also "chorea," in "Lectures chiefly Clinical," Lecture xxx.

Case CCL.—C. J—, aged 26, an accountant, came to me on February 5, 1867, in much shame-faced fear, under the idea that his state of health was the delayed retribution for nasty habits of secret lust practised when he was a boy. He was troubled with involuntary twitches of the facial muscles, low spirits and causeless fears. His breath felt hot to him; his scalp and anus itched; and the abdomen seemed to burn as he lay in bed. His sleep was disturbed by terrible dreams. The spasmodic movements of the face were, however, what troubled him most as a man of business, for they made him ridiculous, and the more he endeavored to control them, the worse they were.

The urine was natural. Indeed, I could detect nothing contrary to nature in the functions of the body, except in those of the alimentary canal, whose excretions were reported very fetid at times.

He had taken Zinc and other specifics without advantage.

I gave him Valerian for two days with some improvement of the sleep (but it is possible that improvement may have been due to his mind being set at rest respecting his youthful nastiness) and then put him on Quinine and Oil of Rosemary, which he continues to take with some benefit.

Symptoms like these are, again, engines by which the advertising quacks extort money. It is difficult to avoid having their obscene literature thrust into one's hands, and it often leaves torturing scars on the mind for life.

Minor degree of reflex manifestations of nervous action assume more familiar forms. Thus we have very commonly what is known as "STOMACH COUGH," that is to say, cough without any bronchial secretion or other morbid condition of the lungs, and connected with, aggravated by, and yielding simultaneously with, catarrhal relaxation of the mucous coats of the stomach. Usually we may infer this gastric derangement from the symptoms; sometimes we have the confirmation (afforded in the case I will quote as an illustration) of a similar catarrhal condition appearing at the visible extremity of the alimentary canal.

Case CCLI.—Henry L—, a manufacturer, aged about 30 when he first consulted me in January, 1861, complained of a constant hacking cough, without expectoration generally, but still aggravated by damp chilly weather, and of pain in the left mammary region. His face was pale and flabby, and he had a tendency to grow fat. The chest was quite healthy as far as could be ascertained by the ear. He got better under the use of Quinine, and after a few visits I did not see him again till November, 1863, when he came to me with a recurrence of his former symptoms. He in addition complained that "his chest" (pointing to his epigastrium) "gets stuffy and feels too large." The uvula was much relaxed, and on looking into the throat, it seemed to be redder as you go down deeper. Acting on this hint, I have since accompanied the tonics by gargles of Oak-bark and Alum, and subsequent attacks have got better the quicker for them. On leaving England, in December, 1864, I commended him to another doctor; but he came to me in October, 1866, saying that he had been quite well till then, when exposure in the country during the wet autumn had brought on an unusually bad attack. I never saw his throat so red, and he said he could feel his œsophagus all the way down to the pit of the stomach. A course of Quinine restored him to his usual health. But again in April he has come again with a sore throat and hoarseness, and this time, as in 1861, he has got a hacking cough, and is convinced, as is also his wife, that he is going into a consumption. The chest is quite normal on examination, but the uvula is much lengthened and the throat red. There is great weight at the epigastrium after food, and a sensation in the œsophagus leading him to hawk mucus continually.

Apropos of the relaxation of the uvula, I may mention that he made it much worse at the first by violently causticing the back of the fauces with Nitrate of Silver.

These cases are best treated by astringents, but changes of weather must be guarded against with especial care. Without the aid of the stethoscope they might readily be mistaken for pulmonary consumption.

The following is a very fair typical example of symptoms such as these occurring, as they often do, in a gouty constitution.

CASE CCLII.—S—, a burly country gentleman of 50, used to have regular attacks of acute gout in the small joints three times a year till 1863, when they ceased. But their place is taken by a worse enemy. For as October comes round annually his throat gets sore and livid red, or else he has a frequent hacking cough, and sometimes both evils together. There is a feeling of weight at the epigastrium after food, and a discomfort scarcely to be called pain, on pressure of the lower part of the stomach. The bowels are apt to be loose, but he usually restrains the looseness from proceeding to diarrhœa by Opium. (*Dec. 19th,* 1866.)

In the last case the pains of gout alternated with the stomach cough and sore throat, but in some unfortunates they may both occur together.

CASE CCLIII.—G. G—, aged 44, is the brother of a patient who has been under my care for gout, and has himself been bald from an early age. He reports also that former generations of the family have been gouty. He has been under my care since November last (1866) for intermittent sciatica and stiff-neck, much dependent on the condition of the stomach, and which I con-

sider therefore to have a gouty character. He was desirous of reporting improvement in this as time went on, but I cannot say I thought it any better
in January, 1867, when he began to have cough and then a red swollen throat.
The lower lobes of the lungs are apt also to become congested occasionally;
but he never expectorates with his cough, which seems to vary much with
the weather and state of the stomach. I treated him for some time with
Quinine, Iodide of Potassium, Baths. &c., without any advantage. In February I gave some unpalatable, or impracticable, advice as to change of
climate and lost sight of him.

Another form of the influence of gastric derangement upon
the nervous system is the production of morbid ANÆSTHESIA.
As I remarked respecting headache the most frequent instances
are found amongst those who have already made the nervous
system susceptible of disease by overstraining it.

CASE CCLIV.—An American speculator, aged 48, was sent to me in June,
1866, by Dr. Forsyth Meigs, of Philadelphia. He had lost a fortune of ten
thousand a year by the civil war, so that he had to begin life over again—an
ordeal not so severe in the United States as in England, but still an ordeal
anywhere; he had worked energetically to recover his position; he had thrown
himself into the turbulent, rather than the quiet joys of life; and he had also
gone through certain matrimonial difficulties not unscathed. In August,
1864, he was taken with vomiting and loss of appetite, general debility, deficient sleep, and occasional flatulence. These ordinary digestive derangements
were the only trouble till the end of September in the same year, when he
found gradually creeping over his hands and feet a peculiar sensation of numbness; not what is commonly called "pins and needles," but a bluntness of
perception especially in the finger-tips, so that he did not know when he was
touching a small object, unless he saw it; and he often tripped from not de-
tecting a small impediment in walking. On resuming matrimonial privileges,
after an interval of abstinence on account of his health, he found that emission
occurred immediately on entrance, or even before entrance was effected. His
head had been bald since the age of twenty, but in general respects he is a
young-looking man. The specific gravity of the urine before breakfast is
1.025, after breakfast 1.015. Under the use of nutritious diet, abstinence
from alcohol, and from over-much anxiety in business, and of Nux Vomica
and Quinia, prescribed by Dr. Meigs, his digestion had strengthened and the
sensation was returning by degrees to his extremities. I thought he could
not do better than take a course of the same drugs for ten days in each
month, and follow strictly the plan of life laid down for him by his first-rate
physician. It is proposed that he shall spend a few years in taking his
daughter round the chief cities of Europe.

He called on me early this spring of 1867 to report improvement.

Cases such as this last receive all sorts of names, according
to the prevailing theories of the period; which nomenclature

does the patient no harm so long as the theory does not influence the treatment. The principal danger is lest he may fall into the hands of a counter-irritator, who should depress the vital powers by making sore places over the spots where he supposes chronic inflammation to exist. Electricity probably does no harm, but it will be observed that recovery is always coincident with an improvement in the digestion, and I think attention to this function is our leading duty; and alone it may be followed by a cure, so that there is no need of additional treatment.

The loss of nervous function in some cases is manifested in the nerves of motion principally, or even solely. We must presume that it depends on the specialities of the nervous system itself in each individual which portion of it be affected, and that the influence of the alimentary canal is general; otherwise we should be able to map out certain tracts as ruled over by certain viscera, and not find motor and sensory fibres indifferently injured by imperfection of the stomach.

CASE CCLV.—Colonel B—, aged 43, applied for my opinion on the 30th of July, 1866, about a loss of power in the legs. He has always had a "weak stomach," feeling a weight at the epigastrium if he takes liberties with his diet, or is exposed to a damp cold. In the summer of 1865 his stomach difficulties were particularly bad, and then he began to notice what he called a "fidgetiness" of the legs, inducing him to kick and stretch them about. Then he found he was less and less able to walk, and then there was pain in the legs felt, especially after any exertion. A mile was the utmost he was able to walk when I saw him. He had been galvanized and had tried a variety of medical treatments, without any advantage that he could discover; the only improvement he could ever notice, was when he was in the bracing air of Scotland. On this hint I sent him thither, with a prescription for Quinine and Strychnine. I heard from him in August that he digested better and walked better, but that the pains were bad in the legs at night. I added therefore to his mixture four grains per dose of Iodide of Potassium.

In this instance the sensory tracts do not seem to be injured at all. And the paralysis is not sufficiently complete to cause atrophy from deficient use of the muscles. Indeed, in gastric paralysis (if I may so call it) I have never seen the loss of power so complete as to deprive the muscles of that amount of motion which is conducive to its welfare. The patients can always get about a little, are willing to do so, and very often disposed to exert themselves too much.

This is an important point in the treatment; for if what I have remarked is true universally, we shall be doing harm by following the common practice in telling the patients to employ and exercise the muscles as much as possible; we ought rather to impress upon them the necessity of avoiding such an amount of motion as nature warns us against by the sensation of consequent fatigue.

When atrophy of the muscles has any connection with derangement of the digestive organs, it is usually to be traced to overwork rather than to underwork. Of this I will quote an instance which I have previously made use of in the later editions of my Clinical Lectures.

CASE CCLVI.—(*Clinical, St. Mary's, June 13th*, 1863.) Nathaniel B— is a top-sawyer by trade, aged 45, and was always a hearty fellow, able to do a good day's work, till ten months ago; when, after violent exertion in turning over a mass of timber, he got what he calls "a wrench" in the pit of the stomach, and " has never been the same man since." The appetite failed, and therewith the strength; the muscles wasted, and the whole body grew emaciated. The loss of appetite then became entire, and then increased to an utter loathing of food. He went into Guy's Hospital three months ago, but left apparently dissatisfied and ungrateful. On gaining admission to St. Mary's, May 22d, he seemed much cast down, expecting never to get any better. He was able to walk about, and the chief loss of power seemed in the shoulder-muscles, the deltoid and biceps; and when he tried to "put up" the latter, that is to throw into it the contractile nervous force, it felt quite soft, without any of the corky elasticity which distinguishes a sawyer's arm. He is the father of thirteen children, but since the commencement of his present illness he has entirely lost virile power. He states himself to be a perfectly sober moderate man, and has a good character on that score from his employer.

It is scarcely necessary to say that the epigastrium and hepatic region were carefully examined for evidence of cancerous degeneration, and none was found. The lungs also were thoroughly auscultated, and nothing abnormal was detected, beyond a suspicion of slight comparative dulness in the right apex. He had not suffered from habitual cough or had any diarrhœa.

He was at first kept in bed and given milk and beef-tea every two hours, with ten grains of Boudault's Pepsine powders three times daily. In a few days his excessive nausea and lowness of spirits had abated, and he was ordered six grains of Quinine and three drachms of Cod-liver Oil daily in addition. In a few days more he was tried with half a mutton-chop, digested it well, and on the 6th of June was able to take our whole ordinary diet. a, pint of milk, and a pint of beef-tea, and a pint of porter. On the 12th (yesterday) he was so much better that I thought it was scarcely justifiable to let him occupy a place in the hospital any longer. and I trust he will be able to get on as an out-patient.

As he was confined to bed at first, it was not convenient for some days to put him in the scales; but on May the 24th we found his weight 8 stone 5½ pounds; on the 30th, 8 stone 7½ pounds; on June 6th, 8 stone 10 pounds; on June 12th, 8 stone 10½ pounds; his height being 5 feet 6 inches.

The only day on which he did not take Pepsine was May 29th, when our stock was accidentally exhausted. He then complained of pain at the epigastrium, and attributed that to the omission of his powders.

A somewhat similar case is related page 37 (CASE VIII).

Now, had this sawyer been a gentleman in easy circumstances, the excessive waste would not have been habitual, and he would not have had muscular atrophy of the limbs. The "wrench" would have been confined to the stomach, and he would probably have suffered only from imperfect indigestion, like the sporting-man (CASE XIII) whose partial paralysis of the stomach dated from hallooing and running on Derby-day. As he went on in life, any extra exertion would have induced flatulence, as in the old fox-hunter (CASE XII). Or if he had been in vigorous bloom, the paralysis would have been only temporary, as in CASES IX, X, XI. The proneness to muscular atrophy may have long existed; but no harm happened so long as the stomach was able to go on supplying nutriment enough to compensate the extreme waste of the violent exertions. No sooner does its debilitated condition fall below a certain point than atrophy is exhibited suddenly and proceeds at a frightful pace.

The only cases in which the easy classes are likely to be similarly affected is when injudicious friends urge them to exertions for the good of their health, or where brain-work occupies the hostile post held by body-work, as in the instance before us.

Another symptom of gastric weakness which obscene advertisers are wont to turn to their profit, as well as vertigo and chorea, is FLUSHING OF THE FACE, or causeless blushing. This fact would be sufficient to show that it is very common in healthy persons; but rarely does it grow bad enough to consult a physician about.

Case CCLVII.—Mrs. R—, aged 56, put herself under my charge in June, 1866, stating that for the last half-dozen years—in fact, since the cessation of the catamenia—she had suffered from flushings of the face at irregular times, accompanied by palpitations of the heart, so severe that she thought that organ must be organically diseased. On examination, I found it healthy in all respects.

On inquiry, I elicited that she suffered excessively from intestinal flatulence, especially of an evening, and not uncommonly had heartburn at night, if she ate pastry or took sugar in her tea.

When she came to me she had been latterly much worse than usual, and this I traced to annoyance about a love difficulty of her favorite son.

A month's course of Quinine and Strychnine twice a day dissipated gradually the inconvenience she suffered; but she remained, and probably considering her age always will remain, a very nervous subject.

A higher degree of the same phenomenon constitutes NETTLE-RASH, which is always partially traceable to the stomach, as is shown by the relief afforded by a change of diet, a change sometimes seemingly insignificant.

Case CCLVIII.—Miss C. R—, aged 35 or so, was a patient of mine in the summer of 1866 for nettlerash, which for the last six years had made her mornings miserable to her, coming on in her legs directly they are put out of her warm bed. She had tried all sorts of treatment in vain; nothing seemed to do her any good. The chief things which make her worse are being worried, and getting wet through when riding.

I have given her Soda, Liquor Ammoniæ, &c., without any apparent effect; but what really seems to afford the greatest relief is leaving off all alcoholic liquids, tea, and fruit.

Case CCLIX.—R. V. E—, aged 50, a commercial man, requested my advice in February, 1850, about a peculiar itching papular eruption. Observing that it ran in straight lines, and was only in the front parts of the body commanded by the hands, I questioned him further, and found that it only appeared when scratched, and in the morning assumed the form of "wheals." He confessed that eating pickles and drinking hard beer used to bring on these "wheals," but that he had for some time carefully avoided all acids. The only unnecessary that I could detect among his "non-naturals" was the use of tobacco. He allowed that strong shag, in which he indulged, certainly did somewhat upset his stomach and make his hand shake.

Leaving off smoking at night, and using only light cheroots by day, cured him, with the help of a little Liquor Ammoniæ. His nettlerash got well, when his nerves and his stomach returned to their duty together.

That nettlerash is a phenomenon connected with imperfect innervation I have the evidence of the following personal experience. A few years ago I was unfortunate enough to have a popliteal aneurism. When this came to press habitually upon the nerves of the ham, of which my sensations gave me due notice, I remarked that bathing in cold water brought on nettle-rash, invariably confined to the affected leg only, and never since suffered from.

Now, when an individual has a nervous system less powerful than the rest of the body, let them be females, feminine men, or men with habits debilitating to innervation, any gastric derangement is liable to bring on nettlerash among other symptoms, just as the cold water brought on nettlerash in the leg whose innervation was partially cut off. That is my theory, and that is why I introduce nettlerash in this somewhat unusual connection.

In the next case the eruption on the skin, though clearly dependent on the digestive organs, was not of such a character as to be correctly called Nettlerash; it was rather an "Erythema circumscriptum." (I presume in cutaneous pathology one may be allowed to invent a nomenclature that best suits the case in hand, inasmuch as no one treats the patient in obedience to the nomenclature.) The depression of spirits and causeless fears point out that connection of the pathology of the case with the nervous system, which I have alluded to above.

CASE CCLX.—W. L—, a country gentleman of 52, of robust jovial aspect, square set and muscular, comes of a gouty family; but he has never developed in his own person his *damnosa hereditas* in its typical form, having never felt any swelling or pain in any of the joints or tendons. He placed himself under my care on the 15th of May, 1867, on account of attacks of which the following is the usual course. About every three or four weeks he finds a dreadful depression of spirits comes over him, without any cause or previous excitement; the urine becomes excessively copious and pale colored, and is passed with great frequency. Then commences gastric flatulence, the air bursting up from the stomach in such quantities as to wake him up at night; and at the same period the heart thumps against the ribs with extraordinary force and frequency. The appetite is quite lost, and disgust to food takes its place. But there is considerable thirst, and the tongue is furred, clammy, and dry. The urine now gets scanty, and as high colored sometimes as porter, but never thick. Then he begins to bring up clear fluid from the mouth nearly every hour in gushes of several ounces at a time. The fluid is always sour, and after meals is intensely so to the taste. These attacks have ordinarily ceased in about four days, and the convalescence from them is almost always announced by a peculiar eruption on the skin, especially on that of the

belly and legs. Sometimes they last longer, and notably on the occasion of
his first coming to my house he had been ill off and on for three weeks, occasionally having a remission but never quite free. When I saw him the tongue
was much furred, he had no appetite, and had been throwing up the acid fluid
already that morning. In spite of a round red face he really had the aspect
of great mental distress, quite incommensurate with the degree of pain experienced. In spite of the depression of spirits his mind was quite clear, and
capable of business (if he had any), and there was no vertigo or headache.
During the three weeks he had lost fifteen pounds in weight. The epigastrium was very drummy on percussion, the resonance stretching to an abnormal extent from side to side, so as to show a very large stomach. The belly
however was flat, though he is a broad dumpy man. I prescribed (May 17th,
1867)—

> R.—Quiniæ Sulphatis, gr. ij,
> Strychniæ Hydrochloratis, gr. $_2^1{}_0$,
> Succi Limonum quant. suff., ad illa sol venda,
> Aquæ ad fl̃3j,
> Potassii Iodidi, gr. iij ; bis die.

> R.—Pilulæ Alöes cum Myrrhâ, gr. iv,
> Strychniæ Hydrochloratis, $_2^1{}_5$; omni nocte ;

and desired him to call again in four days. He was advised to take no stimulants at all.

On the second visit I had an opportunity of seeing what he had described
as the convalescence from his attacks. He reported that with each dose of
the medicine from the time of beginning its use he had felt relief. The flatulence and ejection of sour fluid had ceased ; and on the second day his appetite returned. He had celebrated the return of it by eating whitebait and
spitch-cock eels, and drinking some weak brandy and water at dinner. The
consequence was one attack of the sour waterbrash ; but since then he had
been free. The urine had become nearly natural. On the skin of the arms
and abdomen he showed me spots as large as sixpences of a red erythematous
eruption, closely resembling measles. He said it did not itch, but pricked,
like what Indians call " the prickly heat."

The next day it was still more like measles, for the centre of the spots had
begun to fade while the circumference had spread, like fairy-rings on the green
sward.

ANALYSIS.

THE cases in this volume form a sort of skeleton, which I have articulated together by argument, and tried to make muscularly active by practical observations. A list of them, indexed according to the subjects they profess to illustrate, will serve as a memorandum, just as the dry bones recall succinctly the structure of the animal frame to the anatomical student.

20

CASE XV.—Chronic condition following one mental shock. Flatulence, impotence, red nose.

CASE XVI.—Chronic condition from continued anxiety. Flatulence, emaciation, weight after meals.

CASE XVII.—Flatulence, spasmodic pain, costiveness, from bronchial catarrh.

CASES XVIII, XIX, and XX.—Flatulence, heartburn, waterbrash, diarrhœa, from an attack of epidemic cholera. Adhesion of the poison to the locality. Evil of purgatives.

CASE XXI.—Indigestion from the depressing effects of rheumatic fever. Pain at epigastrium after food.

CASE XXII.—Flatulence and pain in abdomen after child-bearing.

CASE XXIII.—Weight at epigastrium and vertigo, flatulence and acidity, from climatic influences.

CASE XXIV.—The same producing hysteria.

CASE XXV.—The deleterious climatic influences are changes, rather than continued depressants. Resistance gained by use of Iron.

CASE XXVI.—The dependence of indigestion on change of weather a diagnostic sign.

CASE XXVII.—Influence of locality. Flatulence, nausea, acidity, waterbrash, emaciation, when in a malarious district, and not elsewhere.

CASE XXVIII.—Soft air.

CASE XXIX.—Sea air and bathing.

CASE XXX.—Chronic flatulence after tropical dysentery. Remarks on digestion in hot countries.

CASE XXXI.—Effects on flatulence of abstinence from sugar.

§ 2. *Indigestion of albumen and fibrin.*

CASE XXXII.—Pain in epigastrium after meat during the debility induced by an acute fever.

CASE XXXIII.—Pain, nausea, and diarrhœa after meat in an aphthous condition of mucous membrane.

CASE XXXIV.—Phthisis pulmonalis. Weight at the epigastrium after animal food, diarrhœa, emaciation, relieved by Quinine and Iron.

CASE XXXV.—Fatal consequence of a mental shock in a consumptive who could not digest meat.

CASE XXXVI.—Vomiting from œsophageal lesion alleviated by anæsthetics.

CASE XXXVII.—It is the form rather than the chemical constitution which makes meat difficult of digestion in œsophageal cases. Fish easier digested from its friability.

CASE XXXVIII.—Yet in some stomach cases even the softest animal food is rejected.

§ 3. *Indigestion of fat.*

CASE XXXIX.—Consumption fatal from loss of power of assimilating fat, though the amount of tubercle moderate.

CASE XL.—Consumption, though advanced, relieved by great power of assimilating fat.

CASE XLI.—Moderate amount of consumption and moderate amount of assimilating power balancing one another.

CASE XLII and XLIII.—Renewal of fat assimilation overcoming tubercular diarrhœa.

CASE XLIV.—Strumous dyspepsia consists in the non-assimilation of fat.

CASE XLV.—Disgust to fat joined to defective assimilation. The disease induced by starvation in childhood.

CASE XLVI.—Importance of fat at the period of puberty in girls.

CASE XLVII.—The disease induced by over-exertion of intellect.

CASE XLVIII.—Cutaneous symptoms of non-assimilation of fat.

§ 4. *Digestion of water.*

CASE XLIX.—Osmosis of water through membranes defective from the watery condition of the blood in induced anæmia.

CASE L.—Osmosis defective from retarded blood-stream in diseased heart.

CASE LI.—Osmosis defective from retarded blood-stream through emphysematous lungs.

CASE LII.—Illustrations of the variations in the specific gravity and alkalinity of the urine through the normal absorption of fluid.

§ 5. *Treatment of indigestion based on the article of food not digested.*

CASE LIII.—It must not consist in entire omission of the indigested article from the dietary. Scorbutus from leaving off vegetables by medical advice.

CASE LIV.—Purpura from insufficient vegetable diet in a hard-working man.

CASE LV.—Pain at the epigastrium, costiveness and debility, from omitting meat. Amenorrhœa a stomach disease.

HINTS ON DIETETICS.

§ 6. *Treatment based on general pathological condition of indigestion, as used in the foregoing cases. Principally Quinine and Strychnine.*

CHAPTER III.—*Habits of Social Life leading to Indigestion.*
(The treatment is based on their removal.)

§ 1. *Eating too little.*

CASE LVI.—Flatulence in a man over-temperate on false medical theory.

CASE LVII.—Emaciation, flatulence, hysteria, irregular pulse, from omitting meat in consequence of gastralgia.

CASE LVIII.—Debility, flatulence, palpitation of the heart, intermittent pulse, from omitting meat on false medical theory.

CASE LIX.—Mental depression and pain at epigastrium from religious ascetism.

§ 2. Eating too much.

CASE LX.—Eating too much not necessarily a vice. Symptoms in a robust woman of active intellect.

CASE LXI.—Symptoms in a woman of weaker intellect. Obesity. Hysteria.

CASE LXII.—Melancholy in a man from over-eating and not caring.

CASE LXIII.—Caution against overloading the stomach afforded by the sudden death of an old man with diseased heart.

CASE LXIV.—Excess producing sleeplessness and emaciation.

CASE LXV.—Excess producing corpulence in spite of indigestion.

§ 3. Sedentary habits.

CASE LXVI.—Sedentary life not productive of indigestion if proper dietetic habits be adopted.

CASE LXVII.—Violent exercise after eating causes indigestion.

CASE LXVIII.—Even moderate exercise does so in elderly persons.

§ 4. Tight lacing.

CASE LXIX.—Indigestion in a growing girl from not getting a new pair of stays to fit her growth.

CASE LXX.—Chronic vomiting in an adult from bandaging herself too tight to preserve the form after child-bearing.

§ 5. Compression of epigastrium by shoemakers.

CASE LXXI.—Incipient stage. Pain in epigastrium on pressure and soon after food.

CASE LXXII.—A more advanced stage. Pain in epigastrium increased by pressure and immediately after food, eructation, emaciation, debility, broken sleep.

CASE LXXIII.—Deformity of thoracic parietes.

CASE LXXIV.—Final blow to the stomach. Ulceration and hæmatemesis.

§ 6. Sexual excesses.

They do not appear to produce indigestion, though accused of doing so.

CASE LXXV.—But indigestion produces a perversion of the sexual instinct.

§ 7. Solitude.

CASE LXXVI.—Flatulence and confusion of intellect when dining alone, however temperately.

CASE LXXVII.—Vomiting after solitary meals.

§ 8. *Intellectual exertion.*

CASE LXXVIII.—Nocturnal flatulence, nightmare, and seminal emissions from unwonted exertion of a moderate intellect.

CASE LXXIX.—This does not happen in tough-brained men: usually some cause they do not like to own is the origin of their indigestion.

§ 9. *Want of employment.*

CASE LXXX.—The concentration of the mind upon itself in blind people exaggerates internal sensations.

CASE LXXXI.—The same happens in uneducated persons deprived of accustomed employment.

CASE LXXXII.—It may be prevented by simple occupation.

CASE LXXXIII.—Peculiar suspiciousness of insane persons a diagnostic mark to detect imaginary pains.

CASE LXXXIV.—Association with invalids deleterious.

§ 10. *Abuse of purgatives.*

CASE LXXXV.—The habit may be commenced from mere imitation.

CASE LXXXVI.—Gradual omission of purgative habits in those willing to confess their folly.

CASE LXXXVII.—Physiological error of medical practitioners respecting mercurials.

§ 11. *Abuse of alcohol.*

CASE LXXXVIII.—Symptoms in obese persons (male).

CASE LXXXIX.—Ditto (female).

CASE XC.—Occasional effect of a sudden change of habits in elderly persons.

CASE XCI.—Extreme effects of chronic dram-drinking on the stomach.

§ 12. *Tobacco.*

CASES XCII and XCIII.—Effects of this drug on the nervous system.

CASES XCIV and XCV.—Occasional effects on the digestive organs.

§ 13. *Tea.*

CASE XCVI.—Induction of hysterical dyspepsia.

§ 14. *Opium.*

CASE XCVII.—Occasionally induces vomiting.

CASE XCVIII.—Chronically affects the digestion of meat and fat.

CASE XCIX.—Ease with which Opium may be left off by a resolute person.

§ 4. *Spasms.*

CASES CXXVII, CXXVIII, CXXIX, CXXX.—Specimens of spasmodic stomach-ache from insoluble food taken at irregular hours.

CASE CXXXI.—Occurring the morning after an indigestible dinner.

CASE CXXXII.—Accompanied by muscular cramps.

CASE CXXXIII.—Arising from flatulence in a dram-drinker.

CASE CXXXIV.—Abdominal neuralgia from malarious poison.

§ 5. *Gripes evacuating the stomach.*

CASE CXXXV.—In pulmonary consumption a very bad symptom.

CASES CXXXVI and CXXXVII.—With chronic lesion and healthy lungs not so bad.

CASE CXXXVIII.—The relation of the bowels in these cases not always immediate.

§ 6. *Weight at the stomach* (sometimes called " oppression," " distension," and "tightness at the chest").

CASE CXXXIX.—Case referring its origin to excess of mucus on the gastric parietes. (Adherent pericardium.)

CASE CXL.—(Valvular lesion of heart.)

CASE CXLI.—(Functional disturbance of heart.)

CASE CXLII.—The dependence of this collection of mucus on climatic influences.

CASE CXLIII.—Its occasional connection with flatulence.

CASE CXLIV.—With costiveness.

CASE CXLV.—Leads to hypochondriasis and oxaluria.

CASE CXLVI.—The hypochondriasis is apt to take a form suggested by the situation of the discomfort.

CASE CXLVII.—Distinction drawn between weight and heartburn.

CASE CXLVIII.—They may occur together.

§ 7. *Wearing pain.*

CASE CXLIX.—From structural lesion caused by exposure to high temperature.

CASE CL.—The same caused by poor living. Use of Opium.

CASE CLI.—The same from external adhesions probably.

CASE CLII.—Effects of hot food.

CASE CLIII.—Effects of damp climate. Pain increased by motion.

CASE CLIV.—Pain increased by motion and by tonic drugs.

Chapter V.— *Vomiting.*

§ 1. *General remarks on the physiology of the process.*

§ 2. *Vomiting of pus.*

§ 3. *Vomiting of mucus.*

§ 4. *Vomiting of blood.*

Various aspect of the blood thrown out:

In *streaks*, CASE CLXXI;

In a *gush*, CASES CLXXIV, CLXXIX;

In *coagulated masses*, CASE CLXXXIII;

Partially digested into a brown fluid, CASE CLXXXIV;

Green, CASE CLXXXV;

As black stools, CASE CLXXXVI.

CASE CLXXXVII.—Risk of the blood draining away by the bowels instead of by stomach.

CASE CLXXXVIII.—The two symptoms usually joined.

CASE CLXXXIX.—Conjunction of waterbrash and hæmatemesis.

CASE CXC.—Illustration of the slight violence needful to break the gastric bloodvessels.

CASE CXCI.—Hæmatemesis from congestion of the neighboring viscera.

§ 5. *Acid fermentation of vomit.*

CASES CXCII to CXCV.—Cases illustrative of the symptoms accompanying it.

§ 6. *Fecal vomiting.*

The discussion of this is considered not fairly to belong to the subject of indigestion.

§ 7. *Vomiting of unchanged food, and hysterical vomiting.*

CASE CXCVI.—Chewing the cud partially voluntary.

CASE CXCVII.—Vomiting induced by the catamenial period.

CASE CXCVIII.—Vomiting arrested by the occurrence of the catamenia.

CASE CXCIX.—Vomiting without hysteria or any uterine disturbance cut short by Valerian.

CASE CC.—Arrest of acute functional vomiting by withdrawal of all food.

CASE CCI.—The same in a more chronic instance.

CASE CCII.—The same, blood being also thrown up.

CASE CCIII.—Effects of the shower-baths. Hereditary nature of this disease.

CASE CCIV.—Effects of strong mental impression in causing and curing the disease.

CASE CCV.—Association of functional vomiting with functional paralysis of other parts illustrates the true physiology of the act.

CASE CCVI.—Throwing up of food from mechanical impediments to its descent.

CHAPTERS VII and VIII, on *Diarrhœa, Constipation* and *Costiveness, are not illustrated by Cases.*

CHAPTER IX.—*Nerve Disorders connected with Indigestion.*

CASE CCXXXI.—Sick-headache arising from mental causes cured by leaving off purgatives and taking Quinine and Tarnxacum, in a gouty man.

CASE CCXXXII.—Sick-headache in a weakly woman.

CASE CCXXXIII.—The same with erythema nodosum.

CASE CCXXXIV.—Headache causing alarm of apoplexy.

CASE CCXXXV.—Long-ingrained headache cured by Strychnine and Quinine.

CASE CCXXXVI.—Hemicrania in such cases an evidence of improvement.

CASE CCXXXVII.—Occupation of the whole head an evidence of the contrary.

CASE CCXXXVIII.—Purgatives aggravate the disease.

CASES CCXXXIX and CCXL.—Gastric vertigo also aggravated by purgatives.

CASES CCXLI and CCXLII.—Difference of this vertigo and that produced by alcohol.

CASE CCXLIII.—Addled brain and addled stomach.

CASE CCXLIV.—Vertigo and mental helplessness.

CASE CCXLV.—Strong epileptic fit from undigested food in a healthy child.

CASE CCXLVI.—Mild chronic epilepsy from chronic morbid condition of alimentary canal.

CASES CCXLVII and CCXLVIII.—The same. Cure by dietetic discipline.

CASE CCXLIX.—Incurable epileptic fits preceded by flatulence and distension.

CASE CCL.—Choreic spasms.

CASE CCLI.—Stomach cough in phlegmatic temperament.

CASE CCLII.—Stomach cough in a gouty diathesis, alternating with the pains.

CASE CCLIII.—The same coinciding with the pains.

CASE CCLIV.—Anæsthesia.

CASE CCLV.—Paralysis.

CASE CCLVI.—Atrophy of muscles.

CASE CCLVII.—Flushing of face.

CASES CCLVIII and CCLIX.—Nettlerash.

CASE CCLX.—Erythema circumscriptum from indigestion in a gouty person.

ALPHABETICAL INDEX

TO

THE COMMENTARY.

HENRY C. LEA'S
(LATE LEA & BLANCHARD'S)
CLASSIFIED CATALOGUE
OF
MEDICAL AND SURGICAL PUBLICATIONS.

In asking the attention of the profession to the works contained in the following pages, the publisher would state that no pains are spared to secure a continuance of the confidence earned for the publications of the house by their careful selection and accuracy and finish of execution.

The printed prices are those at which books can generally be supplied by booksellers throughout the United States, who can readily procure for their customers any works not kept in stock. Where access to bookstores is not convenient, books will be sent by mail post-paid on receipt of the price, but no risks are assumed either on the money or the books, and no publications but my own are supplied. Gentlemen will therefore in most cases find it more convenient to deal with the nearest bookseller.

An ILLUSTRATED CATALOGUE, of 64 octavo pages, handsomely printed, will be forwarded by mail, postpaid, on receipt of ten cents.

HENRY C. LEA.

Nos. 706 and 708 SANSOM ST., PHILADELPHIA, November, 1868.

ADDITIONAL INDUCEMENT FOR SUBSCRIBERS TO

THE AMERICAN JOURNAL OF THE MEDICAL SCIENCES.
THREE MEDICAL JOURNALS, containing over 2000 LARGE PAGES,
Free of Postage, for SIX DOLLARS Per Annum.

TERMS FOR 1869—IN ADVANCE:

THE AMERICAN JOURNAL OF THE MEDICAL SCIENCES, and } Five Dollars per annum,
THE MEDICAL NEWS AND LIBRARY, both free of postage, } in advance.

OR,

THE AMERICAN JOURNAL OF THE MEDICAL SCIENCES, published quar- } Six Dollars
quarterly (1150 pages per annum), with }
THE MEDICAL NEWS AND LIBRARY, monthly (384 pp. per annum), and } per annum
RANKING'S ABSTRACT OF THE MEDICAL SCIENCES, published half- } in advance.
yearly (600 pages per annum), all free of postage. }

SEPARATE SUBSCRIPTIONS TO

THE AMERICAN JOURNAL OF THE MEDICAL SCIENCES, subject to postage when not paid for in advance, Five Dollars.

THE MEDICAL NEWS AND LIBRARY, free of postage, in advance, One Dollar.

RANKING'S HALF-YEARLY ABSTRACT, Two Dollars and a Half per annum in advance. Single numbers One Dollar and a Half.

In thus offering at a price unprecedentedly low, this vast amount of valuable practical matter, the publisher felt that he could only be saved from loss by an extent of circulation hitherto unknown in the annals of medical journalism. It is therefore with much gratification that he is enabled to state that the marked approbation of the profession, as evinced in the steady increase of the subscription list, promises to render the enterprise a permanent one. He is happy to acknowledge the

1

valuable aid rendered by subscribers who have kindly made known among their friends the advantages thus offered, and he confidently anticipates a continuance of the same friendly interest, which, by enlarging the circulation of these periodicals, will enable him to maintain them at the unexampled rate at which they are now supplied. Arrangements have been made in London by which "RANKING'S ABSTRACT" will be issued here almost simultaneously with its appearance in England: and, with the cöoperation of the profession, the publisher trusts to succeed in his endeavor to lay on the table of every reading practitioner in the United States a monthly, a quarterly, and a half-yearly periodical at the comparatively trifling cost of SIX DOLLARS *per annum.*

These periodicals are universally known for their high professional standing in their several spheres.

I.

THE AMERICAN JOURNAL OF THE MEDICAL SCIENCES,

EDITED BY ISAAC HAYS, M.D.,

is published Quarterly, on the first of January, April, July, and October. Each number contains nearly three hundred large octavo pages, appropriately illustrated, wherever necessary. It has now been issued regularly for over FORTY years, during nearly the whole of which time it has been under the control of the present editor. Throughout this long period, it has maintained its position in the highest rank of medical periodicals both at home and abroad, and has received the cordial support of the entire profession in this country. Among its Collaborators will be found a large number of the most distinguished names of the profession in every section of the United States, rendering the department devoted to

ORIGINAL COMMUNICATIONS

full of varied and important matter, of great interest to all practitioners. Thus, during 1868, contributions have appeared in its pages from more than one hundred gentlemen of the highest standing in the profession throughout the United States.*

Following this is the "REVIEW DEPARTMENT," containing extended and impartial reviews of all important new works, together with numerous elaborate "ANALYTICAL AND BIBLIOGRAPHICAL NOTICES" of nearly all the publications of the day.

This is followed by the "QUARTERLY SUMMARY OF IMPROVEMENTS AND DISCOVERIES IN THE MEDICAL SCIENCES," classified and arranged under different heads, presenting a very complete digest of all that is new and interesting to the physician, abroad as well as at home.

Thus, during the year 1868, the "JOURNAL" furnished to its subscribers One Hundred and Twenty-two Original Communications, Eighty-six Reviews and Bibliographical Notices, and Two Hundred and Forty-nine articles in the Quarterly Summaries, making a total of over FOUR HUNDRED AND FIFTY articles emanating from the best professional minds in America and Europe.

To old subscribers, many of whom have been on the list for twenty or thirty years, the publisher feels that no promises for the future are necessary; but gentlemen who may now propose for the first time to subscribe may rest assured that no exertion will be spared to maintain the "JOURNAL" in the high position which it has so long occupied as a national exponent of scientific medicine, and as a medium of intercommunication between the profession of Europe and America—in the words of the "London Medical Times" (Sept. 5th, 1868) "almost the only one that circulates everywhere, all over the Union and in Europe"—to render it, in fact, necessary to every practitioner who desires to keep on a level with the progress of his science.

The subscription price of the "AMERICAN JOURNAL OF THE MEDICAL SCIENCES" has never been raised, during its long career. It is still FIVE DOLLARS per annum; and when paid in advance, the subscriber receives in addition the "MEDICAL NEWS AND LIBRARY," making in all about 1500 large octavo pages per annum, free of postage.

II.

THE MEDICAL NEWS AND LIBRARY

is a monthly periodical of Thirty-two large octavo pages, making 384 pages per annum. Its "NEWS DEPARTMENT" presents the current information of the day, with Clinical Lectures and Hospital Gleanings; while the "LIBRARY DEPARTMENT" is devoted to publishing standard works on the various branches of medical science, paged

---

* Communications are invited from gentlemen in all parts of the country   All elaborate articles inserted by the Editor are paid for by the Publisher.

separately, so that they can be removed and bound on completion. In this manner subscribers have received, without expense, such works as "WATSON'S PRACTICE," "TODD AND BOWMAN'S PHYSIOLOGY," "WEST ON CHILDREN," "MALGAIGNE'S SURGERY," &c. &c. The work now appearing in its pages, Dr. HUDSON'S valuable "LECTURES ON THE STUDY OF FEVER," will be completed in the number for December, 1868. With January, 1869, another work of similar practical value will be begun, rendering this a very eligible period for the commencement of new subscriptions.

As stated above, the subscription price of the "MEDICAL NEWS AND LIBRARY" is ONE DOLLAR per annum in advance; and it is furnished without charge to all advance paying subscribers to the "AMERICAN JOURNAL OF THE MEDICAL SCIENCES."

III.
RANKING'S ABSTRACT OF THE MEDICAL SCIENCES

is issued in half-yearly volumes, which will be delivered to subscribers about the first of February, and first of August. Each volume contains about 300 closely printed octavo pages, making about six hundred pages per annum.

"RANKING'S ABSTRACT" has now been published in England regularly for more than twenty years, and has acquired the highest reputation for the ability and industry with which the essence of medical literature is condensed into its pages. It purports to be "*A Digest of British and Continental Medicine, and of the progress of Medicine and the Collateral Sciences*," and it presents an abstract of all that is important or interesting in European Medical Literature. Each article is carefully condensed, so as to present its substance in the smallest possible compass, thus affording space for the very large amount of information laid before its readers. The volumes of 1868, for instance, have thus contained

> FORTY-TWO ARTICLES ON GENERAL QUESTIONS IN MEDICINE.
> ONE HUNDRED AND TWENTY-NINE ARTICLES ON SPECIAL QUESTIONS IN MEDICINE.
> SEVENTEEN ARTICLES ON FORENSIC MEDICINE.
> EIGHTY-EIGHT ARTICLES ON THERAPEUTICS.
> FORTY-ONE ARTICLES ON GENERAL QUESTIONS IN SURGERY.
> ONE HUNDRED AND FIFTY SIX ARTICLES ON SPECIAL QUESTIONS IN SURGERY.
> ONE HUNDRED AND TWELVE ARTICLES ON MIDWIFERY AND DISEASES OF WOMEN AND CHILDREN.
> SIX ARTICLES IN APPENDIX.

Making in all nearly six hundred articles in a single year. Each volume, moreover, is systematically arranged, with an elaborate Table of Contents and a very full Index, thus facilitating the researches of the reader in pursuit of particular subjects, and enabling him to refer without loss of time to the vast amount of information contained in its pages.

The subscription price of the "ABSTRACT," mailed free of postage, is Two DOLLARS AND A HALF per annum, payable in advance. Single volumes, $1 50 each.

As stated above, however, it will be supplied in conjunction with the "AMERICAN JOURNAL OF THE MEDICAL SCIENCES" and the "MEDICAL NEWS AND LIBRARY," the whole *free of postage*, for SIX DOLLARS PER ANNUM IN ADVANCE.

For this small sum the subscriber will therefore receive three periodicals costing separately Eight Dollars and a Half, each of them enjoying the highest reputation in its class, containing in all over TWO THOUSAND PAGES of the choicest reading, and presenting a complete view of medical progress throughout the world.

In this effort to bring so large an amount of practical information within the reach of every member of the profession, the publisher confidently anticipates the friendly aid of all who are interested in the dissemination of sound medical literature. He trusts, especially, that the subscribers to the "AMERICAN MEDICAL JOURNAL" will call the attention of their acquaintances to the advantages thus offered, and that he will be sustained in the endeavor to permanently establish medical periodical literature on a footing of cheapness never heretofore attempted.

∗ Gentlemen desiring to avail themselves of the advantages thus offered will do well to forward their subscriptions at an early day, in order to insure the receipt of complete sets for the year 1869.

☞ The safest mode of remittance is by postal money order, drawn to the order of the undersigned. Where money order post-offices are not accessible, remittances for the "JOURNAL" may be made at the risk of the publisher, by forwarding in REGISTERED letters. Address,

HENRY C. LEA,
Nos. 706 and 708 SANSOM ST., PHILADELPHIA, PA.

DUNGLISON (ROBLEY), M. D.,

Professor of Institutes of Medicine in Jefferson Medical College, Philadelphia.

MEDICAL LEXICON; A DICTIONARY OF MEDICAL SCIENCE: Containing a concise explanation of the various Subjects and Terms of Anatomy, Physiology, Pathology, Hygiene, Therapeutics, Pharmacology, Pharmacy, Surgery; Obstetrics, Medical Jurisprudence, and Dentistry. Notices of Climate and of Mineral Waters; Formulæ for Officinal, Empirical, and Dietetic Preparations; with the Accentuation and Etymology of the Terms, and the French and other Synonymes; so as to constitute a French as well as English Medical Lexicon. Thoroughly Revised, and very greatly Modified and Augmented In one very large and handsome royal octavo volume of 1048 double-columned pages, in small type; strongly done up in extra cloth, $6 00; leather, raised bands, $6 75.

The object of the author from the outset has not been to make the work a mere lexicon or dictionary of terms, but to afford, under each, a condensed view of its various medical relations, and thus to render the work an epitome of the existing condition of medical science. Starting with this view, the immense demand which has existed for the work has enabled him, in repeated revisions, to augment its completeness and usefulness, until at length it has attained the position of a recognized and standard authority wherever the language is spoken. The mechanical execution of this edition will be found greatly superior to that of previous impressions. By enlarging the size of the volume to a royal octavo, and by the employment of a small but clear type, on extra fine paper, the additions have been incorporated without materially increasing the bulk of the volume, and the matter of two or three ordinary octavos has been compressed into the space of one not unhandy for consultation and reference.

It would be a work of supererogation to bestow a word of praise upon this Lexicon. We can only wonder at the labor expended, for whenever we refer to its pages for information we are seldom disappointed in finding all we desire, whether it be in accentuation, etymology, or definition of terms.—*New York Medical Journal*, November, 1865.

It would be mere waste of words in us to express our admiration of a work which is so universally and deservedly appreciated. The most admirable work of its kind in the English language. As a book of reference it is invaluable to the medical practitioner, and in every instance that we have turned over its pages for information we have been charmed by the clearness of language and the accuracy of detail with which each abounds. We can most cordially and confidently commend it to our readers.—*Glasgow Medical Journal*, January, 1866.

A work to which there is no equal in the English language.—*Edinburgh Medical Journal*.

It is something more than a dictionary, and something less than an encyclopædia. This edition of the well-known work is a great improvement on its predecessors. The book is one of the very few of which it may be said with truth that every medical man should possess it.—*London Medical Times*, Aug. 26, 1865.

Few works of the class exhibit a grander monument of patient research and of scientific lore. The extent of the sale of this lexicon is sufficient to testify to its usefulness, and to the great service conferred by Dr. Robley Dunglison on the profession, and indeed on others, by its issue.—*London Lancet*, May 13, 1865.

The old edition, which is now superseded by the new, has been universally looked upon by the medical profession as a work of immense research and great value. The new has increased usefulness; for medicine, in all its branches, has been making such progress that many new terms and subjects have recently been introduced; all of which may be found fully defined in the present edition. We know of no other dictionary in the English language that can bear a comparison with it in point of completeness of subjects and accuracy of statement.—*N. Y. Druggists' Circular*, 1865.

For many years Dunglison's Dictionary has been the standard book of reference with most practitioners in this country, and we can certainly commend this work to the renewed confidence and regard of our readers.—*Cincinnati Lancet*, April, 1865.

It is undoubtedly the most complete and useful medical dictionary hitherto published in this country.—*Chicago Med. Examiner*, February, 1865.

What we take to be decidedly the best medical dictionary in the English language. The present edition is brought fully up to the advanced state of science. For many a long year "Dunglison" has been at our elbow, a constant companion and friend, and we greet him in his replenished and improved form with especial satisfaction.—*Pacific Med. and Surg. Journal*, June 27, 1865.

This is, perhaps, the book of all others which the physician or surgeon should have on his shelves. It is more needed at the present day than a few years back.—*Canada Med. Journal*, July, 1865.

It deservedly stands at the head, and cannot be surpassed in excellence.—*Buffalo Med. and Surg. Journal*, April, 1865.

We can sincerely commend Dr. Dunglison's work as most thorough, scientific, and accurate. We have tested it by searching its pages for new terms, which have abounded so much of late in medical nomenclature, and our search has been successful in every instance. We have been particularly struck with the fulness of the synonymy and the accuracy of the derivation of words. It is as necessary a work to every enlightened physician as Worcester's English Dictionary is to every one who would keep up his knowledge of the English tongue to the standard of the present day. It is, to our mind, the most complete work of the kind with which we are acquainted.—*Boston Med. and Surg. Journal*, June 22, 1865.

We are free to confess that we know of no medical dictionary more complete; no one better, if so well adapted for the use of the student; no one that may be consulted with more satisfaction by the medical practitioner.—*Am. Jour. Med. Sciences*, April, 1865.

The value of the present edition has been greatly enhanced by the introduction of new subjects and terms, and a more complete etymology and accentuation, which renders the work not only satisfactory and desirable, but indispensable to the physician.—*Chicago Med. Journal*, April, 1865.

No intelligent member of the profession can or will be without it.—*St. Louis Med. and Surg. Journal*, April, 1865.

It has the rare merit that it certainly has no rival in the English language for accuracy and extent of references.—*London Medical Gazette*.

HOBLYN (RICHARD D.), M. D.

A DICTIONARY OF THE TERMS USED IN MEDICINE AND THE COLLATERAL SCIENCES. A new American edition, revised, with numerous additions, by ISAAC HAYS, M. D., Editor of the "American Journal of the Medical Sciences." In one large royal 12mo. volume of over 500 double-columned pages; extra cloth, $1 50; leather, $2 00.

It is the best book of definitions we have, and ought always to be upon the student's table.—*Southern Med. and Surg. Journal*.

*N*EILL (*JOHN*), *M. D.*, and *S*MITH (*FRANCIS G.*), *M. D.*,
Prof. of the Institutes of Medicine in the Univ. of Penna.

AN ANALYTICAL COMPENDIUM OF THE VARIOUS
BRANCHES OF MEDICAL SCIENCE; for the Use and Examination of Students. A new edition, revised and improved. In one very large and handsomely printed royal 12mo. volume, of about one thousand pages, with 374 wood cuts, extra cloth, $4; strongly bound in leather, with raised bands, $4 75.

The Compend of Drs. Neill and Smith is incomparably the most valuable work of its class ever published in this country. Attempts have been made in various quarters to squeeze Anatomy, Physiology, Surgery, the Practice of Medicine, Obstetrics, Materia Medica, and Chemistry into a single manual; but the operation has signally failed in the hands of all up to the advent of "Neill and Smith's" volume, which is quite a miracle of success. The outlines of the whole are admirably drawn and illustrated, and the authors are eminently entitled to the grateful consideration of the student of every class.—*N. O. Med. and Surg. Journal.*

This popular favorite with the student is so well known that it requires no more at the hands of a medical editor than the annunciation of a new and improved edition. There is no sort of comparison between this work and any other on a similar plan, and for a similar object.—*Nash. Journ. of Medicine.*

There are but few students or practitioners of medicine unacquainted with the former editions of this unassuming though highly instructive work. The whole science of medicine appears to have been sifted, as the gold-bearing sands of El Dorado, and the precious facts treasured up in this little volume. A complete portable library so condensed that the student may make it his constant pocket companion.—*Western Lancet.*

To compress the whole science of medicine in less

than 1,000 pages is an impossibility, but we think that the book before us approaches as near to it as is possible. Altogether, it is the best of its class, and has met with a deserved success. As an elementary text-book for students, it has been useful, and will continue to be employed in the examination of private classes, whilst it will often be referred to by the country practitioner.—*Va. Med. Journal.*

As a handbook for students it is invaluable, containing in the most condensed form the established facts and principles of medicine and its collateral sciences.—*N. H. Journal of Medicine.*

In the rapid course of lectures, where work for the students is heavy, and review necessary for an examination, a compend is not only valuable, but it is almost a sine qua non. The one before us is, in most of the divisions, the most unexceptionable of all books of the kind that we know of. The newest and soundest doctrines and the latest improvements and discoveries are explicitly, though concisely, laid before the student. Of course it is useless for us to recommend it to all last course students, but there is a class to whom we very sincerely recommend this cheap book as worth its weight in silver—that class is the graduates in medicine of more than ten years' standing, who have not studied medicine since. They will perhaps find out from it that the science is not exactly now what it was when they left it off.—*The Stethoscope.*

*H*ARTSHORNE (*HENRY*), *M. D.*,
Professor of Hygiene in the University of Pennsylvania.

A CONSPECTUS OF THE MEDICAL SCIENCES; containing
Handbooks on Anatomy, Physiology, Chemistry, Materia Medica, Practical Medicine, Surgery, and Obstetrics. In one large royal 12mo. volume of nearly 1000 closely printed pages, with about 390 illustrations on wood. (*Nearly Ready.*)

The ability of the author, and his practical skill in condensation, give assurance that this work will prove valuable not only to the student preparing for examination, but also to the practitioner desirous of obtaining within a moderate compass, a view of the existing condition of the various departments of science connected with medicine.

*L*UDLOW (*J. L.*), *M. D.*,

A MANUAL OF EXAMINATIONS upon Anatomy, Physiology,
Surgery, Practice of Medicine, Obstetrics, Materia Medica, Chemistry, Pharmacy, and Therapeutics. To which is added a Medical Formulary. Third edition, thoroughly revised and greatly extended and enlarged. With 370 illustrations. In one handsome royal 12mo. volume of 816 large pages, extra cloth, $3 25; leather, $8 75.

The arrangement of this volume in the form of question and answer renders it especially suitable for the office examination of students, and for those preparing for graduation.

We know of no better companion for the student during the hours spent in the lecture-room, or to refresh, at a glance, his memory of the various topics crammed into his head by the various professors to whom he is compelled to listen.—*Western Lancet.*

As it embraces the whole range of medical studies it is necessarily voluminous, containing 816 large duodecimo pages. After a somewhat careful examination of its contents, we have formed a much more favorable opinion of it than we are wont to regard such works. Although well adapted to meet the wants

of the student in preparing for his final examination, it might be profitably consulted by the practitioner also, who is most apt to become rusty in the very kind of details here given, and who, amid the hurry of his daily routine, is but too prone to neglect the study of more elaborate works. The possession of a volume of this kind might serve as an inducement for him to seize the moment of excited curiosity to inform himself on any subject, and which is otherwise too often allowed to pass unimproved.—*St. Louis Med. and Surg. Journal.*

*T*ANNER (*THOMAS HAWKES*), *M. D.*,

A MANUAL OF CLINICAL MEDICINE AND PHYSICAL DIAG-
NOSIS. Third American, from the second enlarged and revised English edition. To which is added The Code of Ethics of the American Medical Association. In one handsome volume 12mo. (*Preparing for early publication.*)

GRAY (HENRY), F.R.S.,

Lecturer on Anatomy at St. George's Hospital, London.

ANATOMY, DESCRIPTIVE AND SURGICAL.   The Drawings by
H. V. CARTER, M. D., late Demonstrator on Anatomy at St. George's Hospital; the Dissections jointly by the AUTHOR and DR. CARTER. ; Second American, from the second revised and improved London edition. In one magnificent imperial octavo volume, of over 800 pages, with 388 large and elaborate engravings on wood. Price in extra cloth, $6 00; leather, raised bands, $7 00.

The author has endeavored in this work to cover a more extended range of subjects than is customary in the ordinary text-books, by giving not only the details necessary for the student, but also the application of those details in the practice of medicine and surgery, thus rendering it both a guide for the learner, and an admirable work of reference for the active practitioner. The engravings form a special feature in the work, many of them being the size of nature, nearly all original, and having the names of the various parts printed on the body of the cut, in place of figures of reference, with descriptions at the foot. They thus form a complete and splendid series, which will greatly assist the student in obtaining a clear idea of Anatomy, and will also serve to refresh the memory of those who may find in the exigencies of practice the necessity of recalling the details of the dissecting room; while combining, as it does, a complete Atlas of Anatomy, with a thorough treatise on systematic, descriptive, and applied Anatomy, the work will be found of essential use to all physicians who receive students in their offices, relieving both preceptor and pupil of much labor in laying the groundwork of a thorough medical education.

Notwithstanding its exceedingly low price, the work will be found, in every detail of mechanical execution, one of the handsomest that has yet been offered to the American profession; while the careful scrutiny of a competent anatomist has relieved it of whatever typographical errors existed in the English edition.

Thus it is that book after book makes the labor of the student easier than before, and since we have seen Blanchard & Lea's new edition of Gray's Anatomy, certainly the finest work of the kind now extant, we would fain hope that the bugbear of medical students will lose half its horrors, and this necessary foundation of physiological science will be much facilitated and advanced.—*N. O. Med. News.*

The various points illustrated are marked directly on the structure; that is, whether it be muscle, process, artery, nerve, valve, etc. etc.—we say each point is distinctly marked by lettered engravings, so that the student perceives at once each point described as readily as if pointed out on the subject by the demonstrator. Most of the illustrations are thus rendered exceedingly satisfactory, and to the physician they serve to refresh the memory with great readiness.

and with scarce a reference to the printed text. The surgical application of the various regions is also presented with force and clearness, impressing upon the student at each step of his research all the important relations of the structure demonstrated.—*Cincinnati Lancet.*

This is, we believe, the handsomest book on Anatomy as yet published in our language, and bids fair to become in a short time THE standard text-book of our colleges and studies. Students and practitioners will alike appreciate this book. We predict for it a bright career, and are fully prepared to endorse the statement of the *London Lancet*, that "We are not acquainted with any work in any language which can take equal rank with the one before us." Paper, printing, binding, all are excellent, and we feel that a grateful profession will not allow the publishers to go unrewarded.—*Nashville Med. and Surg. Journal.*

SMITH (HENRY H.), M.D.,   and   HORNER (WILLIAM E.), M.D.,

Prof. of Surgery in the Univ. of Penna., &c.   *Late Prof. of Anatomy in the Univ. of Penna., &c.*

AN ANATOMICAL ATLAS, illustrative of the Structure of the Human Body. In one volume, large imperial octavo, extra cloth, with about six hundred and fifty beautiful figures. $4 50.

The plan of this Atlas, which renders it so peculiarly convenient for the student, and its superb artistical execution, have been already pointed out. We must congratulate the student upon the completion of this Atlas, as it is the most convenient work of

the kind that has yet appeared; and we must add, the very beautiful manner in which it is "got up" is so creditable to the country as to be flattering to our national pride.—*American Medical Journal.*

HORNER (WILLIAM E.), M.D.,

SPECIAL ANATOMY AND HISTOLOGY. Eighth edition, extensively revised and modified. In two large octavo volumes of over 1000 pages, with more than 300 wood-cuts; extra cloth, $6 00.

SHARPEY (WILLIAM), M.D.,   and   QUAIN (JONES & RICHARD).

HUMAN ANATOMY. Revised, with Notes and Additions, by JOSEPH LEIDY, M. D., Professor of Anatomy in the University of Pennsylvania. Complete in two large octavo volumes, of about 1300 pages, with 511 illustrations; extra cloth, $6 00.

The very low price of this standard work, and its completeness in all departments of the subject, should command for it a place in the library of all anatomical students.

ALLEN (J. M.), M.D.

THE PRACTICAL ANATOMIST; OR, THE STUDENT'S GUIDE IN THE DISSECTING ROOM. With 266 illustrations. In one very handsome royal 12mo. volume, of over 600 pages; extra cloth, $2 00.

One of the most useful works upon the subject ever written.—*Medical Examiner*

WILSON (ERASMUS), F. R. S.

A SYSTEM OF HUMAN ANATOMY, General and Special. A new

and revised American, from the last and enlarged English edition. Edited by W. H. Go-
BRECHT, M. D., Professor of General and Surgical Anatomy in the Medical College of Ohio.
Illustrated with three hundred and ninety-seven engravings on wood. In one large and
handsome octavo volume, of over 600 large pages; extra cloth, $4 00; leather, $5 00.
    The publisher trusts that the well-earned reputation of this long-established favorite will be
more than maintained by the present edition. Besides a very thorough revision by the author, it
has been most carefully examined by the editor, and the efforts of both have been directed to introducing everything which increased experience in its use has suggested as desirable to render it
a complete text-book for those seeking to obtain or to renew an acquaintance with Human Anatomy. The amount of additions which it has thus received may be estimated from the fact that
the present edition contains over one-fourth more matter than the last, rendering a smaller type
and an enlarged page requisite to keep the volume within a convenient size. The author has not
only thus added largely to the work, but he has also made alterations throughout, wherever there
appeared the opportunity of improving the arrangement or style, so as to present every fact in its
most appropriate manner, and to render the whole as clear and intelligible as possible. The editor
has exercised the utmost caution to obtain entire accuracy in the text, and has largely increased
the number of illustrations, of which there are about one hundred and fifty more in this edition
than in the last, thus bringing distinctly before the eye of the student everything of interest or
importance.

BY THE SAME AUTHOR.

THE DISSECTOR'S MANUAL; OR, PRACTICAL AND SURGICAL ANA-

TOMY. Third American, from the last revised and enlarged English edition. Modified and
rearranged by WILLIAM HUNT, M. D., late Demonstrator of Anatomy in the University of
Pennsylvania. In one large and handsome royal 12mo. volume, of 582 pages, with 154
illustrations; extra cloth, $2 00.

HODGES, (RICHARD M.), M. D.,
    Late Demonstrator of Anatomy in the Medical Department of Harvard University.

PRACTICAL DISSECTIONS. Second Edition, thoroughly revised. In

one neat royal 12mo. volume, half-bound, $2 00. (*Just Issued.*)
    The object of this work is to present to the anatomical student a clear and concise description
of that which he is expected to observe in an ordinary course of dissections. The author has
endeavored to omit unnecessary details, and to present the subject in the form which many years'
experience has shown him to be the most convenient and intelligible to the student. In the
revision of the present edition, he has sedulously labored to render the volume more worthy of
the favor with which it has heretofore been received.

MACLISE (JOSEPH).

SURGICAL ANATOMY. By JOSEPH MACLISE, Surgeon. In one

volume, very large imperial quarto; with 68 large and splendid plates, drawn in the best
style and beautifully colored, containing 190 figures, many of them the size of life; together
with copious explanatory letter-press. Strongly and handsomely bound in extra cloth.
Price $14 00.
    As no complete work of the kind has heretofore been published in the English language, the
present volume will supply a want long felt in this country of an accurate and comprehensive
Atlas of Surgical Anatomy, to which the student and practitioner can at all times refer to ascertain the exact relative positions of the various portions of the human frame towards each other
and to the surface, as well as their abnormal deviations. The importance of such a work to the
student, in the absence of anatomical material, and to practitioners, either for consultation in
emergencies or to refresh their recollections of the dissecting room, is evident. Notwithstanding
the large size, beauty and finish of the very numerous illustrations, it will be observed that the
price is so low as to place it within the reach of all members of the profession.

We know of no work on surgical anatomy which
can compete with it.—*Lancet.*

The work of Maclise on surgical anatomy is of the
highest value. In some respects it is the best publication of its kind we have seen, and is worthy of a
place in the library of any medical man, while the
student could scarcely make a better investment than
this.—*The Western Journal of Medicine and Surgery.*

No such lithographic illustrations of surgical regions have hitherto, we think, been given. While
the operator is shown every vessel and nerve where
an operation is contemplated, the exact anatomist is

refreshed by those clear and distinct dissections,
which every one must appreciate who has a particle
of enthusiasm. The English medical press has quite
exhausted the words of praise, in recommending this
admirable treatise. Those who have any curiosity
to gratify, in reference to the perfectibility of the
lithographic art in delineating the complex mechanism of the human body, are invited to examine our
specimen copy. If anything will induce surgeons
and students to patronize a book of such rare value
and everyday importance to them, it will be a survey
of the artistical skill exhibited in these fac-similes of
nature.—*Boston Med. and Surg. Journal.*

PEASLEE (E. R.), M. D.,
    Professor of Anatomy and Physiology in Dartmouth Med. College, N. H.

HUMAN HISTOLOGY, in its relations to Anatomy, Physiology, and

Pathology; for the use of medical students. With four hundred and thirty-four illustrations. In one handsome octavo volume of over 600 pages, extra cloth. $3 75.

*M*ARSHALL (JOHN), F. R. S.
   Professor of Surgery in University College, London, &c.

OUTLINES OF PHYSIOLOGY, HUMAN AND COMPARATIVE.
With Additions by FRANCIS GURNEY SMITH, M. D., Professor of the Institutes of Medicine in the University of Pennsylvania, &c. With numerous illustrations. In one large and handsome octavo volume, of 1026 pages, extra cloth, $6 50; leather, raised bands, $7 50. (*Now Ready.*)

We may now congratulate him on having completed the latest as well as the best summary of modern physiological science, both human and comparative, with which we are acquainted. To speak of this work in the terms ordinarily used on such occasions would not be agreeable to ourselves, and would fail to do justice to its author. To write such a book requires a varied and wide range of knowledge, considerable power of analysis, correct judgment, skill in arrangement, and conscientious spirit. It must have entailed great labor, but now that the task has been fulfilled, the book will prove not only invaluable to the student of medicine and surgery, but serviceable to all candidates in natural science examinations, to teachers in schools, and to the lover of nature generally. In conclusion, we can only express the conviction that the merits of the work will command for it that success which the ability and vast labor displayed in its production so well deserve.—*London Lancet*, Feb. 22, 1868.

If the possession of knowledge, and peculiar apti-

tude and skill in expounding it, qualify a man to write an educational work, Mr. Marshall's treatise might be reviewed favorably without even opening the covers. There are few, if any, more accomplished anatomists and physiologists than the distinguished professor of surgery at University College; and he has long enjoyed the highest reputation as a teacher of physiology, possessing remarkable powers of clear exposition and graphic illustration. It is only remarkable that Mr. Marshall has allowed so long a time to elapse before producing a text-book after his own heart. The plan of this book differs in many respects from that of existing educational books; the science of human physiology being treated in wider and more constant reference to chemistry, physics, and comparative anatomy and physiology. There can be no question, we think, that this is the most satisfactory, philosophic, and fruitful mode of teaching physiology. We have rarely the pleasure of being able to recommend a text-book so unreservedly as this.—*British Med. Journal*, Jan. 25, 1868.

*C*ARPENTER (WILLIAM B.), M. D., F. R. S.,
   Examiner in Physiology and Comparative Anatomy in the University of London.

PRINCIPLES OF HUMAN PHYSIOLOGY; with their chief applications to Psychology, Pathology, Therapeutics, Hygiene and Forensic Medicine. A new American from the last and revised London edition. With nearly three hundred illustrations. Edited, with additions, by FRANCIS GURNEY SMITH, M. D., Professor of the Institutes of Medicine in the University of Pennsylvania, &c. In one very large and beautiful octavo volume, of about 900 large pages, handsomely printed; extra cloth, $5 50; leather, raised bands, $6 50.

The highest compliment that can be extended to this great work of Dr. Carpenter is to call attention to this, another new edition, which the favorable regard of the profession has called for. Carpenter is the standard authority on physiology, and no physician or medical student will regard his library as complete without a copy of it.—*Cincinnati Med. Observer.*

With Dr. Smith, we confidently believe "that the present will more than sustain the enviable reputation already attained by former editions, of being one of the fullest and most complete treatises on the subject in the English language." We know of none from the pages of which a satisfactory knowledge of the physiology of the human organism can be as well obtained, none better adapted for the use of such as take up the study of physiology in its reference to the institutes and practice of medicine.—*Am. Jour. Med. Sciences.*

We doubt not it is destined to retain a strong hold on public favor, and remain the favorite text-book in our colleges.—*Virginia Medical Journal.*

We have so often spoken in terms of high commendation of Dr. Carpenter's elaborate work on human physiology that, in announcing a new edition, it is unnecessary to add anything to what has heretofore been said, and especially is this the case since every intelligent physician is as well aware of the character and merits of the work as we ourselves are.—*St. Louis Med. and Surg. Journal.*

The above is the title of what is emphatically the great work on physiology; and we are conscious that it would be a useless effort to attempt to add anything to the reputation of this invaluable work, and can only say to all with whom our opinion has any influence, that it is our *authority.—Atlanta Med. Journal.*

*B*Y THE SAME AUTHOR.

PRINCIPLES OF COMPARATIVE PHYSIOLOGY. New American, from the Fourth and Revised London Edition. In one large and handsome octavo volume, with over three hundred beautiful illustrations. Pp. 752. Extra cloth, $5 00.

As a complete and condensed treatise on its extended and important subject, this work becomes a necessity to students of natural science, while the very low price at which it is offered places it within the reach of all.

*B*Y THE SAME AUTHOR.

THE MICROSCOPE AND ITS REVELATIONS. Illustrated by four hundred and thirty-four beautiful engravings on wood. In one large and very handsome octavo volume, of 724 pages, extra cloth, $5 25.

*K*IRKES (WILLIAM SENHOUSE), M. D.,

A MANUAL OF PHYSIOLOGY. A new American from the third and improved London edition. With two hundred illustrations. In one large and handsome royal 12mo. volume. Pp. 586. Extra cloth, $2 25; leather, $2 75.

It is at once convenient in size, comprehensive in design, and concise in statement, and altogether well adapted for the purpose designed.—*St. Louis Med. and Surg. Journal.*

The physiological reader will find it a most excel-

lent guide in the study of physiology in its most advanced and perfect form. The author has shown himself capable of giving details sufficiently ample in a condensed and concentrated shape, on a science in which it is necessary at once to be correct and not lengthened.—*Edinburgh Med. and Surg. Journal.*

*D*ALTON (*J. C.*), *M. D.,*
Professor of Physiology in the College of Physicians and Surgeons, New York, &c.

A TREATISE ON HUMAN PHYSIOLOGY, Designed for the use of Students and Practitioners of Medicine. Fourth edition, revised, with nearly three hundred illustrations on wood. In one very beautiful octavo volume, of about 700 pages, extra cloth, $5 25; leather, $6 25. (*Just Issued.*)

From the Preface to the New Edition.

"The progress made by Physiology and the kindred Sciences during the last few years has required, for the present edition of this work, a thorough and extensive revision. This progress has not consisted in any very striking single discoveries, nor in a decided revolution in any of the departments of Physiology; but it has been marked by great activity of investigation in a multitude of different directions, the combined results of which have not failed to impress a new character on many of the features of physiological knowledge. . . . In the revision and correction of the present edition, the author has endeavored to incorporate all such improvements in physiological knowledge with the mass of the text in such a manner as not essentially to alter the structure and plan of the work, so far as they have been found adapted to the wants and convenience of the reader. . . . Several new illustrations are introduced, some of them as additions, others as improvements or corrections of the old. Although all parts of the book have received more or less complete revision, the greatest number of additions and changes were required in the Second Section, on the Physiology of the Nervous System."

The advent of the first edition of Prof. Dalton's Physiology, about eight years ago, marked a new era in the study of physiology to the American student. Under Dalton's skilful management, physiological science threw off the long, loose, ungainly garments of probability and surmise, in which it had been arrayed by most artists, and came among us smiling and attractive, in the beautifully tinted and closely-fitting dress of a demonstrated science. It was a stroke of genius, as well as a result of erudition and talent, that led Prof. Dalton to present to the world a work on physiology at once brief, pointed. and comprehensive, and which exhibited plainly in letter and drawings the basis upon which the conclusions arrived at rested. It is no disparagement of the many excellent works on physiology, published prior to that of Dalton, to say that none of them, either in plan of arrangement or clearness of execution, could be compared with his for the use of students or general practitioners of medicine. For this purpose his book has no equal in the English language.—*Western Journal of Medicine*, Nov. 1867.

A capital text-book in every way. We are, therefore, glad to see it in its fourth edition. It has already been examined at full length in these columns, so that we need not now further advert to it beyond remarking that both revision and enlargement have been most judicious.—*London Med. Times and Gazette*, Oct. 19, 1867.

No better proof of the value of this admirable work could be produced than the fact that it has already reached a fourth edition in the short space of eight years. Possessing in an eminent degree the

merits of clearness and condensation, and being fully brought up to the present level of Physiology, it is undoubtedly one of the most reliable text-books upon this science that could be placed in the hands of the medical student.—*Am. Journal Med. Sciences*, Oct. 1867.

Prof. Dalton's work has such a well-established reputation that it does not stand in need of any recommendation. Ever since its first appearance it has become the highest authority in the English language; and that it is able to maintain the enviable position which it has taken, the rapid exhaustion of the different successive editions is sufficient evidence. The present edition, which is the fourth, has been thoroughly revised, and enlarged by the incorporation of all the many important advances which have lately been made in this rapidly progressing science.—*N. Y. Med. Record*, Oct. 15, 1867.

As it stands, we esteem it the very best of the physiological text-books for the student, and the most concise reference and guide-book for the practitioner.—*N. Y. Med. Journal*, Oct. 1867.

The present edition of this now standard work fully sustains the high reputation of its accomplished author. It is not merely a reprint, but has been faithfully revised, and enriched by such additions as the progress of physiology has rendered desirable. Taken as a whole, it is unquestionably the most reliable and useful treatise on the subject that has been issued from the American press.—*Chicago Med. Journal*, Sept. 1867.

*D*UNGLISON (*ROBLEY*), *M. D.,*
Professor of Institutes of Medicine in Jefferson Medical College, Philadelphia.

HUMAN PHYSIOLOGY. Eighth edition. Thoroughly revised and extensively modified and enlarged, with five hundred and thirty-two illustrations. In two large and handsomely printed octavo volumes of about 1500 pages, extra cloth. $7 00.

*L*EHMANN (*C. G.*)
PHYSIOLOGICAL CHEMISTRY. Translated from the second edition by GEORGE E. DAY, M. D., F. R. S., &c., edited by R. E. ROGERS, M. D., Professor of Chemistry in the Medical Department of the University of Pennsylvania, with illustrations selected from Funke's Atlas of Physiological Chemistry, and an Appendix of plates. Complete in two large and handsome octavo volumes, containing 1200 pages, with nearly two hundred illustrations, extra cloth. $6 00.

BY THE SAME AUTHOR.

MANUAL OF CHEMICAL PHYSIOLOGY. Translated from the German, with Notes and Additions, by J CHESTON MORRIS, M. D., with an Introductory Essay on Vital Force, by Professor SAMUEL JACKSON, M D., of the University of Pennsylvania. With illustrations on wood. In one very handsome octavo volume of 336 pages, extra cloth. $2 25.

*T*ODD (*ROBERT B.*), *M. D., F. R. S., and* *B*OWMAN (*W.*), *F. R. S.*

THE PHYSIOLOGICAL ANATOMY AND PHYSIOLOGY OF MAN. With about three hundred large and beautiful illustrations on wood. Complete in one large octavo volume of 950 pages, extra cloth. Price $4 75.

*B*RANDE (*WM. T.*), *D. C. L.*, and *T*AYLOR (*ALFRED S.*), *M. D., F. R. S.*

CHEMISTRY. Second American edition, thoroughly revised by Dr.
TAYLOR. In one handsome 8vo. volume of 764 pages, extra cloth, $5 00; leather, $6 00.
(*Just Issued.*)

FROM DR. TAYLOR'S PREFACE.

"The revision of the second edition, in consequence of the death of my lamented colleague,
has devolved entirely upon myself. Every chapter, and indeed every page, has been revised,
and numerous additions made in all parts of the volume. These additions have been restricted
chiefly to subjects having some practical interest, and they have been made as concise as possible,
in order to keep the book within those limits which may retain for it the character of a Student's
Manual "—*London*, June 29, 1867.

A book that has already so established a reputation, as has Brande and Taylor's Chemistry, can
hardly need a notice, save to mention the additions
and improvements of the edition. Doubtless the
work will long remain a favorite text-book in the
schools, as well as a convenient book of reference for
all.—*N. Y. Medical Gazette*, Oct. 12, 1867.

For this reason we hail with delight the republication, in a form which will meet with general approval
and command public attention, of this really valuable standard work on chemistry—more particularly
as it has been adapted with such care to the wants of
the general public. The well known scholarship of
its authors, and their extensive researches for many
years in experimental chemistry, have been long appreciated in the scientific world, but in this work they
have been careful to give the largest possible amount
of information with the most sparing use of technical
terms and phraseology, so as to furnish the reader,
"whether a student of medicine, or a man of the
world, with a plain introduction to the science and
practice of chemistry."—*Journal of Applied Chemistry*, Oct. 1867.

This second American edition of an excellent treatise on chemical science is not a mere republication
from the English press, but is a revision and enlargement of the original, under the supervision of
the surviving author, Dr. Taylor. The favorable
opinion expressed on the publication of the former
edition of this work is fully sustained by the present
revision, in which Dr. T. has increased the size of
the volume, by an addition of sixty-eight pages.—*Am.
Journ. Med. Sciences*, Oct. 1867.

THE HANDBOOK IN CHEMISTRY OF THE STUDENT.—
For clearness of language, accuracy of description,
extent of information, and freedom from pedantry
and mysticism, no other text-book comes into competition with it.—*The Lancet*.

The authors set out with the definite purpose of
writing a book which shall be intelligible to any
educated man. This conceived, and worked out in
the most sturdy, common-sense method, this book
gives in the clearest and most summary method
possible all the facts and doctrines of chemistry.—
Medical Times.

*B*OWMAN (*JOHN E.*), *M. D.*

PRACTICAL HANDBOOK OF MEDICAL CHEMISTRY. Edited
by C. L. BLOXAM, Professor of Practical Chemistry in King's College, London. Fourth
American, from the fourth and revised English Edition. In one neat volume, royal 12mo.,
pp. 351, with numerous illustrations, extra cloth. $2 25.

The fourth edition of this invaluable text-book of
Medical Chemistry was published in England in October of the last year. The Editor has brought down
the Handbook to that date, introducing, as far as was
compatible with the necessary conciseness of such a
work, all the valuable discoveries in the science

which have come to light since the previous edition
was printed. The work is indispensable to every
student of medicine or enlightened practitioner. It
is printed in clear type, and the illustrations are
numerous and intelligible.—*Boston Med. and Surg.
Journal.*

*B*Y THE SAME AUTHOR.

INTRODUCTION TO PRACTICAL CHEMISTRY, INCLUDING
ANALYSIS. Fourth American, from the fifth and revised London edition. With numerous illustrations. In one neat vol., royal 12mo., extra cloth. $2 25. (*Just Issued.*)

One of the most complete manuals that has for a
long time been given to the medical student.—
Athenæum.

We regard it as realizing almost everything to be
desired in an introduction to Practical Chemistry.

It is by far the best adapted for the Chemical student
of any that has yet fallen in our way.—*British and
Foreign Medico-Chirurgical Review.*

The best introductory work on the subject with
which we are acquainted.—*Edinburgh Monthly Jour.*

*G*RAHAM (*THOMAS*), *F. R. S.*

THE ELEMENTS OF INORGANIC CHEMISTRY, including the
Applications of the Science in the Arts. New and much enlarged edition, by HENRY
WATTS and ROBERT BRIDGES, M. D. Complete in one large and handsome octavo volume,
of over 800 very large pages, with two hundred and thirty-two wood-cuts, extra cloth.
$5 50.

Part II., completing the work from p. 431 to end, with Index, Title Matter, &c., may be had
separate, cloth backs and paper sides. Price $3 00.

From Prof. E. N. Horsford, Harvard College.

It has, in its earlier and less perfect editions, been
familiar to me, and the excellence of its plan and
the clearness and completeness of its discussions,
have long been my admiration.

No reader of English works on this science can

afford to be without this edition of Prof. Graham's
Elements.—*Silliman's Journal*, March, 1858.

From Prof. Wolcott Gibbs, N. Y. Free Academy.

The work is an admirable one in all respects, and
its republication here cannot fail to exert a positive
influence upon the progress of science in this country.

FOWNES (GEORGE), Ph. D.

A MANUAL OF ELEMENTARY CHEMISTRY; Theoretical and

Practical. With one hundred and ninety-seven illustrations. Edited by ROBERT BRIDGES, M. D. In one large royal 12mo. volume, of 600 pages, extra cloth, $2 00; leather, $2 50.

We know of no treatise in the language so well calculated to aid the student in becoming familiar with the numerous facts in the intrinsic science on which it treats, or one better calculated as a text-book for those attending Chemical lectures. * * * The best text-book on Chemistry that has issued from our press.—*American Medical Journal.*

We again most cheerfully recommend it as the best text-book for students in attendance upon Chemical lectures that we have yet examined.—*Ill. and Ind. Med. and Surg. Journal.*

A first-rate work upon a first-rate subject.—*St. Louis Med. and Surg. Journal.*

No manual of Chemistry which we have met comes so near meeting the wants of the beginner.—*Western Journal of Medicine and Surgery.*

We know of none within the same limits which has higher claims to our confidence as a college class-book, both for accuracy of detail and scientific arrangement.—*Augusta Medical Journal.*

We know of no text-book on chemistry that we would sooner recommend to the student than this edition of Prof. Fownes' work.—*Montreal Medical Chronicle.*

A new and revised edition of one of the best elementary works on chemistry accessible to the American and English student.—*N. Y. Journal of Medical and Collateral Science.*

We unhesitatingly recommend it to medical students.—*N. W. Med. and Surg. Journal.*

This is a most excellent text-book for class instruction in chemistry, whether for schools or colleges.—*Silliman's Journal.*

ABEL AND BLOXAM'S HANDBOOK OF CHEMISTRY, Theoretical, Practical, and Technical. In one vol. 8vo. of 662 pages, extra cloth, $4 50.

GARDNER'S MEDICAL CHEMISTRY. 1 vol. 12mo., with wood-cuts, pp. 396, extra cloth, $1 00.

KNAPP'S TECHNOLOGY; or Chemistry Applied to the Arts, and to Manufactures. With American additions, by Prof. WALTER R. JOHNSON. In two very handsome octavo volumes, with 500 wood engravings, extra cloth, $6 00.

PARRISH (EDWARD),

Professor of Materia Medica in the Philadelphia College of Pharmacy.

A TREATISE ON PHARMACY. Designed as a Text-Book for the

Student, and as a Guide for the Physician and Pharmaceutist. With many Formulæ and Prescriptions. Third Edition, greatly improved. In one handsome octavo volume, of 850 pages, with several hundred illustrations, extra cloth. $5 00.

The immense amount of practical information condensed in this volume may be estimated from the fact that the Index contains about 4700 items. Under the head of Acids there are 312 references; under Emplastrum, 36; Extracts, 159; Lozenges, 25; Mixtures, 55; Pills, 56; Syrups, 131; Tinctures, 138; Unguentum, 57, &c.

We have examined this large volume with a good deal of care, and find that the author has completely exhausted the subject upon which he treats; a more complete work, we think, it would be impossible to find. To the student of pharmacy the work is indispensable; indeed, so far as we know, it is the only one of its kind in existence, and even to the physician or medical student who can spare five dollars to purchase it, we feel sure the practical information he will obtain will more than compensate him for the outlay.—*Canada Med. Journal,* Nov. 1864.

The medical student and the practising physician will find the volume of inestimable worth for study and reference.—*San Francisco Med. Press,* July, 1864.

When we say that this book is in some respects the best which has been published on the subject in the English language for a great many years, we do not wish it to be understood as very extravagant praise. In truth, it is not so much the best as the only book.—*The London Chemical News.*

An attempt to furnish anything like an analysis of Parrish's very valuable and elaborate *Treatise on Practical Pharmacy* would require more space than we have at our disposal. This, however, is not so much a matter of regret, inasmuch as it would be difficult to think of any point, however minute and apparently trivial, connected with the manipulation of pharmaceutic substances or appliances which has

not been clearly and carefully discussed in this volume. Want of space prevents our enlarging further on this valuable work, and we must conclude by a simple expression of our hearty appreciation of its merits.—*Dublin Quarterly Jour. of Medical Science,* August, 1864.

We have in this able and elaborate work a fair exposition of pharmaceutical science as it exists in the United States; and it shows that our transatlantic friends have given the subject most elaborate consideration, and have brought their art to a degree of perfection which, we believe, is scarcely to be surpassed anywhere. The book is, of course, of more direct value to the medicine maker than to the physician; yet Mr. PARRISH has not failed to introduce matter in which the prescriber is quite as much interested as the compounder of remedies. In conclusion, we can only express our high opinion of the value of this work as a guide to the pharmaceutist, and in many respects to the physician, not only in America, but in other parts of the world.—*British Med. Journal,* Nov. 12th, 1864.

The former editions have been sufficiently long before the medical public to render the merits of the work well known. It is certainly one of the most complete and valuable works on practical pharmacy to which the student, the practitioner, or the apothecary can have access.—*Chicago Medical Examiner,* March, 1864.

DUNGLISON (ROBLEY), M.D.,

Professor of Institutes of Medicine in Jefferson Medical College, Philadelphia.

GENERAL THERAPEUTICS AND MATERIA MEDICA; adapted

for a Medical Text-Book. With Indexes of Remedies and of Diseases and their Remedies. Sixth edition, revised and improved. With one hundred and ninety-three illustrations. In two large and handsomely printed octavo vols. of about 1100 pages, extra cloth. $6 50.

BY THE SAME AUTHOR.

NEW REMEDIES, WITH FORMULÆ FOR THEIR PREPARA-

TION AND ADMINISTRATION. Seventh edition, with extensive additions. In one very large octavo volume of 770 pages, extra cloth. $4 00.

STILLÉ (ALFRED), M. D.,
Professor of Theory and Practice of Medicine in the University of Penna,

THERAPEUTICS AND MATERIA MEDICA; a Systematic Treatise
on the Action and Uses of Medicinal Agents, including their Description and History.
Third edition, revised and enlarged. In two large and handsome octavo volumes of about
1700 pages, extra cloth, $10; leather, $12.   (*Now Ready.*)

That two large editions of a work of such magnitude should be exhausted in a few years, is
sufficient evidence that it has supplied a want generally felt by the profession, and the unanimous commendation bestowed upon it by the medical press, abroad as well as at home, shows
that the author has successfully accomplished his object in presenting to the profession a systematic treatise suited to the wants of the practising physician, and unincumbered with details
interesting only to the naturalist or the dealer. Notwithstanding its enlargement, the present
edition has been kept at the former very moderate price.

From the Preface to the Third Edition.

Although the second edition of this work had for many months been out of print, the author
preferred to delay a new issue of it, rather than omit anything which appeared to be substantially
valuable among the recent advances of the science and art of Therapeutics.   The subjects now
treated of for the first time, are : CHROMIC ACID; PERMANGANATE OF POTASSA; THE SULPHITES
OF SODA, ETC.; CARBOLIC ACID; NITROUS OXIDE; RHIGOLENE; and CALABAR BEAN.   The
article on BROMINE has been prepared entirely anew; and that on ELECTRICITY very materially
enlarged by an account of the most recent improvements in electrical apparatus, and in the application of this agent to the cure of disease.   The additions which have been mentioned, with
much new matter besides, which will be found under the more important titles, occupy nearly
one hundred pages.
     April, 1868.
A few notices of former editions are subjoined.

We have placed first on the list Dr. Stillé's great | hardly find a work written in a style more clear and
work on Therapeutics. When the first edition of this | simple, conveying forcibly the facts taught, and yet
work made its appearance nearly five years ago, we | free from turgidity and redundancy. There is a fas-
expressed our high sense of its value as containing a | cination in its pages that will insure to it a wide po-
full and philosophical account of the existing state | pularity and attentive perusal, and a degree of use-
of Therapeutics. From the opinion expressed at that | fulness not often attained through the influence of a
time we have nothing to retract; we have, on the | single work. The author has much enhanced the
contrary, to state that the introduction of numerous | practical utility of his book by passing briefly over
additions has rendered the work even more complete | the physical, botanical, and commercial history of
than formerly. We can cordially recommend to those | medicines, and directing attention chiefly to their
of our readers who are interested in Therapeutics a | physiological action, and their application for the
careful perusal of Dr. Stillé's work.—*Edinburgh Med.* | amelioration or cure of disease.—*Chicago Med. Jour-*
Journal, 1865. | *nal,* March, 1860.

An admirable digest of our present knowledge of | It has held from its appearance in 1860, the place
Materia Medica and Therapeutics.—*Am. Journ. Med.* | it so well deserves, that of the best treatise on Thera-
Sciences, July, 1860. | peutics in the English language. A considerable
Dr. Stillé's splendid work on therapeutics and ma- | amount of new matter has been added to this edition
teria medica.—*London Med. Times,* April 8, 1865. | without increasing its bulk, and its general appear-
We think this work will do much to obviate the | ance is all that could be desired of a work which
reluctance to a thorough investigation of this branch | should find a place in the hand library of every stu-
of scientific study, for in the whole range of medical | dent and physician.—*Boston Med. and Surg. Jour-*
literature treasured in the English tongue, we shall | *nal,* Dec. 15, 1864.

GRIFFITH (ROBERT E.), M.D.

A UNIVERSAL FORMULARY, Containing the Methods of Pre-
paring and Administering Officinal and other Medicines. The whole adapted to Physicians
and Pharmaceutists. Second edition, thoroughly revised, with numerous additions, by
ROBERT P. THOMAS, M.D., Professor of Materia Medica in the Philadelphia College of
Pharmacy. In one large and handsome octavo volume of 650 pages, double-columns.
Extra cloth, $4 00; leather, $5 00.

In this volume, the Formulary proper occupies over 400 double-column pages, and contains
about 5000 formulas, among which, besides those strictly medical, will be found numerous valuable
receipts for the preparation of essences, perfumes, inks, soaps, varnishes, &c. &c.   In addition to
this, the work contains a vast amount of information indispensable for daily reference by the practising physician and apothecary, embracing Tables of Weights and Measures, Specific Gravity,
Temperature for Pharmaceutical Operations, Hydrometrical Equivalents, Specific Gravities of some
of the Preparations of the Pharmacopœias, Relation between different Thermometrical Scales,
Explanation of Abbreviations used in Formulæ, Vocabulary of Words used in Prescriptions, Observations on the Management of the Sick Room, Doses of Medicines, Rules for the Administration
of Medicines, Management of Convalescence and Relapses, Dietetic Preparations not included in
the Formulary, List of Incompatibles, Posological Table, Table of Pharmaceutical Names which
differ in the Pharmacopœias, Officinal Preparations and Directions, and Poisons.

Three complete and extended Indexes render the work especially adapted for immediate consultation.   One, of DISEASES AND THEIR REMEDIES, presents under the head of each disease the
remedial agents which have been usefully exhibited in it, with reference to the formulæ containing
them—while another of PHARMACEUTICAL and BOTANICAL NAMES, and a very thorough GENERAL
INDEX afford the means of obtaining at once any information desired.   The Formulary itself is
arranged alphabetically, under the heads of the leading constituents of the prescriptions.

This is one of the most useful books for the prac- | We know of none in our language, or any other, so
tising physician which has been issued from the press | comprehensive in its details.—*London Lancet.*
of late years, containing a vast variety of formulas |
for the safe and convenient administration of medi- | One of the most complete works of the kind in any
cines, all arranged upon scientific and rational prin- | language.—*Edinburgh Med. Journal.*
ciples, with the quantities stated in full, without |
signs or abbreviations.—*Memphis Med. Recorder.* | We are not cognizant of the existence of a parallel
| work.—*London Med. Gazette.*

*P*EREIRA (*JONATHAN*), *M.D.*, *F.R.S.* and *L.S.*

MATERIA MEDICA AND THERAPEUTICS; being an Abridgment of the late Dr. Pereira's Elements of Materia Medica, arranged in conformity with the British Pharmacopœia, and adapted to the use of Medical Practitioners, Chemists and Druggists, Medical and Pharmaceutical Students, &c. By F. J. FARRE, M.D., Senior Physician to St. Bartholomew's Hospital, and London Editor of the British Pharmacopœia; assisted by ROBERT BENTLEY, M.R.C.S., Professor of Materia Medica and Botany to the Pharmaceutical Society of Great Britain; and by ROBERT WARINGTON, F.R.S., Chemical Operator to the Society of Apothecaries. With numerous additions and references to the United States Pharmacopœia, by HORATIO C. WOOD, M.D., Professor of Botany in the University of Pennsylvania. In one large and handsome octavo volume of 1040 closely printed pages, with 236 illustrations, extra cloth, $7 00; leather, raised bands, $8 00. (*Just Issued.*)

The task of the American editor has evidently been no sinecure, for not only has he given to us all that is contained in the abridgment useful for our purposes, but by a careful and judicious embodiment of over a hundred new remedies has increased the size of the former work fully one-third, besides adding many new illustrations, some of which are original. We unhesitatingly say that by so doing he has proportionately increased the value, not only of the condensed edition, but has extended the applicability of the great original, and has placed his medical countrymen under lasting obligations to him. The American physician now has all that is needed in the shape of a complete treatise on materia medica, and the medical student has a text-book which, for practical utility and intrinsic worth, stands unparalleled. Although of considerable size, it is none too large for the purposes for which it has been intended, and every medical man should, in justice to himself, spare a place for it upon his book-shelf, resting assured that the more he consults it the better he will be satisfied of its excellence.—*N. Y. Med. Record*, Nov. 15, 1866.

It will fill a place which no other work can occupy in the library of the physician, student, and apothecary.—*Boston Med. and Surg. Journal*, Nov. 8, 1866.

Of the many works on Materia Medica which have appeared since the issuing of the British Pharmaco-

pœia, none will be more acceptable to the student and practitioner than the present. Pereira's Materia Medica had long ago asserted for itself the position of being the most complete work on the subject in the English language. But its very completeness stood in the way of its success. Except in the way of reference, or to those who made a special study of Materia Medica, Dr. Pereira's work was too full, and its perusal required an amount of time which few had at their disposal. Dr. Farre has very judiciously availed himself of the opportunity of the publication of the new Pharmacopœia, by bringing out an abridged edition of the great work. This edition of Pereira is by no means a mere abridged re-issue, but contains many improvements, both in the descriptive and therapeutical departments. We can recommend it as a very excellent and reliable text-book.—*Edinburgh Med. Journal*, February, 1866.

The reader cannot fail to be impressed, at a glance, with the exceeding value of this work as a compend of nearly all useful knowledge on the materia medica. We are greatly indebted to Professor Wood for his adaptation of it to our meridian. Without his emendations and additions it would lose much of its value to the American student. With them it is an American book.—*Pacific Medical and Surgical Journal*, December, 1866.

*E*LLIS (*BENJAMIN*), *M.D.*

THE MEDICAL FORMULARY: being a Collection of Prescriptions derived from the writings and practice of many of the most eminent physicians of America and Europe. Together with the usual Dietetic Preparations and Antidotes for Poisons. The whole accompanied with a few brief Pharmaceutic and Medical Observations. Twelfth edition, carefully revised and much improved by ALBERT H. SMITH, M.D. In one volume 8vo. of 376 pages, extra cloth, $3 00. (*Now Ready.*)

We endorse the favorable opinion which the book has so long established for itself, and take this occasion to commend it to our readers as one of the convenient handbooks of the office and library.—*Cincinnati Lancet*, Feb. 1864.

The work has long been before the profession, and its merits are well known. The present edition contains many valuable additions, and will be found to be an exceedingly convenient and useful volume for reference by the medical practitioner.—*Chicago Medical Examiner*, March, 1864.

The work is now so well known, and has been so

frequently noticed in this Journal as the successive editions appeared, that it is sufficient, on the present occasion, to state that the editor has introduced into the eleventh edition a large amount of new matter, derived from the current medical and pharmaceutical works, as well as a number of valuable prescriptions furnished from private sources. A very comprehensive and extremely useful index has also been supplied, which facilitates reference to the particular article the prescriber may wish to administer; and the language of the Formulary has been made to correspond with the nomenclature of the new national Pharmacopœia.—*Am. Jour. Med. Sciences*, Jan. 1864.

*C*ARSON (*JOSEPH*), *M.D.*,
Professor of Materia Medica and Pharmacy in the University of Pennsylvania, &c.

SYNOPSIS OF THE COURSE OF LECTURES ON MATERIA MEDICA AND PHARMACY, delivered in the University of Pennsylvania. With three Lectures on the Modus Operandi of Medicines. Fourth and revised edition, extra cloth, $3 00. (*Now Ready.*)

ROYLE'S MATERIA MEDICA AND THERAPEUTICS; including the Preparations of the Pharmacopœias of London, Edinburgh, Dublin, and of the United States. With many new medicines. Edited by JOSEPH CARSON, M.D. With ninety-eight illustrations. In one large octavo volume of about 700 pages, extra cloth. $4 00.

CHRISTISON'S DISPENSATORY; OR, COMMENTARY on the Pharmacopœias of Great Britain and the United States. With copious additions, and 213 large wood-engravings. By R. EGLESFELD GRIFFITH, M.D. In one very handsome octavo volume of over 1000 pages, extra cloth. $4 00.

CARPENTER'S PRIZE ESSAY ON THE USE OF ALCOHOLIC LIQUORS IN HEALTH AND DISEASE. New edition, with a Preface by D. F. CONDIE, M.D., and explanations of scientific words. In one neat 12mo. volume, pp. 178, extra cloth. 60 cents.

DE JONGH ON THE THREE KINDS OF COD-LIVER Oil, with their Chemical and Therapeutic Properties. 1 vol. 12mo., cloth. 75 cents.

MAYNE'S DISPENSATORY AND THERAPEUTICAL REMEMBRANCER. With every Practical Formula contained in the three British Pharmacopœias. Edited, with the addition of the Formulæ of the U. S. Pharmacopœia, by R. E. GRIFFITH, M.D. In one 12mo. volume, 300 pp., extra cloth. 75 cents.

14   HENRY C. LEA'S PUBLICATIONS—(*Pathology*).

*G*ROSS (SAMUEL D.), M. D.,
    Professor of Surgery in the Jefferson Medical College of Philadelphia.

ELEMENTS OF PATHOLOGICAL ANATOMY. Third edition, thoroughly revised and greatly improved. In one large and very handsome octavo volume of nearly 800 pages, with about three hundred and fifty beautiful illustrations, of which a large number are from original drawings; extra cloth. $4 00.

The very beautiful execution of this valuable work, and the exceedingly low price at which it is offered, should command for it a place in the library of every practitioner.

To the student of medicine we would say that we know of no work which we can more heartily commend than Gross's Pathological Anatomy.—*Southern Med. and Surg. Journal.*

The volume commends itself to the medical student; it will repay a careful perusal, and should be upon

the book-shelf of every American physician.—*Charleston Med. Journal.*

It contains much new matter, and brings down our knowledge of pathology to the latest period.—*London Lancet.*

---

*J*ONES (C. HANDFIELD), F. R. S., and SIEVEKING (ED. H.), M. D.,
    Assistant Physicians and Lecturers in St. Mary's Hospital.

A MANUAL OF PATHOLOGICAL ANATOMY. First American edition, revised. With three hundred and ninety-seven handsome wood engravings. In one large and beautifully printed octavo volume of nearly 750 pages, extra cloth, $3 50.

Our limited space alone restrains us from noticing more at length the various subjects treated of in this interesting work; presenting, as it does, an excellent summary of the existing state of knowledge in relation to pathological anatomy, we cannot too strongly urge upon the student the necessity of a thorough acquaintance with its contents.—*Medical Examiner.*

We have long had need of a hand-book of pathological anatomy which should thoroughly reflect the present state of that science. In the treatise before us this desideratum is supplied. Within the limits of a moderate octavo, we have the outlines of this great department of medical science accurately defined,

and the most recent investigations presented in sufficient detail for the student of pathology. We cannot at this time undertake a formal analysis of this treatise, as it would involve a separate and lengthy consideration of nearly every subject discussed; nor would such analysis be advantageous to the medical reader. The work is of such a character that every physician ought to obtain it, both for reference and study.—*N. Y. Journal of Medicine.*

Its importance to the physician cannot be too highly estimated, and we would recommend our readers to add it to their library as soon as they conveniently can.—*Montreal Med. Chronicle.*

---

*R*OKITANSKY (CARL), M. D.,
    Curator of the Imperial Pathological Museum, and Professor at the University of Vienna.

A MANUAL OF PATHOLOGICAL ANATOMY. Translated by W. E. SWAINE, EDWARD SIEVEKING, C. H. MOORE, and G. E. DAY. Four volumes octavo, bound in two, of about 1200 pages, extra cloth. $7 50.

GLUGE'S ATLAS OF PATHOLOGICAL HISTOLOGY. Translated, with Notes and Additions, by JOSEPH LEIDY, M. D. In one volume, very large imperial quarto, with 320 copper-plate figures, plain and colored, extra cloth. $4 00.

SIMON'S GENERAL PATHOLOGY, as conducive to the Establishment of Rational Principles for the Prevention and Cure of Disease. In one octavo volume of 212 pages, extra cloth. $1 25.

---

*W*ILLIAMS (CHARLES J. B.), M. D.,
    Professor of Clinical Medicine in University College, London.

PRINCIPLES OF MEDICINE. An Elementary View of the Causes, Nature, Treatment, Diagnosis, and Prognosis of Disease; with brief remarks on Hygienics, or the preservation of health. A new American, from the third and revised London edition. In one octavo volume of about 500 pages, extra cloth. $3 50.

The unequivocal favor with which this work has been received by the profession, both in Europe and America, is one among the many gratifying evidences which might be adduced as going to show that there is a steady progress taking place in the science as well as in the art of medicine.—*St. Louis Med. and Surg. Journal.*

No work has ever achieved or maintained a more deserved reputation.— *Virginia Med. and Surg. Journal.*

One of the best works on the subject of which it treats in our language.

It has already commended itself to the high regard of the profession; and we may well say that we know of no single volume that will afford the source of so thorough a drilling in the principles of practice as this. Students and practitioners should make themselves intimately familiar with its teachings; they will find their labor and study most amply repaid.—*Cincinnati Med. Observer.*

There is no work in medical literature which can fill the place of this one. It is the *Primer* of the young practitioner, the *Koran* of the scientific one.—*Stethoscope.*

A text-book to which no other in our language is comparable.—*Charleston Med. Journal.*

The lengthened analysis we have given of Dr. Williams's Principles of Medicine will, we trust, clearly prove to our readers his perfect competency for the task he has undertaken—that of imparting to the student, as well as to the more experienced practitioner, a knowledge of those general principles of pathology on which alone a correct practice can be founded. The absolute necessity of such a work must be evident to all who pretend to more than mere empiricism. We must conclude by again expressing our high sense of the immense benefit which Dr. Williams has conferred on medicine by the publication of this work. We are certain that in the present state of our knowledge his Principles of Medicine could not possibly be surpassed. While we regret the loss which many of the rising generation of practitioners have sustained by his resignation of the Chair at University College, it is comforting to feel that his writings must long continue to exert a powerful influence on the practice of that profession for the improvement of which he has so assiduously and successfully labored, and in which he holds so distinguished a position.—*London Jour. of Medicine.*

*F*LINT (AUSTIN), M. D.,
Professor of the Principles and Practice of Medicine in Bellevue Med. College, N. Y.

A TREATISE ON THE PRINCIPLES AND PRACTICE OF MEDICINE;

designed for the use of Students and Practitioners of Medicine. Third edition, revised and enlarged. In one large and closely printed octavo volume of 1002 pages; handsome extra cloth, $6 00; or strongly bound in leather, with raised bands, $7 00. (*Just Ready.*)

From the Preface to the Third Edition.

Since the publication, in December, 1866, of the second edition of this treatise, much time has been devoted to its revision. Recognizing in the favor with which it has been received a proportionate obligation to strive constantly to increase its worthiness, the author has introduced in the present edition additions, derived from his clinical studies, and from the latest contributions in medical literature, which, it is believed, will enhance considerably the practical utility of the work. A slight modification in the typographical arrangement has accommodated these additions without materially increasing the bulk of the volume.

NEW YORK, October, 1868.

At the very low price affixed, the profession will find this to be one of the cheapest volumes within their reach. A few notices of former editions are subjoined.

We are happy in being able once more to commend this work to the students and practitioners of medicine who seek for accurate information conveyed in language at once clear, precise, and expressive.—*Amer. Journ. Med. Sciences*, April, 1867.

Dr. Flint, who has been known in this country for many years, both as an author and teacher, who has discovered truth, and pointed it out clearly and distinctly to others, investigated the symptoms and natural history of disease and recorded its language and facts, and devoted a life of incessant study and thought to the doubtful or obscure in his profession, has at length, in his ripe scholarship, given this work to the profession as a crowning gift. If we have a spoken highly of its value to the profession and world; if we have said, all considered, it is the very best work upon medical practice in any language; if we have spoken of its excellences in detail, and given points of special value, we have yet failed to express in any degree our present estimate of its value as a guide in the practice of medicine. It does not contain too much or too little; it is not positive where doubt should be expressed, or hesitate where truth is known. It is philosophical and speculative where philosophy and speculation are all that can at present be obtained, but nothing is admitted to the elevation of established truth, without the most thorough investigation. It is truly remarkable with what even hand this work has been written, and how it all shows the most careful thought and untiring study. We conclude that, though it may yet be susceptible of improvement, it still constitutes the very best which human knowledge can at present produce. "When knowledge is increased," the work will doubtless be again revised; meanwhile we shall accept it as the rule of practice. —*Buffalo Med. and Surg. Journal*, Feb. 1867.

Dr. Flint's book is the only one on the practice of medicine that can benefit the young practitioner.—*Nashville Med. Journal*, Aug. 1866.

We consider the book, in all its essentials, as the best adapted to the student of any of our numerous text-books on this subject.—*N. Y. Med. Journ.*, Jan. '67.

Its terse conciseness fully redeems it from being ranked among heavy and common-place works, while the unmistakable way in which Dr Flint gives his own views is quite refreshing, and far from common. It is a book of enormous research; the writer is evidently a man of observation and large experience; his views are practically sound and theoretically moderate, and we have no hesitation in commending his *magnum opus* to our readers.—*Dublin Medical Press and Circular*, May 16, 1866.

In the plan of the work and the treatment of individual subjects there is a freshness and an originality which make it worthy of the study of practitioners as well as students. It is, indeed, an admirable book, and highly creditable to American medicine. For clearness and conciseness in style, for careful reasoning upon what is known, for lucid distinction between what we know and what we do not know, between what nature does in disease and what the physician can do and should, for richness in good clinical observation, for independence of statement and opinion on great points of practice, and for general sagacity and good judgment, the work is most meritorious. It is singularly rich in good qualities, and free from faults.—*London Lancet*, June 23, 1866.

In following out such a plan Dr. Flint has succeeded most admirably, and gives to his readers a work that is not only very readable, interesting, and concise, but in every respect calculated to meet the requirements of professional men of every class. The student has presented to him, in the plainest possible manner, the symptoms of disease, the principles which should guide him in its treatment, and the difficulties which have to be surmounted in order to arrive at a correct diagnosis. The practitioner, besides having such aids, has offered to him the conclusion which the experience of the professor has enabled him to arrive at in reference to the relative merits of different therapeutical agents, and different methods of treatment. This new work will add not a little to the well-earned reputation of Prof. Flint as a medical teacher.—*N. Y. Med. Record*, April 2, 1866.

*D*UNGLISON, FORBES, TWEEDIE, AND CONOLLY.

THE CYCLOPÆDIA OF PRACTICAL MEDICINE:

comprising Treatises on the Nature and Treatment of Diseases, Materia Medica and Therapeutics, Diseases of Women and Children, Medical Jurisprudence, &c. &c. In four large super-royal octavo volumes, o 13254 double-columned pages, strongly and handsomely bound in leather, $15; extra cloth, $11.

⁎⁎⁎ This work contains no less than four hundred and eighteen distinct treatises, contributed by sixty-eight distinguished physicians.

The most complete work on practical medicine extant, or at least in our language.—*Buffalo Medical and Surgical Journal.*

For reference, it is above all price to every practitioner.—*Western Lancet.*

One of the most valuable medical publications of the day. As a work of reference it is invaluable.—*Western Journal of Medicine and Surgery.*

It has been to us, both as learner and teacher, a work for ready and frequent reference, one in which modern English medicine is exhibited in the most advantageous light.—*Medical Examiner.*

BARLOW'S MANUAL OF THE PRACTICE OF MEDICINE. With Additions by D. F. CONDIE, M. D. 1 vol. 8vo., pp. 600, cloth. $2 50.

HOLLAND'S MEDICAL NOTES AND REFLECTIONS. From the third and enlarged English edition. In one handsome octavo volume of about 500 pages, extra cloth. $3 50.

*H*ARTSHORNE (HENRY), M. D.,
Professor of Hygiene in the University of Pennsylvania.

ESSENTIALS OF THE PRINCIPLES AND PRACTICE OF MEDICINE.
A handy-book for Students and Practitioners. In one handsome royal 12mo. volume of 418 pages, clearly printed on small type, cloth, $2 38; half bound, $2 63. (*Just Issued.*)

The very cordial reception with which this work has met shows that the author has fully succeeded in his attempt to condense within a convenient compass the essential points of scientific and practical medicine, so as to meet the wants not only of the student, but also of the practitioner who desires to acquaint himself with the results of recent advances in medical science.

As a strikingly terse, full, and comprehensive embodiment in a condensed form of the essentials in medical science and art, we hazard nothing in saying that it is incomparably in advance of any work of the kind of the past, and will stand long in the future without a rival. A mere glance will, we think, impress others with the correctness of our estimate. Nor do we believe there will be found many who, after the most cursory examination, will fail to possess it. How one could be able to crowd so much that is valuable, especially to the student and young practitioner, within the limits of so small a book, and yet embrace and present all that is important in a well-arranged, clear form, convenient, satisfactory for reference, with so full a table of contents, and extended general index, with nearly three hundred formulas and recipes, is a marvel.—*Western Journal of Medicine*, Aug. 1867.

The little book before us has this quality, and we can therefore say that all students will find it an invaluable guide in their pursuit of clinical medicine. Dr. Hartshorne speaks of it as "an unambitious effort to make useful the experience of twenty years of private and hospital medical practice, with its attendant study and redaction." That the effort will prove successful we have no doubt, and in his study, and at the bedside, the student will find Dr Hartshorne a safe and accomplished companion. We speak thus highly of the volume, because it approaches more

nearly than any similar manual lately before us the standard at which all such books should aim—of teaching much, and suggesting more. To the student we can heartily recommend the work of our transatlantic colleague, and the busy practitioner, we are sure, will find in it the means of solving many a doubt, and will rise from the perusal of its pages, having gained clearer views to guide him in his daily struggle with disease.—*Dub. Med. Press*, Oct. 2, 1867.

Pocket handbooks of medicine are not desirable, even when they are as carefully and elaborately compiled as this, the latest, most complete, and most accurate which we have seen.—*British Med. Journal*, Sept. 21, 1867.

This work of Dr. Hartshorne must not be confounded with the medical manuals so generally to be found in the hands of students, serving them at best but as blind guides, better adapted to lead them astray than to any useful and reliable knowledge. The work before us presents a careful synopsis of the essential elements of the theory of diseased action, its causes, phenomena, and results, and of the art of healing, as recognized by the most authoritative of our professional writers and teachers. A very careful and candid examination of the volume has convinced us that it will be generally recognized as one of the best manuals for the use of the student that has yet appeared.—*American Journal Med. Sciences*, Oct. 1867.

*W*ATSON (THOMAS), M. D., &c.

LECTURES ON THE PRINCIPLES AND PRACTICE OF PHYSIC,
Delivered at King's College, London. A new American, from the last revised and enlarged English edition, with Additions, by D. FRANCIS CONDIE, M. D., author of "A Practical Treatise on the Diseases of Children," &c. With one hundred and eighty-five illustrations on wood. In one very large and handsome volume, imperial octavo, of over 1200 closely printed pages in small type; extra cloth, $6 50; strongly bound in leather, with raised bands, $7 50.

Believing this to be a work which should lie on the table of every physician, and be in the hands of every student, every effort has been made to condense the vast amount of matter which it contains within a convenient compass, and at a very reasonable price, to place it within reach of all. In its present enlarged form, the work contains the matter of at least three ordinary octavos, rendering it one of the cheapest works now offered to the American profession, while its mechanical execution makes it an exceedingly attractive volume.

DICKSON'S ELEMENTS OF MEDICINE; a Compendious View of Pathology and Therapeutics, or the History and Treatment of Diseases. Second edition, revised. 1 vol. 8vo. of 750 pages, extra cloth. $4 00.

WHAT TO OBSERVE AT THE BEDSIDE AND AFTER DEATH IN MEDICAL CASES. Published under the authority of the London Society for Medical Obser-

vation. From the second London edition. 1 vol. royal 12mo., extra cloth. $1 00.

LAYCOCK'S LECTURES ON THE PRINCIPLES AND METHODS OF MEDICAL OBSERVATION AND RESEARCH. For the use of advanced students and junior practitioners. In one very neat royal 12mo. volume, extra cloth. $1 00.

*B*ARCLAY (A. W.), M. D.

A MANUAL OF MEDICAL DIAGNOSIS; being an Analysis of the Signs and Symptoms of Disease.
Third American from the second and revised London edition. In one neat octavo volume of 451 pages, extra cloth. $3 50.

A work of immense practical utility.—*London Med. Times and Gazette.*

The book should be in the hands of every practical man.—*Dublin Med. Press.*

*F*ULLER (HENRY WILLIAM), M. D.,
Physician to St. George's Hospital, London.

ON DISEASES OF THE LUNGS AND AIR-PASSAGES. Their
Pathology, Physical Diagnosis, Symptoms, and Treatment. From the second and revised English edition. In one handsome octavo volume of about 500 pages, extra cloth, $3 50. (*Just Issued.*)

Dr. Fuller's work on diseases of the chest was so favorably received, that so many who did not know the extent of his engagements, it was a matter of wonder that it should be allowed to remain three years out of print. Determined, however, to improve it, Dr. Fuller would not consent to a mere reprint, and

accordingly we have what might be with perfect justice styled an entirely new work from his pen, the portion of the work treating of the heart and great vessels being excluded. Nevertheless, this volume is of almost equal size with the first.—*London Medical Times and Gazette*, July 20, 1867.

FLINT (AUSTIN), M. D.,

Professor of the Principles and Practice of Medicine in Bellevue Hospital Med. College, N. Y.

A PRACTICAL TREATISE ON THE PHYSICAL EXPLORA-
TION OF THE CHEST AND THE DIAGNOSIS OF DISEASES AFFECTING THE RESPIRATORY ORGANS. Second and revised edition. In one handsome octavo volume of 595 pages, extra cloth, $4 50. (*Just Issued.*)

Premising this observation of the necessity of each student and practitioner making himself acquainted with auscultation and percussion, we may state our honest opinion that Dr. Flint's treatise is one of the most trustworthy guides which he can consult, the style is clear and distinct, and is also concise, being free from that tendency to over-refinement and unnecessary minuteness which characterizes many works on the same subject.—*Dublin Medical Press,* Feb. 6, 1867.

Id the invaluable work before us, we have a book of *facts* of nearly 600 pages, admirably arranged, clear, thorough, and lucid on all points, without prolixity; exhausting every point and topic touched; a monument of patient and long-continued observation, which does credit to its author, and reflects honor on

American medicine.—*Atlanta Med. and Surg. Journal,* Feb. 1867.

The chapter on Phthisis is replete with interest; and his remarks on the diagnosis, especially in the early stages, are remarkable for their acumen and great practical value. Dr. Flint's style is clear and elegant, and the tone of freshness and originality which pervades his whole work lend an additional force to its thoroughly practical character, which cannot fail to obtain for it a place as a standard work on diseases of the respiratory system. — *London Lancet,* Jan. 19, 1867.

This is an admirable book. Excellent in detail and execution, nothing better could be desired by the practitioner. Dr. Flint enriches his subject with much solid and not a little original observation.— *Ranking's Abstract,* Jan. 1867.

BY THE SAME AUTHOR.

A PRACTICAL TREATISE ON THE DIAGNOSIS, PATHOLOGY,
AND TREATMENT OF DISEASES OF THE HEART. In one neat octavo volume of nearly 500 pages, with a plate; extra cloth, $3 50.

We question the fact of any recent American author in our profession being more extensively known, or more deservedly esteemed in this country than Dr. Flint. We willingly acknowledge his success, more particularly in the volume on diseases of the heart, in

making an extended personal clinical study available for purposes of illustration, in connection with cases which have been reported by other trustworthy observers.—*Brit. and For. Med.-Chir. Review.*

CHAMBERS (T. K.), M. D.,

Consulting Physician to St. Mary's Hospital, London, &c.

THE INDIGESTIONS; or, Diseases of the Digestive Organs Functionally
Treated. Second American, from the second and revised English Edition. In one handsome octavo volume of over 300 pages, extra cloth, $3 00. (*Now Ready.*)

He is perhaps the most vivid and brilliant of living medical writers; and here he supplies, in a graphic series of illustrations, bright sketches from his well-stored portfolio. His is an admirable clinical book, like all that he publishes, original, brilliant, and interesting. Everywhere he is graphic, and his work supplies numerous practical hints of much value.— *Edinburgh Med. and Surg. Journal,* Nov. 1867.

Associate with this the rare faculty which Dr. Chambers has of infusing an enthusiasm in his subject, and we have in this little work all the elements which make it a model of its sort. We have perused it carefully; have studied every page; our interest in the subject has been intensified as we proceeded, and we are enabled to lay it down with unqualified praise.—*N. Y. Med. Record,* April 15, 1867.

It is in the combination of these qualities—clear and vivid expression, with thorough scientific knowledge

and practical skill—that his success as a teacher or literary expositor of the medical art consists; and the volume before us is a better illustration than its author has yet produced of the rare degree in which those combined qualities are at his command. Next to the diseases of children, there is no subject on which the young practitioner is oftener consulted, or on which the public are more apt to form their opinions of his professional skill, than the various phenomena of indigestion. Dr. Chambers comes most opportunely and effectively to his assistance. In fact, there are few situations in which the conscientious practitioner can place himself in which Dr. Chambers' conclusions on digestion will not be of service.— *London Lancet,* February 23, 1867.

This is one of the most valuable works which it has ever been our good fortune to receive.—*London Med. Mirror,* Feb. 1867.

BRINTON (WILLIAM), M. D., F. R. S.

LECTURES ON THE DISEASES OF THE STOMACH; with an
Introduction from its Anatomy and Physiology. From the second and enlarged London edition. With illustrations on wood. In one handsome octavo volume of about 300 pages, extra cloth. $3 25. (*Just issued.*)

Nowhere can be found a more full, accurate, plain, and instructive history of these diseases, or more rational views respecting their pathology and therapeutics.—*Am. Journ. of the Med. Sciences,* April, 1865.

The most complete work in our language upon the diagnosis and treatment of these puzzling and important diseases.—*Boston Med. and Surg. Journal,* Nov. 1865.

HABERSHON (S. O.), M. D.

PATHOLOGICAL AND PRACTICAL OBSERVATIONS ON DIS-
EASES OF THE ALIMENTARY CANAL, ŒSOPHAGUS, STOMACH, CÆCUM, AND INTESTINES. With illustrations on wood. In one handsome octavo volume of 312 pages, extra cloth. $2 50.

HUDSON (A.), M. D., M. R. I. A.,

Physician to the Meath Hospital.

· LECTURES ON THE STUDY OF FEVER. In one vol. 8vo., extra cloth. (*Ready in November.*)

*R*OBERTS (*WILLIAM*), M. D.,
  Lecturer on Medicine in the Manchester School of Medicine, &c.

A PRACTICAL TREATISE ON URINARY AND RENAL DIS-
EASES, including Urinary Deposits. Illustrated by numerous cases and engravings. In
one very handsome octavo volume of 516 pp., extra cloth. $4 50. (*Just Issued.*)

In carrying out this design, he has not only made
good use of his own practical knowledge, but has
brought together from various sources a vast amount
of information, some of which is not generally pos-
sessed by the profession in this country. We must
now bring our notice of this book to a close, regretting only that we are obliged to resist the temptation of giving further extracts from it. Dr. Roberts
has already on several occasions placed before the
profession the results of researches made by him on
various points connected with the urine, and had thus
led us to expect from him something good—in which
expectation we have been by no means disappointed.
The book is, beyond question, the most comprehen-

sive work on urinary and renal diseases, considered
in their strictly practical aspect, that we possess in
the English language.—*British Medical Journal*,
Dec. 9, 1865.

We have read this book with much satisfaction.
It will take its place beside the best treatises in our
language upon urinary pathology and therapeutics.
Not the least of its merits is that the author, unlike
some other book-makers, is contented to withhold
much that he is well qualified to discuss in order to
impart to his volume such a strictly practical character as cannot fail to render it popular among British
readers.—*London Med. Times and Gazette*, March
17, 1866.

.*. "Bird on Urinary Deposits," being for the present out of print, gentlemen will find in the
above work a trustworthy substitute.

MORLAND ON RETENTION IN THE BLOOD OF
THE ELEMENTS OF THE URINARY SECRE-
TION. 1 vol. 8vo., extra cloth. 75 cents.
BLOOD AND URINE (MANUALS ON). By J. W.

GRIFFTH, G. O. REESE, and A. MARKWICK. 1 vol.
12mo., extra cloth, with plates. pp. 460. $1 25.
BUDD ON DISEASES OF THE LIVER. Third edition.
1 vol. 8vo., extra cloth, with four beautifully colored
plates, and numerous wood-cuts. pp. 500. $4 00.

*J*ONES (*C. HANDFIELD*), M. D.,
  Physician to St. Mary's Hospital, &c.

CLINICAL OBSERVATIONS ON FUNCTIONAL NERVOUS
DISORDERS. Second American Edition. In one handsome octavo volume of 348 pages,
extra cloth, $3 25. (*Just Issued.*)

Taken as a whole, the work before us furnishes a
short but reliable account of the pathology and treatment of a class of very common but certainly highly
obscure disorders. The advanced student will find it
a rich mine of valuable facts, while the medical practitioner will derive from it many a suggestive hint to
aid him in the diagnosis of "nervous cases," and in
determining the true indications for their amelioration or cure.—*Amer. Journ. Med. Sci.*, Jan. 1867.

We must cordially recommend it to the profession
of this country as supplying, in a great measure, a
deficiency which exists in the medical literature of
the English language.—*New York Med. Journ.*, April,
1867.

The volume is a most admirable one—full of hints
and practical suggestions.—*Canada Med. Journal*,
April, 1867.

HARRISON'S ESSAY TOWARDS A CORRECT
THEORY OF THE NERVOUS SYSTEM. In one
octavo volume of 292 pp. $1 50.
SOLLY ON THE HUMAN BRAIN; its Structure, Physiology, and Diseases. From the Second and much
enlarged London edition. In one octavo volume of
500 pages, with 120 wood-cuts; extra cloth. $2 50.

BUCKNILL AND TUKE'S MANUAL OF PSYCHO-
LOGICAL MEDICINE; containing the History,
Nosology, Description, Statistics, Diagnosis, Patho-
logy, and Treatment of Insanity. With a Plate.
In one handsome octavo volume, of 536 pages, extra
cloth. $4 25.

*S*LADE (*D. D.*), M.D.

DIPHTHERIA; its Nature and Treatment, with an account of the History of its Prevalence in various Countries. Second and revised edition. In one neat
royal 12mo. volume, extra cloth. $1 25. (*Just issued.*)

SMITH ON CONSUMPTION; ITS EARLY AND RE-
MEDIABLE STAGES. In one neat octavo volume
of 254 pages, extra cloth. $2 25.
SALTER ON ASTHMA; its Pathology, Causes, Con-
sequences, and Treatment. In one volume octavo,
extra cloth. $2 50.
BUCKLER ON FIBRO-BRONCHITIS AND RHEU-
MATIC PNEUMONIA. In one octavo vol., extra
cloth, pp. 150. $1 25.
FISKE FUND PRIZE ESSAYS.—LEE ON THE EF-
FECTS OF CLIMATE ON TUBERCULOUS DIS-
EASE. AND WARREN ON THE INFLUENCE OF

PREGNANCY ON THE DEVELOPMENT OF TU-
BERCLES. Together in one neat octavo volume
extra cloth, $1 00.
HUGHES' CLINICAL INTRODUCTION TO AUS-
CULTATION AND OTHER MODES OF PHYSICAL
DIAGNOSIS. Second edition. One volume royal
12mo., extra cloth, pp. 304. $1 25.
WALSHE'S PRACTICAL TREATISE ON THE DIS-
EASES OF THE HEART AND GREAT VESSELS.
Third American, from the third revised and much
enlarged London edition. In one handsome octavo
volume of 420 pages, extra cloth. $3 00.

*L*YONS (*ROBERT D.*), K. C. C.

A TREATISE ON FEVER; or, Selections from a Course of Lectures
on Fever. Being part of a Course of Theory and Practice of Medicine. In one neat octavo
volume, of 362 pages, extra cloth. $2 25.

CLYMER ON FEVERS; THEIR DIAGNOSIS, PA-
THOLOGY AND TREATMENT. In one octavo volume
of 600 pages, leather. $1 75.
TODD'S CLINICAL LECTURES ON CERTAIN ACUTE
DISEASES. In one neat octavo volume, of 320 pages,
extra cloth. $2 50
LA ROCHE ON YELLOW FEVER, considered in its
Historical, Pathological, Etiological, and Therapeu-
tical Relations. Including a Sketch of the Disease
as it has occurred in Philadelphia from 1699 to 1854,

with an examination of the connections between it
and the fevers known under the same name in other
parts of temperate as well as in tropical regions.
In two large and handsome octavo volumes, of
nearly 1500 pages, extra cloth. $7 00.
LA ROCHE ON PNEUMONIA; its Supposed Connection, Pathological, and Etiological, with Autumnal
Fevers, including an Inquiry into the Existence and
Morbid Agency of Malaria. In one handsome octavo volume, extra cloth, of 500 pages. Price $3 00.

*B*UMSTEAD (*FREEMAN J.*), M.D.,
Professor of Venereal Diseases at the Col. of Phys: and Surg., New York, &c.

THE PATHOLOGY AND TREATMENT OF VENEREAL DIS-
EASES. Including the results of recent investigations upon the subject. A new and revised edition, with illustrations. In one large and handsome octavo volume of 640 pages,
extra cloth, $5 00. (*Lately Issued.*)

Well known as one of the best authorities of the
present day on the subject.—*British and For. Med.-
Chirurg. Review*, April, 1866.

A regular store-house of special information.—
London Lancet, Feb. 24, 1866.

A remarkably clear and full systematic treatise on
the whole subject.—*Lond. Med. Times and Gazette.*

The best, completest, fullest monograph on this
subject in our language.—*British American Journal.*

Indispensable in a medical library.—*Pacific Med.
and Surg. Journal.*

We have no doubt that it will supersede in America
every other treatise on Venereal.—*San Francisco
Med. Press*, Oct. 1861.

A perfect compilation of all that is worth knowing
on venereal diseases in general. It fills up a gap

which has long been felt in English medical literature.
—*Brit. and Foreign Med.-Chirurg. Review*, Jan., '65.

We have not met with any which so highly merits
our approval and praise as the second edition of Dr.
Bumstead's work.—*Glasgow Med. Journal*, Oct. 1864.

We know of no treatise in any language which is
its equal in point of completeness and practical simplicity.—*Boston Medical and Surgical Journal*,
Jan. 30, 1864.

The book is one which every practitioner should
have in his possession, and, we may further say, the
only book upon the subject which he should acknowledge as competent authority.—*Buffalo Medical and
Surgical Journal*, July, 1864.

The best work with which we are acquainted, and
the most convenient hand-book for the busy practitioner.—*Cincinnati Lancet*, July, 1864.

*C*ULLERIER (A.),   and
Surgeon to the Hôpital du Midi.

*B*UMSTEAD (*FREEMAN J.*),
*Professor of Venereal Diseases in the College of
Physicians and Surgeons, N. Y.*

AN ATLAS OF VENEREAL DISEASES. Translated and Edited by
FREEMAN J. BUMSTEAD. In one large imperial 4to. volume of 328 pages, double-columns,
with 26 plates, containing about 150 figures, beautifully colored, many of them the size of
life; strongly bound in extra cloth, $17 00; also, in five parts, stout wrappers for mailing, at
$3 per part. (*Just Ready.*)

As the successor of Ricord in the great Venereal Hospital of Paris, M. Cullerier has enjoyed
special advantages for the present undertaking, and his series of illustrations, though only recently
finished, is already recognized as the most complete and comprehensive that has yet appeared on
this subject. In reproducing these plates every care has been had to preserve their artistic finish
and accuracy, and they are confidently presented as equal to anything that has yet been produced
in this country. The reputation of Dr. Bumstead as a writer and syphilographer is too well known
to require other guarantee for the fidelity of the translation or the value of the additions introduced.
Anticipating a very large sale for this work, it is offered at the very low price of THREE DOL-
LARS a Part, thus placing it within the reach of all who are interested in this department of practice. Gentlemen desiring early impressions of the plates would do well to order it without delay.
   ** A specimen of the plates and text sent free by mail, on receipt of 25 cents.

This is a very handsome edition in English of a
well-known and highly valued French publication.
That Dr. Bumstead, the author of by far the best and
most popular treatise on venereal diseases in the
English language, should think it proper to translate
and edit this one, speaks more highly in its favor than
anything that can be said. It is a judgment ex ca-
thedra. The translation is an excellent one. The
plates in the first Part represent blennorrhagia and
its complications, swelled testicle, and gonorrhoeal
ophthalmia. They are admirably and artistically executed. Indeed they are superior to any illustrations
of the kind hitherto executed in this country. The
notes added by Dr. Bumstead enhance the value of
the work. The whole getting-up of this publication
is of rare excellence, and most creditable to all concerned.—*Am. Journ. of Med. Sciences*, April, 1865.

A magnificent work, in the best style of artistic
illustration.—*Chicago Med. Journal*, April, 1868.

We desire now especially to call the attention of
the profession to the appearance of this magnificent
work. The plates in chromo-lithography are most
admirably executed, and compare very favorably in
distinctness and brilliancy with the originals, as we

know from personal examination. The appearance of
the work in parts places it within the reach of all, and
when completed it will be a most valuable accession
to our literature.—*N. Y. Med. Journal*, April, 1868.

This is probably the handsomest work of its class
ever published in this country.—*Boston Med. and
Surg. Journal*, April 15, 1868.

The two parts thus published are illustrated by
plates, than which none superior have been issued
from the press; in fact, in this country, no more
magnificent work has ever been published. Infinite
credit is due the publisher and translator that they
should have placed before the profession, in such a
style, so valuable a production.—*St. Louis Med. Reporter*, May, 1868.

This is one of the most elegantly published and
valuable works that have been reprinted and edited
in this country, relating to the loathsome, though important, class of venereal diseases. The author and
editor are alike men of experience and eminent ability; and we freely commend the product of their
labors to the general patronage of the profession.—
Chicago Med. Examiner, May, 1868.

*L*ALLEMAND AND WILSON.

A PRACTICAL TREATISE ON THE CAUSES, SYMPTOMS,
AND TREATMENT OF SPERMATORRHŒA. By M. LALLEMAND. Translated and
edited by HENRY J. McDOUGALL. Fifth American edition. To which is added ———— ON
DISEASES OF THE VESICULÆ SEMINALES, AND THEIR ASSOCIATED ORGANS. With
special reference to the Morbid Secretions of the Prostatic and Urethral Mucous Membrane.
By MARRIS WILSON, M.D. In one neat octavo volume, of about 400 pp., extra cloth, $2 75.

*H*ILL (*BERKELEY*),
Surgeon to the Lock Hospital, London.

ON SYPHILIS AND LOCAL CONTAGIOUS DISORDERS. In
one handsome octavo volume. (*Nearly Ready.*)

WILSON (ERASMUS), F.R.S.,

ON DISEASES OF THE SKIN. With Illustrations on wood. Seventh American, from the sixth and enlarged English edition. In one large octavo volume of over 800 pages, $5. (*Now Ready.*)

A SERIES OF PLATES ILLUSTRATING "WILSON ON DISEASES OF THE SKIN;" consisting of twenty beautifully executed plates, of which thirteen are exquisitely colored, presenting the Normal Anatomy and Pathology of the Skin, and embracing accurate representations of about one hundred varieties of disease, most of them the size of nature. Price, in extra cloth, $5 50.
Also, the Text and Plates, bound in one handsome volume. Extra cloth, $10.

From the Preface to the Sixth English Edition.

The present edition has been carefully revised, in many parts rewritten, and our attention has been specially directed to the practical application and improvements of treatment. And, in conclusion, we venture to remark that if an acute and friendly critic should discover any difference between our present opinions and those announced in former editions, we have only to observe that science and knowledge are progressive, and that we have done our best to move onward with the times.

The industry and care with which the author has revised the present edition are shown by the fact that the volume has been enlarged by more than a hundred pages. In its present improved form it will therefore doubtless retain the position which it has acquired as a standard and classical authority, while at the same time it has additional claims on the attention of the profession as the latest and most complete work on the subject in the English language.

Such a work as the one before us is a most capital and acceptable help. Mr. Wilson has long been held as high authority in this department of medicine, and his book on diseases of the skin has long been regarded as one of the best text-books extant on the subject. The present edition is carefully prepared, and brought up in its revision to the present time. In this edition we have also included the beautiful series of plates illustrative of the text, and to the last edition published separately. There are twenty of these plates, nearly all of them colored to nature, and exhibiting with great fidelity the various groups of diseases treated of in the body of the work.—*Cincinnati Lancet*, June, 1863.

No one treating skin diseases should be without a copy of this standard work.— *Canada Lancet.* August, 1863.

We can safely recommend it to the profession as the best work on the subject now in existence in the English language.—*Medical Times and Gazette.*

Mr. Wilson's volume is an excellent digest of the actual amount of knowledge of cutaneous diseases; it includes almost every fact or opinion of importance connected with the anatomy and pathology of the skin.—*British and Foreign Medical Review.*

These plates are very accurate, and are executed with an elegance and taste which are highly creditable to the artistic skill of the American artist who executed them.—*St. Louis Med. Journal.*

The drawings are very perfect, and the finish and coloring artistic and correct; the volume is an indispensable companion to the book it illustrates and completes.—*Charleston Medical Journal.*

BY THE SAME AUTHOR.

THE STUDENT'S BOOK OF CUTANEOUS MEDICINE and DISEASES OF THE SKIN. In one very handsome royal 12mo. volume. $3 50. (*Lately Issued.*)

BY THE SAME AUTHOR.

HEALTHY SKIN; a Popular Treatise on the Skin and Hair, their Preservation and Management. One vol. 12mo., pp. 291, with illustrations, cloth. $1 00.

NELIGAN (J. MOORE), M.D., M.R.I.A.,

A PRACTICAL TREATISE ON DISEASES OF THE SKIN. Fifth American, from the second and enlarged Dublin edition by T. W. Belcher, M.D. In one neat royal 12mo. volume of 462 pages, extra cloth. $2 25. (*Just Issued.*)

Of the remainder of the work we have nothing beyond unqualified commendation to offer. It is so far the most complete one of its size that has appeared, and for the student there can be none which can compare with it in practical value. All the late discoveries in Dermatology have been duly noticed, and their value justly estimated; in a word, the work is fully up to the times, and is thoroughly stocked with most valuable information.—*New York Med. Record*, Jan. 15, 1867.

This instructive little volume appears once more. Since the death of its distinguished author, the study of skin diseases has been considerably advanced, and the results of these investigations have been added by the present editor to the original work of Dr. Neligan. This, however, has not so far increased its bulk as to destroy its reputation as the most convenient manual of diseases of the skin that can be procured by the student.—*Chicago Med. Journal*, Dec. 1866.

BY THE SAME AUTHOR.

ATLAS OF CUTANEOUS DISEASES. In one beautiful quarto volume, with exquisitely colored plates, &c., presenting about one hundred varieties of disease. Extra cloth, $5 50.

The diagnosis of eruptive disease, however, under all circumstances, is very difficult. Nevertheless, Dr. Neligan has certainly, "as far as possible," given a faithful and accurate representation of this class of diseases, and there can be no doubt that these plates will be of great use to the student and practitioner in drawing a diagnosis as to the class, order, and species to which the particular case may belong. While looking over the "Atlas" we have been induced to examine also the "Practical Treatise," and we are

inclined to consider it a very superior work, combining accurate verbal description with sound views of the pathology and treatment of eruptive diseases. —*Glasgow Med. Journal.*

A compend which will very much aid the practitioner, in this difficult branch of diagnosis. Taken with the beautiful plates of the Atlas, which are remarkable for their accuracy and beauty of coloring, it constitutes a very valuable addition to the library of a practical man.—*Buffalo Med. Journal.*

HILLIER (THOMAS), M.D.,
Physician to the Skin Department of University College Hospital, &c.

HAND-BOOK OF SKIN DISEASES, for Students and Practitioners. In one neat royal 12mo. volume of about 300 pages, with two plates; extra cloth, $2 25. (*Just Issued.*)

CONDIE (D. FRANCIS), M. D.

A PRACTICAL TREATISE ON THE DISEASES OF CHILDREN.
Sixth edition, revised and augmented. In one large octavo volume of nearly 800 closely-printed pages, extra cloth, $5 25; leather, $6 25. (*Now Ready.*)

From the Author's Preface.

In preparing for the press this sixth edition of his treatise on the Diseases of Children, the great aim of the author has been to present a complete and faithful exposition of the pathology and therapeutics of the maladies incident to the earlier stages of existence. The entire work has undergone a careful and thorough revision ; while in the different sections has been incorporated every important observation in reference to the diseases of which they treat, that has been recorded since the appearance of the last edition. Every effort has been made, and every available source of information sought after, to render the treatise a reliable and useful guide to the actual state of medical knowledge in reference to all those diseases which either exclusively or most usually occur between birth and puberty—diseases which form, in some degree, a class apart from those of the adult—and demand for their cure a particular plan of treatment.

Dr. Condie has been one of those who have performed such a service satisfactorily, and, as a result, his popular, comprehensive, and practical work has received that high compliment of approval on the part of his brethren, which several editions incontestably set forth. The present edition, which is the sixth, is fully up to the times in the discussion of all those points in the pathology and treatment of infantile diseases which have been brought forward by the German and French teachers. As a whole, however, the work is the best American one that we have, and in its special adaptation to American practitioners it certainly has no equal.— *New York Med. Record,* March 2, 1868.

No other treatise on this subject is better adapted to the American physician. Dr. Condie has long stood before his countrymen as one peculiarly pre-eminent in this department of medicine. His work has been so long a standard for practitioners and medical students that we do no more now than refer to the fact that it has reached its sixth edition. We are glad once more to refresh the impressions of our earlier days by wandering through its pages, and at the same time to be able to recommend it to the youngest members of the profession, as well as to those who have the older editions on their shelves.—*St. Louis Med. Reporter,* Feb. 15, 1868.

The work of Dr. Condie is unquestionably a very able one. It is practical in its character, as its title imports ; but the practical precepts recommended in it are based, as all practice should be, upon a familiar knowledge of disease. The opportunities of Dr. Condie for the practical study of the diseases of children have been great, and his work is a proof that they have not been thrown away. He has read much, but observed more ; and we think that we may safely say that the American student cannot find, in his own language, a better book upon the subject of which it treats.—*Am. Journal Medical Sciences.*

We pronounced the first edition to be the best work on the diseases of children in the English language, and, notwithstanding all that has been published, we still regard it in that light.—*Medical Examiner.*

SMITH (J. LEWIS), M. D.,
Professor of Morbid Anatomy in the Bellevue Hospital Med. College, N. Y.

A COMPLETE PRACTICAL TREATISE ON THE DISEASES OF CHILDREN. In one large octavo volume of about 650 pages. (*Nearly Ready.*)

The object of the author has been to present within a moderate compass a complete but concise account of the pathology and therapeutics of the diseases incident to infancy and childhood; as viewed from the standpoint of the most advanced condition of medical science. The unusual opportunities which he has enjoyed of clinical observation in public and private practice cannot fail to render the work one of great practical value.

WEST (CHARLES), M. D.,
Physician to the Hospital for Sick Children, &c.

LECTURES ON THE DISEASES OF INFANCY AND CHILDHOOD. Fourth American from the fifth revised and enlarged English edition. In one large and handsome octavo volume of 650 closely-printed pages. Extra cloth, $4 50; leather, $5 50. (*Just issued.*)

Of all the English writers on the diseases of children, there is no one so entirely satisfactory to us as Dr. West. For years we have held his opinion as judicial, and have regarded him as one of the highest living authorities in the difficult department of medical science in which he is most widely known. His writings are characterized by a sound, practical common sense, at the same time that they bear the marks of the most laborious study and investigation. We commend it to all as a most reliable adviser on many occasions when many treatises on the same subjects will utterly fail to help us. It is supplied with a very copious general index, and a special index to the formulæ scattered throughout the work.—*Boston Med. and Surg. Journal,* April 26, 1866.

Dr. West's volume is, in our opinion, incomparably the best authority upon the maladies of children that the practitioner can consult. Withal, too—a minor matter, truly, but still not one that should be neglected—Dr. West's composition possesses a peculiar charm, beauty and clearness of expression, thus

affording the reader much pleasure, even independent of that which arises from the acquisition of valuable truths.—*Cincinnati Jour. of Medicine,* March, 1866.

We have long regarded it as the most scientific and practical book on diseases of children which has yet appeared in this country.—*Buffalo Medical Journal.*

Dr. West's book is the best that has ever been written in the English language on the diseases of infancy and childhood.—*Columbus Review of Med. and Surgery.*

There is no part of the volume, no subject on which it treats which does not exhibit the keen perception, the clear judgment, and the sound reasoning of the author. It will be found a most useful guide to the young practitioner, directing him in his management of children's diseases in the clearest possible manner, and enlightening him on many a dubious pathological point, while the older one will find in it many a suggestion and practical hint of great value.—*Brit. Am. Med. Journal.*

DEWEES (WILLIAM P.), M. D.,
Late Professor of Midwifery, &c., in the University of Pennsylvania, &c.

A TREATISE ON THE PHYSICAL AND MEDICAL TREATMENT OF CHILDREN. Eleventh edition, with the author's last improvements and corrections. In one octavo volume of 548 pages. $2 80.

THOMAS (T. GAILLARD), M. D..
  Professor of Obstetrics, &c. in the College of Physicians and Surgeons, N. Y., &c.

A PRACTICAL TREATISE ON THE DISEASES OF WOMEN. In
one large and handsome octavo volume of over 600 pages, with 219 illustrations, extra
cloth, $5; leather, $6. (*Now Ready.*)

From the Preface.

"This work was undertaken with the conviction that a treatise, such as that which the Author
has aimed to prepare, was needed as a text-book for the American student, and as a book of
reference for the busy practitioner.

"No department of medicine has made greater advances within the last few years than Gynecology; yet the record of its progress is, for the most part, to be found only in special monographs,
journals, transactions of societies, &c., and is thus inaccessible to the mass of the profession in
this country. It has, therefore, seemed to the author that a volume which should, within a
limited space, present the latest aspect of the subject in a systematic form, could scarcely fail to
prove useful, while his position for the last thirteen years as a teacher in this department has
encouraged him, in the hope that his familiarity with the needs of the student may, to some extent,
have fitted him to undertake the task."

The best text-book for students, and a perfect *vade
mecum* to the gynecologist.—*N. Y. Journal of Obstetrics*, May, 1868.

We have recently been almost flooded by works
from American sources on the allied subjects of obstetrics and gynecology. Nor has quantity alone been
the most noticeable feature connected with them, for
their quality has invariably been of the highest. Dr.
Thomas's work is no exception to this rule; it is pre-
eminently sound and practical, evincing much research and great clinical experience. It is further
illustrated by upwards of two hundred engravings,
some of them of great value.—*London Med. Times
and Gaz.*, April 25, 1868.

The book of Prof. Thomas is well calculated to do
away with this latter objection, and to accomplish
more towards the establishment of a rational system
of uterine therapeutics than any other work of its
size in any language. We have rarely read any treatise upon a medical topic that has given us better
satisfaction, or impressed us with the fitness of an author for the proper performance of a most responsible
task. It is perhaps unnecessary for us to commend the
work most heartily to every one who is or may be
liable to treat uterine diseases.—*N. York Medical Record*, April 15, 1868.

A work which will certainly add to the reputation
of its author, and come into general use in this country as a text-book. The style is exceedingly clear
and concise, and the views expressed eminently practical and scientific.—*Quarterly Journ. of Psychol.
Medicine*, April, 1868.

The work is concise and practical, avoiding the discussion of unsettled questions, but giving a judicious
résumé of known facts.—*Chicago Med. Journal*,
April, 1868.

We are led to believe that it is the best work that
has yet appeared on the diseases of women. There
is about it a precision and accuracy, a fulness and

completeness, a clearness and simplicity, to be found
in no other book on the same subject within our
knowledge; its views on uterine pathology are, to
our mind, more consonant with reason and common
sense than any other we have seen. Enjoying fine
opportunities in an extensive field of observation, Dr.
Thomas, as an author, has fully sustained his high
reputation as a teacher. We have no hesitation in
strongly recommending it to the profession as the best
exposition yet published of the subjects of which it
treats.—*Atlanta Med. and Surg. Journal*, April, '68.

In no work with which we are acquainted, is there
to be found so full and complete an exhibit of the improved means of diagnosis of the obscure subjects of
gynecology, or of their more enlightened therapy.
Did our space permit, we should be glad to go into an
extended review of the work before us, giving our
merely an outline of the subjects treated of, but full
extracts from the text itself. As it is, we can only
recommend our readers to buy it, feeling convinced
they will be amply repaid for the outlay.—*Leavenworth Medical Herald*, May, 1868.

Indeed, we do not know a better study in briefer
space, not for medical students merely, but for practitioners, who may really wish to have light upon an
obscure pathway, to be trained in the thorough investigation of diseases peculiar to women, than is
herein to be found.—*Western Journal of Medicine*,
May, 1868.

It is a masterly *résumé* of what is known, by an
experienced and honest observer. The Profession
is indebted to Prof. Thomas for thus tabulating the
much that has been accepted as valuable and reliable
in gynecology, but this is not by any means all the
merit of the work; we have his ample experience,
given in his opinions on pathology and treatment,
which carry conviction by the confidence we feel that
they are the candid opinions of an honest and competent observer.—*Humboldt Med. Archives*, April, 1868.

CHURCHILL (FLEETWOOD), M. D., M. R. I. A.

ON THE DISEASES OF WOMEN; including those of Pregnancy
and Childbed. A new American edition, revised by the Author. With Notes and Additions,
by D. FRANCIS CONDIE, M. D., author of "A Practical Treatise on the Diseases of Children." With numerous illustrations. In one large and handsome octavo volume of 768
pages, extra cloth, $4 00; leather, $5 00.

BY THE SAME AUTHOR.

ESSAYS ON THE PUERPERAL FEVER, AND OTHER DIS-
EASES PECULIAR TO WOMEN. Selected from the writings of British Authors previous to the close of the Eighteenth Century. In one neat octavo volume of about 450
pages, extra cloth. $2 50.

BROWN ON SOME DISEASES OF WOMEN AD-
MITTING OF SURGICAL TREATMENT. With
handsome illustrations. One volume 8vo., extra
cloth, pp. 276. $1 60.

ASHWELL'S PRACTICAL TREATISE ON THE DIS-
EASES PECULIAR TO WOMEN. Illustrated by
Cases derived from Hospital and Private Practice.
Third American, from the Third and revised Lon-
don edition. In one octavo volume, extra cloth,
of 528 pages. $3 50.

RIGBY ON THE CONSTITUTIONAL TREATMENT
OF FEMALE DISEASES. In one neat royal 12mo.
volume, extra cloth, of about 250 pages. $1 00).

DEWEES'S TREATISE ON THE DISEASES OF FE-

MALES. With illustrations. Eleventh Edition,
with the Author's last improvements and corrections. In one octavo volume of 536 pages, with
plates, extra cloth, $3 00.

COLOMBAT DE L'ISERE ON THE DISEASES OF
FEMALES. Translated by C. D. MEIGS, M. D. Second edition. In one vol. 8vo, extra cloth, with
numerous wood-cuts. pp. 720. $3 75.

BENNETT'S PRACTICAL TREATISE ON INFLAM-
MATION OF THE UTERUS, ITS CERVIX AND
APPENDAGES, and on its connection with Uterine
Disease. Sixth American, from the fourth and re-
vised English edition. 1 vol. 8vo., of about 500
pages, extra cloth. $3 75.

HODGE (*HUGH L.*), *M.D.*
Emeritus Professor of Obstetrics, &c., in the University of Pennsylvania.

ON DISEASES PECULIAR TO WOMEN; including Displacements
of the Uterus. With original Illustrations. Second edition, revised and enlarged. In one beautifully printed octavo volume of 531 pages, extra cloth. $4 50. · (*Now Ready*)

In the preparation of this edition the author has spared no pains to improve it with the results of his observation and study during the interval which has elapsed since the first appearance of the work. Considerable additions have thus been made to it, which have been partially accommodated by an enlargement in the size of the page, to avoid increasing unduly the bulk of the volume.

From PROF. W. H. BYFORD, *of the Rush Medical College, Chicago.*

The book bears the impress of a master hand, and must, as its predecessor, prove acceptable to the profession. In diseases of women Dr. Hodge has established a school of treatment that has become worldwide in fame.

Professor Hodge's work is truly an original one from beginning to end, consequently no one can peruse its pages without learning something new. The book, which is by no means a large one, is divided into two grand sections, so to speak: first, that treating of the nervous sympathies of the uterus, and, secondly, that which speaks of the mechanical treatment of displacements of that organ. He is disposed, as a non-believer in the frequency of inflammations of the

uterus, to take strong ground against many of the highest authorities in this branch of medicine, and the arguments which he offers in support of his position are, to say the least, well put. Numerous woodcuts adorn this portion of the work, and add incalculably to the proper appreciation of the variously shaped instruments referred to by our author. As a contribution to the study of women's diseases, it is of great value, and is abundantly able to stand on its own merits.—*N. Y. Medical Record*, Sept. 15, 1868.

In this point of view, the treatise of Professor Hodge will be indispensable to every student in its department. The large, fair type and general perfection of workmanship will render it doubly welcome.—*Pacific Med. and Surg. Journal*, Oct. 1868.

WEST (*CHARLES*), *M.D.*,

LECTURES ON THE DISEASES OF WOMEN. Third American,
from the Third London edition. In one neat octavo volume of about 550 pages, extra cloth. $3 75; leather, $4 75. (*Now Ready.*)

The reputation which this volume has acquired as a standard book of reference in its department, renders it only necessary to say that the present edition has received a careful revision at the hands of the author, resulting in a considerable increase of size. A few notices of previous editions are subjoined.

The manner of the author is excellent, his descriptions graphic and perspicuous, and his treatment up to the level of the time—clear, precise, definite, and marked by strong common sense.—*Chicago Med. Journal*, Dec. 1861.

We cannot too highly recommend this, the second edition of Dr. West's excellent lectures on the diseases of females. We know of no other book on this subject from which we have derived so much pleasure and instruction. Every page gives evidence of the honest, earnest, and diligent searcher after truth. He is not the mere compiler of other men's ideas, but his lectures are the result of ten years' patient investigation in one of the widest fields for women's diseases—St. Bartholomew's Hospital. As a teacher, Dr. West is ample and earnest in his language, clear and comprehensive in his perceptions, and logical in his deductions—*Cincinnati Lancet*, Jan. 1862.

We return the author our grateful thanks for the vast amount of instruction he has afforded us. His valuable treatise needs no eulogy on our part. His graphic diction and truthful pictures of disease all speak for themselves.—*Medico-Chirurg. Review.*

Most justly esteemed a standard work. . . . . It bears evidence of having been carefully revised, and is well worthy of the fame it has already obtained.—*Dub. Med. Quar. Jour.*

As a writer, Dr. West stands, in our opinion, second only to Watson, the "Macaulay of Medicine;" he possesses that happy faculty of clothing instruction in easy garments; combining pleasure with profit, he leads his pupils, in spite of the ancient proverb, along a royal road to learning. His work is one which will not satisfy the extreme on either side, but it is one that will please the great majority who are seeking truth, and hope that will convince the student that he has committed himself to a candid, safe, and valuable guide.—*N. A. Med.-Chirurg Review.*

We must now conclude this hastily written sketch with the confident assurance to our readers that the work will well repay perusal. The conscientious, painstaking, practical physician is apparent on every page.—*N. Y. Journal of Medicine.*

We have to say of it, briefly and decidedly, that it is the best work on the subject in any language, and that it stamps Dr. West as the *facile princeps* of British obstetric authors.—*Edinburgh Med. Journal.*

We gladly recommend his lectures as in the highest degree instructive to all who are interested in obstetric practice.—*London, Lancet.*

We know of no treatise of the kind so complete, and yet so compact.—*Chicago Med. Journal.*

BY THE SAME AUTHOR.

AN ENQUIRY INTO THE PATHOLOGICAL IMPORTANCE OF
ULCERATION OF THE OS UTERI. In one neat octavo volume, extra cloth. $1 25.

MEIGS (*CHARLES D.*), *M.D.*,
Late Professor of Obstetrics, &c. in Jefferson Medical College, Philadelphia.

WOMAN: HER DISEASES AND THEIR REMEDIES. A Series
of Lectures to his Class. Fourth and Improved edition. In one large and beautifully printed octavo volume of over 700 pages, extra cloth, $5 00; leather, $6 00.

BY THE SAME AUTHOR.

ON THE NATURE, SIGNS, AND TREATMENT OF CHILDBED
FEVER. In a Series of Letters addressed to the Students of his Class. In one handsome octavo volume of 365 pages, extra cloth. $2 00.

SIMPSON (*SIR JAMES Y.*), *M.D.*

CLINICAL LECTURES ON THE DISEASES OF WOMEN. With
numerous Illustrations. In one octavo volume of over 500 pages. Second edition, *preparing.*

24   HENRY C. LEA'S PUBLICATIONS—(*Midwifery*).

*H*ODGE (HUGH L.), M. D.,
Emeritus Professor of Midwifery, &c. in the University of Pennsylvania, &c.

THE PRINCIPLES AND PRACTICE OF OBSTETRICS. Illustrated with large lithographic plates containing one hundred and fifty-nine figures from original photographs, and with numerous wood-cuts. In one large and beautifully printed quarto volume of 550 double-columned pages, strongly bound in extra cloth, $14. (*Lately published.*)

The work of Dr. Hodge is something more than a simple presentation of his particular views in the department of Obstetrics; it is something more than an ordinary treatise on midwifery; it is, in fact, a cyclopædia of midwifery. He has aimed to embody in a single volume the whole science and art of Obstetrics. An elaborate text is combined with accurate and varied pictorial illustrations, so that no fact or principle is left unstated or unexplained.—*Am. Med. Times*, Sept. 3, 1864.

We should like to analyze the remainder of this excellent work, but already has this review extended beyond our limited space. We cannot conclude this notice without referring to the excellent finish of the work. In typography it is not to be excelled; the paper is superior to what is usually afforded by our American cousins, quite equal to the best of English books. The engravings and lithographs are most beautifully executed. The work recommends itself for its originality, and is in every way a most valuable addition to those on the subject of obstetrics.—*Canada Med. Journal*, Oct. 1864.

It is very large, profusely and elegantly illustrated, and is fitted to take its place near the works of great obstetricians. Of the American works on the subject it is decidedly the best.—*Edinb. Med. Jour.*, Dec. '64.

We have examined Professor Hodge's work with great satisfaction; every topic is elaborated most fully. The views of the author are comprehensive, and concisely stated. The rules of practice are judicious, and will enable the practitioner to meet every emergency of obstetric complication with confidence.—*Chicago Med. Journal*, Aug. 1864.

More time than we have had at our disposal since we received the great work of Dr. Hodge is necessary to do it justice. It is undoubtedly by far the most original, complete, and carefully composed treatise on the principles and practice of Obstetrics which has ever been issued from the American press.—*Pacific Med. and Surg. Journal*, July, 1864.

We have read Dr. Hodge's book with great pleasure, and have much satisfaction in expressing our commendation of it as a whole. It is certainly highly instructive, and in the main, we believe, correct. The great attention which the author has devoted to the mechanism of parturition, taken along with the conclusions at which he has arrived, point, we think, conclusively to the fact that, in Britain at least, the doctrines of Naegele have been too blindly received.—*Glasgow Med. Journal*, Oct. 1864.

⁎⁎⁎ Specimens of the plates and letter-press will be forwarded to any address, free by mail, on receipt of six cents in postage stamps.

*T*ANNER (THOMAS H.), M. D.,

ON THE SIGNS AND DISEASES OF PREGNANCY. First American from the Second and Enlarged English Edition. With four colored plates and illustrations on wood. In one handsome octavo volume of about 500 pages, extra cloth, $4 25. (*Just Issued.*)

The very thorough revision the work has undergone has added greatly to its practical value, and increased materially its efficiency as a guide to the student and to the young practitioner.—*Am. Journ. Med. Sci.*, April, 1868.

With the immense variety of subjects treated of and the ground which they are made to cover, the impossibility of giving an extended review of this truly remarkable work must be apparent. We have not a single fault to find with it, and most heartily commend it to the careful study of every physician who would not only always be sure of his diagnosis of pregnancy, but always ready to treat all the numerous ailments that are, unfortunately for the civilized women of to-day, so commonly associated with the function.—*N. Y. Med. Record*, March 16, 1868.

We have much pleasure in calling the attention of our readers to the volume produced by Dr. Tanner, the second edition of a work that was, in its original

state even, acceptable to the profession. We recommend obstetrical students, young and old, to have this volume in their collections. It contains not only a fair statement of the signs, symptoms, and diseases of pregnancy, but comprises in addition much interesting relative matter that is not to be found in any other work that we can name.—*Edinburgh Med. Journal*, Jan. 1868.

In its treatment of the signs and diseases of pregnancy it is the most complete book we know of, abounding on every page with matter valuable to the general practitioner.—*Cincinnati Med. Repertory*, March, 1868.

This is a most excellent work, and should be on the table or in the library of every practitioner.—*Humboldt Med. Archives*, Feb. 1868.

A valuable compendium, enriched by his own labors, of all that is known on the signs and diseases of pregnancy.—*St. Louis Med. Reporter*, Feb. 15, 1868.

*M*ONTGOMERY (W. F.), M. D.,
Professor of Midwifery in the King's and Queen's College of Physicians in Ireland.

AN EXPOSITION OF THE SIGNS AND SYMPTOMS OF PREGNANCY. With some other Papers on Subjects connected with Midwifery. From the second and enlarged English edition. With two exquisite colored plates, and numerous wood-cuts In one very handsome octavo volume of nearly 600 pages, extra cloth. $3 75.

*M*ILLER (HENRY), M. D.,
Professor of Obstetrics and Diseases of Women and Children in the University of Louisville.

PRINCIPLES AND PRACTICE OF OBSTETRICS, &c.; including the Treatment of Chronic Inflammation of the Cervix and Body of the Uterus considered as a frequent cause of Abortion. With about one hundred illustrations on wood. In one very handsome octavo volume of over 600 pages, extra cloth. $3 75.

RIGBY'S SYSTEM OF MIDWIFERY. With Notes and Additional Illustrations. Second American edition. One volume octavo, extra cloth, 422 pages. $2 50.

DEWEES'S COMPREHENSIVE SYSTEM OF MIDWIFERY. Twelfth edition, with the author's last improvements and corrections. In one octavo volume, extra cloth, of 600 pages. $3 50.

MEIGS (CHARLES D.), M. D.,
Lately Professor of Obstetrics, &c., in the Jefferson Medical College, Philadelphia.

OBSTETRICS: THE SCIENCE AND THE ART. Fifth edition,
revised. With one hundred and thirty illustrations. In one beautifully printed octavo volume of 760 large pages. Extra cloth, $5 50; leather, $6 50. (*Just Issued.*)

The original edition is already so extensively and favorably known to the profession that no recommendation is necessary; it is sufficient to say, the present edition is very much extended, improved, and perfected. Whilst the great practical talents and unlimited experience of the author render it a most valuable acquisition to the practitioner, it is so condensed as to constitute a most eligible and excellent text-book for the student.—*Southern Med. and Surg. Journal*, July, 1867.

It is to the student that our author has more particularly addressed himself; but to the practitioner we believe it would be equally serviceable as a book of reference. No work that we have met with so thoroughly details everything that falls to the lot of the accoucheur to perform. Every detail, no matter how minute or how trivial, has found a place.—*Canada Medical Journal*, July, 1867.

This very excellent work on the science and art of obstetrics should be in the hands of every student and

practitioner. The rapidity with which the very large editions have been exhausted is the best test of its true merit. Besides, it is the production of an American who has probably had more experience in this branch than any other living practitioner of the country.—*St. Louis Med. and Surg. Journal*, Sept. 1867.

He has also carefully endeavored to be minute and clear in his details, with as little reiteration as possible, and beautifully combines the relations of science to art, as far as the different classifications will admit.—*Detroit Review of Med. and Pharm.*, Aug. 1867.

We now take leave of Dr. Meigs. There are many other and interesting points in his book on which we would fain dwell, but are constrained to bring our observations to a close. We again heartily express our approbation of the labors of Dr. Meigs, extending over many years, and culminating in the work before us, full of practical hints for the inexperienced, and even for those whose experience has been considerable.—*Glasgow Medical Journal*, Sept. 1867.

RAMSBOTHAM (FRANCIS H.), M. D.

THE PRINCIPLES AND PRACTICE OF OBSTETRIC MEDI-
CINE AND SURGERY, in reference to the Process of Parturition. A new and enlarged edition, thoroughly revised by the author. With additions by W. V. KEATING, M. D., Professor of Obstetrics, &c., in the Jefferson Medical College, Philadelphia. In one large and handsome imperial octavo volume of 650 pages, strongly bound in leather, with raised bands; with sixty-four beautiful plates, and numerous wood-cuts in the text, containing in all nearly 200 large and beautiful figures. $7 00.

We will only add that the student will learn from it all he need to know, and the practitioner will find it, as a book of reference, surpassed by none other.—*Stethoscope*.

The character and merits of Dr. Ramsbotham's work are so well known and thoroughly established, that comment is unnecessary and praise superfluous. The illustrations, which are numerous and accurate, are executed in the highest style of art. We cannot too highly recommend the work to our readers.—*St. Louis Med. and Surg. Journal*.

To the physician's library it is indispensable, while to the student, as a text-book, from which to extract the material for laying the foundation of an education on obstetrical science, it has no superior.—*Ohio Med. and Surg. Journal*.

When we call to mind the toil we underwent in acquiring a knowledge of this subject, we cannot but envy the student of the present day the aid which this work will afford him.—*Am. Jour. of the Med. Sciences*.

CHURCHILL (FLEETWOOD), M. D., M. R. I. A.

ON THE THEORY AND PRACTICE OF MIDWIFERY. A new
American from the fourth revised and enlarged London edition. With notes and additions by D. FRANCIS CONDIE, M. D., author of a "Practical Treatise on the Diseases of Children," &c. With one hundred and ninety-four illustrations. In one very handsome octavo volume of nearly 700 large pages. Extra cloth, $4 00; leather, $5 00.

In adapting this standard favorite to the wants of the profession in the United States, the editor has endeavored to insert everything that his experience has shown him would be desirable for the American student, including a large number of illustrations. With the sanction of the author, he has added, in the form of an appendix, some chapters from a little "Manual for Midwives and Nurses," recently issued by Dr. Churchill, believing that the details there presented can hardly fail to prove of advantage to the junior practitioner. The result of all these additions is that the work now contains fully one-half more matter than the last American edition, with nearly one-half more illustrations; so that, notwithstanding the use of a smaller type, the volume contains almost two hundred pages more than before.

These additions render the work still more complete and acceptable than ever; and with the excellent style in which the publishers have presented this edition of Churchill, we can commend it to the profession with great cordiality and pleasure.—*Cincinnati Lancet*.

Few works on this branch of medical science are equal to it, certainly none excel it, whether in regard to theory or practice, and in one respect it is superior to all others, viz., in its statistical information, and therefore, on these grounds a most valuable work for the physician, student, or lecturer, all of whom will find in it the information which they are seeking.—*Brit. Am. Journal*.

The present treatise is very much enlarged and simplified beyond the previous editions but nothing

has been added which could be well dispensed with. An examination of the table of contents shows how thoroughly the author has gone over the ground, and the care he has taken in the text to present the subjects in all their bearings, will render this new edition even more necessary to the obstetric student than were either of the former editions at the date of their appearance. No treatise on obstetrics with which we are acquainted can compare favorably with this, in respect to the amount of material which has been gathered from every source.—*Boston Med. and Surg. Journal*.

There is no better text-book for students, or work of reference and study for the practising physician than this. It should adorn and enrich every medical library.—*Chicago Med. Journal*.

GROSS (SAMUEL D.), M.D.,

Professor of Surgery in the Jefferson Medical College of Philadelphia.

A SYSTEM OF SURGERY: Pathological, Diagnostic, Therapeutic,

and Operative. Illustrated by upwards of Thirteen Hundred Engravings. Fourth edition, carefully revised, and improved. In two large and beautifully printed royal octavo volumes of 2200 pages, strongly bound in leather, with raised bands. $15 00.

The continued favor, shown by the exhaustion of successive large editions of this great work, proves that it has successfully supplied a want felt by American practitioners and students. Though but little over six years have elapsed since its first publication, it has already reached its fourth edition, while the care of the author in its revision and correction has kept it in a constantly improved shape. By the use of a close, though very legible type, an unusually large amount of matter is condensed in its pages, the two volumes containing as much as four or five ordinary octavos. This, combined with the most careful mechanical execution, and its very durable binding, renders it one of the cheapest works accessible to the profession. Every subject properly belonging to the domain of surgery. is treated in detail, so that the student who possesses this work may be said to have in it a surgical library.

It must long remain the most comprehensive work on this important part of medicine.—*Boston Medical and Surgical Journal,* March 23, 1865.

We have compared it with most of our standard works, such as those of Erichsen, Miller, Fergusson, Syme, and others, and we must, in justice to our author, award it the pre-eminence. As a work, complete in almost every detail, no matter how minute or trifling, and embracing every subject known in the principles and practice of surgery, we believe it stands without a rival. Dr. Gross, in his preface, remarks "my aim has been to embrace the whole domain of surgery, and to allot, to every subject its legitimate claim to notice;" and, we assure our readers, he has kept his word. It is a work which we can most confidently recommend to our brethren, for its utility is becoming the more evident the longer it is upon the shelves of our library.—*Canada Med. Journal,* September, 1865.

The first two editions of Professor Gross' System of Surgery are so well known to the profession, and so highly prized, that it would be idle for us to speak in praise of this work.— *Chicago Medical Journal,* September, 1865.

We gladly indorse the favorable recommendation of the work, both as regards matter and style, which we made when noticing its first appearance.—*British and Foreign Medico-Chirurgical Review,* Oct. 1865.

The most complete work that has yet issued from the press on the science and practice of surgery.— *London Lancet.*

This system of surgery is, we predict, destined to take a commanding position in our surgical literature, and be the crowning glory of the author's well earned fame. As an authority on general surgical subjects, this work is long to occupy a pre-eminent place, not only at home, but abroad. We have no hesitation in pronouncing it without a rival in our language, and equal to the best systems of surgery in any language.—*N. Y. Med. Journal.*

Not only by far the best text-book on the subject, as a whole, within the reach of American students, but one which will be much more than ever likely to be resorted to and regarded as a high authority abroad.—*Am. Journal Med. Sciences,* Jan. 1865.

The work contains everything, minor and major, operative and diagnostic, including mensuration and examination, venereal diseases, and uterine manipulations and operations. It is a complete Thesaurus of modern surgery, where the student and practi-

tioner shall not seek in vain for what they desire.— *San Francisco Med. Press,* Jan. 1865.

Open it where we may, we find sound practical information conveyed in plain language. This book is no mere provincial or even national system of surgery, but a work which, while very largely indebted to the past, has a strong claim on the gratitude of the future of surgical science.—*Edinburgh Med. Journal,* Jan. 1865.

A glance at the work is sufficient to show that the author and publisher have spared no labor in making it the most complete "System of Surgery" ever published in any country.—*St. Louis Med. and Surg. Journal,* April, 1865.

The third opportunity is now offered during our editorial life to review, or rather to indorse and recommend this great American work on Surgery. Upon this last edition a great amount of labor has been expended, though to all others except the author the work was regarded in its previous editions as so full and complete as to be hardly capable of improvement. Every chapter has been revised; the text augmented by nearly two hundred pages, and a considerable number of wood-cuts have been introduced. Many portions have been entirely re-written, and the additions made to the text are principally of a practical character. This comprehensive treatise upon surgery has undergone revisions and enlargements, keeping pace with the progress of the art and science of surgery, so that whoever is in possession of this work may consult its pages upon any topic embraced within the scope of its department, and rest satisfied that its teaching is fully up to the present standard of surgical knowledge. It is also so comprehensive that it may truthfully be said to embrace all that is actually known, that is really of any value in the diagnosis and treatment of surgical diseases and accidents. Wherever illustration will add clearness to the subject, or make better or more lasting impression, it is not wanting; in this respect the work is eminently superior.—*Buffalo Med. Journal,* Dec. 1864.

A system of surgery which we think unrivalled in our language, and which will indelibly associate his name with surgical science. And what, in our opinion, enhances the value of the work is that, while the practising surgeon will find all that he requires in it, it is at the same time one of the most valuable treatises which he can put into the hands of the student seeking to know the principles and practice of this branch of the profession which he designs subsequently to follow.—*The Brit. Am. Journ., Montreal.*

BY THE SAME AUTHOR.

A PRACTICAL TREATISE ON THE DISEASES, INJURIES,

AND MALFORMATIONS OF THE URINARY BLADDER, THE PROSTATE GLAND, AND THE URETHRA. Second edition, revised and much enlarged, with one hundred and eighty-four illustrations. In one large and very handsome octavo volume, of over nine hundred pages, extra cloth. $4 00.

Whoever will peruse the vast amount of valuable practical information it contains will, we think, agree with us, that there is no work in the English lan-

guage which can make any just pretensions to be its equal.—*N. Y. Journal of Medicine.*

BY THE SAME AUTHOR.

A PRACTICAL TREATISE ON FOREIGN BODIES IN THE

AIR-PASSAGES. In one handsome octavo volume, extra cloth, with illustrations. pp. 468. $2 75.

*E*RICHSEN (JOHN),
Professor of Surgery in University College, London.

THE SCIENCE AND ART OF SURGERY; being a Treatise on Surgical Injuries, Diseases, and Operations. New and improved American, from the Second enlarged and carefully revised London edition. Illustrated with over four hundred wood engravings. In one large and handsome octavo volume of 1000 closely printed pages; extra cloth, $6; leather, raised bands, $7.

We are bound to state, and we do so without wishing to draw invidious comparisons, that the work of Mr. Erichsen, in most respects, surpasses any that has preceded it. Mr. Erichsen's is a practical work, combining a due proportion of the "Science and Art of Surgery." Having derived no little instruction from it, in many important branches of surgery, we can have no hesitation in recommending it as a valuable book alike to the practitioner and the student.—*Dublin Quarterly.*

Gives a very admirable practical view of the science and art of surgery.—*Edinburgh Med. and Surg. Journal.*

We recommend it as the best compendium of surgery in our language.—*London Lancet.*

It is, we think, the most valuable practical work on surgery in existence, both for young and old practitioners.—*Nashville Med. and Surg. Journal.*

The limited time we have to review this improved edition of a work, the first issue of which we prized

as one of the very best, if not the best text-book of surgery with which we were acquainted, permits us to give it but a passing notice totally unworthy of its merits. It may be confidently asserted, that no work on the science and art of surgery has ever received more universal commendation or occupied a higher position as a general text-book on surgery, than this treatise of Professor Erichsen.—*Savannah Journal of Medicine.*

In fulness of practical detail and perspicuity of style, convenience of arrangement and soundness of discrimination, as well as fairness and completeness of discussion, it is better suited to the wants of both student and practitioner than any of its predecessors. —*Am. Journal of Med. Sciences.*

After careful and frequent perusals of Erichsen's surgery, we are at a loss fully to express our admiration of it. The author's style is eminently didactic, and characterized by a most admirable directness, clearness, and compactness.—*Ohio Med. and Surg. Journal.*

*B*Y THE SAME AUTHOR. (*Ready in June.*)

ON RAILWAY, AND OTHER INJURIES OF THE NERVOUS SYSTEM. In small octavo volume. Extra cloth, $1 00.

We welcome this as perhaps the most practically useful treatise written for many a day.—*Medical Times.*

It will serve as a most useful and trustworthy guide

to the profession in general, many of whom may be consulted in such cases; and it will, no doubt, take its place as a text-book on the subject of which it treats.—*Medical Press.*

*M*ILLER (JAMES),
Late Professor of Surgery in the University of Edinburgh, &c.

PRINCIPLES OF SURGERY. Fourth American, from the third and revised Edinburgh edition. In one large and very beautiful volume of 700 pages, with two hundred and forty illustrations on wood, extra cloth. $3 75.

*B*Y THE SAME AUTHOR.

THE PRACTICE OF SURGERY. Fourth American, from the last Edinburgh edition. Revised by the American editor. Illustrated by three hundred and sixty-four engravings on wood. In one large octavo volume of nearly 700 pages, extra cloth. $3 75.

It is seldom that two volumes have ever made so profound an impression in so short a time as the "Principles" and the "Practice" of Surgery by Mr. Miller, or so richly merited the reputation they have

acquired. The author is an eminently sensible, practical, and well-informed man, who knows exactly what he is talking about and exactly how to talk it.— *Kentucky Medical Recorder.*

*P*IRRIE (WILLIAM), F. R. S. E.,
Professor of Surgery in the University of Aberdeen.

THE PRINCIPLES AND PRACTICE OF SURGERY. Edited by JOHN NEILL, M. D., Professor of Surgery in the Penna. Medical College, Surgeon to the Pennsylvania Hospital, &c. In one very handsome octavo volume of 780 pages, with 316 illustrations, extra cloth. $3 75.

*S*ARGENT (F. W.), M. D.

ON BANDAGING AND OTHER OPERATIONS OF MINOR SURGERY. New edition, with an additional chapter on Military Surgery. One handsome royal 12mo. volume, of nearly 400 pages, with 184 wood-cuts. Extra cloth, $1 75.

Exceedingly convenient and valuable to all members of the profession.—*Chicago Medical Examiner, May, 1862.*

The very best manual of Minor Surgery we have seen.—*Buffalo Medical Journal.*

We cordially commend this volume as one which the medical student should most closely study; and to the surgeon in practice it must prove itself instructive on many points which he may have forgotten.— *Brit. Am. Journal, May, 1862.*

MALGAIGNE'S OPERATIVE SURGERY. With numerous illustrations on wood. In one handsome octavo volume, extra cloth, of nearly 600 pp. $2 50.

SKEY'S OPERATIVE SURGERY. In one very handsome octavo volume, extra cloth, of over 650 pages, with about 100 wood-cuts. $3 25.

*D*RUITT (*ROBERT*), *M. R. C. S., &c.*

THE PRINCIPLES AND PRACTICE OF MODERN SURGERY.

A new and revised American, from the eighth enlarged and improved London edition. Illustrated with four hundred and thirty-two wood-engravings. In one very handsome octavo volume, of nearly 700 large and closely printed pages. Extra cloth, $4 00; leather, $5 00.

Besides the careful revision of the author, this work has had the advantage of very thorough editing on the part of a competent surgeon to adapt it more completely to the wants of the American student and practitioner. Many illustrations have been introduced, and every care has been taken to render the mechanical execution unexceptionable. At the very low price affixed, it will therefore be found one of the most attractive and useful volumes accessible to the American practitioner.

All that the surgical student or practitioner could desire.—*Dublin Quarterly Journal.*

It is a most admirable book. We do not know when we have examined one with more pleasure.—*Boston Med. and Surg. Journal.*

In Mr. Druitt's book, though containing only some seven hundred pages, both the principles and the practice of surgery are treated, and so clearly and perspicuously, as to elucidate every important topic. The fact that twelve editions have already been called for, in these days of active competition, would of itself show it to possess marked superiority. We have examined the book most thoroughly, and can say that this success is well merited. His book, moreover, possesses the inestimable advantages of having the subjects perfectly well arranged and classified, and of being written in a style at once clear and succinct.—*Am. Journal of Med. Sciences.*

Whether we view Druitt's Surgery as a guide to operative procedures, or as representing the latest

theoretical surgical opinions, no work that we are at present acquainted with can at all compare with it. It is a compendium of surgical theory (if we may use the word) and practice in itself, and well deserves the estimate placed upon it.—*Brit. Am. Journal.*

Thus enlarged and improved, it will continue to rank among our best text-books on elementary surgery.—*Columbus Rev. of Med. and Surg.*

We must close this brief notice of an admirable work by recommending it to the earnest attention of every medical student.—*Charleston Medical Journal and Review.*

A text-book which the general voice of the profession in both England and America has commended as one of the most admirable "manuals," or, "*vade mecum*," as its English title runs, which can be placed in the hands of the student. The merits of Druitt's Surgery are too well known to every one to need any further eulogium from us.—*Nashville Med. Journal.*

---

*H*AMILTON (*FRANK H.*), *M. D.,*
Professor of Fractures and Dislocations, &c. in Bellevue Hosp. Med. College, New York.

A PRACTICAL TREATISE ON FRACTURES AND DISLOCA-

TIONS. Third edition, thoroughly revised. In one large and handsome octavo volume of 777 pages, with 294 illustrations, extra cloth, $5 75. (*Just Issued.*)

The demand which has so speedily exhausted two large editions of this work shows that the author has succeeded in supplying a want, felt by the profession at large, of an exhaustive treatise on a frequent and troublesome class of accidents. The unanimous voice of the profession, abroad as well as at home, has pronounced it the most complete work to which the surgeon can refer for information respecting all details of the subject. In the preparation of this new edition, the author has sedulously endeavored to render it worthy a continuance of the favor which has been accorded to it, and the experience of the recent war has afforded a large amount of material which he has sought to turn to the best practical account.

In fulness of detail, simplicity of arrangement, and accuracy of description, this work stands unrivalled. So far as we know, no other work on the subject in the English language can be compared with it. While congratulating our trans-Atlantic brethren on the European reputation which Dr. Hamilton, along with many other American surgeons, has attained, we also may be proud that, in the *mother tongue*, a classical work has been produced which need not fear comparison with the standard treatises of any other nation.—*Edinburgh Med. Journal,* Dec. 1866.

The credit of giving to the profession the only complete practical treatise on fractures and dislocations in our language during the present century, belongs to the author of the work before us, a distinguished

American professor of surgery; and his book adds one more to the list of excellent practical works which have emanated from his country, notices of which have appeared from time to time in our columns during the last few months.—*London Lancet,* Dec. 15, 1866.

These additions make the work much more valuable, and it must be accepted as the most complete monograph on the subject, certainly in our own, if not even in any other language.—*American Journal Med. Sciences,* Jan. 1867.

This is the most complete treatise on the subject in the English language.—*Ranking's Abstract,* Jan. 1867.

A mirror of all that is valuable in modern surgery. *Richmond Med. Journal,* Nov. 1866.

BRODIE'S CLINICAL LECTURES ON SURGERY. 1 vol. 8vo., 350 pp.; cloth, $1 25.

BARWELL'S TREATISE ON DISEASES OF THE JOINTS. With illustrations. 1 vol. 8vo., of about 500 pages; extra cloth, $3 00.

COOPER'S LECTURES ON THE PRINCIPLES AND PRACTICE OF SURGERY. In one very large octavo volume, extra cloth, of 750 pages. $2 00.

GIBSON'S INSTITUTES AND PRACTICE OF SURGERY. Eighth edition, improved and altered. With thirty-four plates. In two handsome octavo volumes, about 1000 pages, leather, raised bands. $6 50

JONES' PRINCIPLES AND PRACTICE OF OPH-

THALMIC MEDICINE AND SURGERY. With one hundred and seventeen illustrations. Third and revised American, with Additions from the second London edition. In one handsome octavo volume of 455 pages, extra cloth. $3 25.

MACKENZIE'S PRACTICAL TREATISE ON DISEASES AND INJURIES OF THE EYE. From the fourth revised and enlarged London edition. With Notes and Additions by ADDINELL HEWSON, M. D., Surgeon to Wills Hospital, &c. &c. In one very large and handsome octavo volume of 1027 pages, extra cloth, with plates and numerous wood-cuts. $6 50.

TOYNBEE (JOSEPH), F. R. S.,
Aural Surgeon to and Lecturer on Surgery at St. Mary's Hospital.

THE DISEASES OF THE EAR: their Nature, Diagnosis, and Treatment. With one hundred engravings on wood. Second American edition. In one very handsomely printed octavo volume of 440 pages; extra cloth, $4.

The appearance of a volume of Mr. Toynbee's, therefore, in which the subject of aural disease is treated in the most scientific manner, and our knowledge is respect to it placed fully on a par with that which we possess respecting most other organs of the body, is a matter for sincere congratulation. We may reasonably hope that henceforth the subject of this treatise will cease to be among the *opprobria* of medical science.—*London Medical Review.*

The work, as was stated at the outset of our notice, is a model of its kind, and every page and paragraph of it are worthy of the most thorough study. Considered all in all—as an original work, well written, philosophically elaborated, and happily illustrated with cases and drawings—it is by far the ablest monograph that has ever appeared on the anatomy and diseases of the ear, and one of the most valuable contributions to the art and science of surgery in the nineteenth century.—*N. Am. Med.-Chirurg. Review.*

LAURENCE (JOHN Z.), F. R. C. S., and MOON (ROBERT C.),
Editor of the Ophthalmic Review, &c. *House Surgeon to the Southwark Ophthalmic Hospital, &c.*

A HANDY-BOOK OF OPHTHALMIC SURGERY, for the use of Practitioners. With numerous illustrations. In one very handsome octavo volume, extra cloth. $2.50. (*Just Issued.*)

No book on ophthalmic surgery was more needed. Designed, as it is, for the wants of the busy practitioner, it is the *ne plus ultra* of perfection. It epitomizes all the diseases incidental to the eye in a clear and masterly manner, not only enabling the practitioner readily to diagnose each variety of disease, but affording him the more important assistance of proper treatment. Altogether this is a work which ought certainly to be in the hands of every general practitioner.—*Dublin Med. Press and Circular, Sept. 12, '66.*

We cordially recommend this book to the notice of our readers, as containing an excellent outline of modern ophthalmic surgery.—*British Med. Journal, October 13, 1866.*

Not only, as its modest title suggests, a "Handy-Book" of Ophthalmic Surgery, but an excellent and well-digested *résumé* of all that is of practical value in the specialty.—*New York Medical Journal, November, 1866.*

This object the authors have accomplished in a highly satisfactory manner, and we know no work we can more highly recommend to the "busy practitioner" who wishes to make himself acquainted with the recent improvements in ophthalmic science. Such a work as this was much wanted at this time, and this want Messrs. Laurence and Moon have now well supplied.—*Am. Journal Med. Sciences, Jan. 1867.*

LAWSON (GEORGE), F. R. C. S., Engl.
Assistant Surgeon to the Royal London Ophthalmic Hospital, Moorfields, &c.

INJURIES OF THE EYE, ORBIT, AND EYELIDS: their Immediate and Remote Effects. With about one hundred illustrations. In one very handsome octavo volume, extra cloth, $3.50. (*Now Ready.*)

This work will be found eminently fitted for the general practitioner. In cases of functional or structural diseases of the eye, the physician who has not made ophthalmic surgery a special study can, in most instances, refer a patient to some competent practitioner. Cases of injury, however, supervene suddenly and usually require prompt assistance, and a work devoted especially to them cannot but prove essentially useful to those who may at any moment be called upon to treat such accidents. The present volume, as the work of a gentleman of large experience, may be considered as eminently worthy of confidence for reference in all such emergencies.

It is an admirable practical book in the highest and best sense of the phrase. Copiously illustrated by excellent woodcuts, and with well-selected, well-described cases, it is written in plain, simple language, and in a style the transparent clearness and frankness, so to speak, of which, add greatly to its value and usefulness. Only a master of his subject could so write; every topic is handled with an ease, decision, and straightforwardness, that show the skilful and highly educated surgeon writing from

fulness of practical knowledge. We predict for Mr. Lawson's work a great and well-merited success. We are confident that the profession, and especially, as we have said, our country brethren, will feel grateful to him for having given them in it a guide and counsellor fully up to the most advanced state of Ophthalmic Surgery, and of whom they can make a trusty and familiar friend.—*London Medical Times and Gazette, May 18, 1867.*

THOMPSON (SIR HENRY).
Surgeon and Professor of Clinical Surgery to University College Hospital.

LECTURES ON DISEASES OF THE URINARY ORGANS. With illustrations on wood. In one neat octavo volume. (*Preparing for early publication.*)

MORLAND (W. W.), M. D.

DISEASES OF THE URINARY ORGANS; a Compendium of their Diagnosis, Pathology, and Treatment. With illustrations. In one large and handsome octavo volume of about 600 pages, extra cloth. $3.50.

Taken as a whole, we can recommend Dr. Morland's compendium as a very desirable addition to the library of every medical or surgical practitioner.—*Brit. and For. Med.-Chir. Review, April, 1859.*

CURLING (T. B.), F. R. S.,
Surgeon to the London Hospital, President of the Hunterian Society, &c.

A PRACTICAL TREATISE ON DISEASES OF THE TESTIS, SPERMATIC CORD, AND SCROTUM. Second American, from the second and enlarged English edition. In one handsome octavo volume, extra cloth, with numerous illustrations. pp. 420. $2.00.

WALES (PHILIP S.), *M. D.*, Surgeon *U. S. N.*

MECHANICAL THERAPEUTICS: a Practical Treatise on Surgical Apparatus, Appliances, and Elementary Operations: embracing Minor Surgery, Bandaging, Orthopraxy, and the Treatment of Fractures and Dislocations. With six hundred and forty-two illustrations on wood. In one large and handsome octavo volume of about 700 pages: extra cloth, $5 75; leather, $6 75. (*Just Issued.*)

A Naval Medical Board directed to examine and report upon the merits of this volume, officially states that "it should in our opinion become a standard work in the hands of every naval surgeon;" and its adoption for use in both the Army and Navy of the United States is sufficient guarantee of its adaptation to the needs of every-day practice.

The title of this book will give a reasonably good idea of its scope, but its merits can only be appreciated by a careful perusal of its text. No one who undertakes such a task will have any reason to complain that the author has not performed his duty, and has not taken every pains to present every subject in a clear, common-sense, and practical light. It is a unique specimen of literature in its way, in that, treating upon such a variety of subjects, it is as a whole so completely up to the wants of the student and the general practitioner. We have never seen any work of its kind that can compete with it in real utility and extensive adaptability. Dr. Wales perfectly understands what may naturally be required of him in the premises, and in the work before us has bridged over a very wide gap which has always heretofore existed between the first rudiments of surgery and practical surgery proper. He has emphatically given us a comprehensive work for the beginner; and when we say of his labors, that in their particular sphere they leave nothing to be desired, we assert a great deal to recommend the book to the attention of those specially concerned. In conclusion, we would state, at the risk of reiteration, that this is the most comprehensive book on the subject that we have seen; is the best that can be placed in the hands of the student in need of a first book on surgery, and the most useful that can be named for such general practitioners who, without any special pretensions to surgery, are occasionally liable to treat surgical cases.—*N. Y. Med. Record*, March 2, 1868.

It is certainly the most complete and thorough work of its kind in the English language. Students and young practitioners of surgery will find it invaluable. It will prove especially useful to inexperienced country practitioners, who are continually required to take charge of surgical cases, under circumstances precluding them from the aid of experienced surgeons. —*Pacific Med. and Surg. Journal*, Feb. 1868.

This is a most complete and elegant work of 673 pages, and is certainly well deserving of the commendation of every American surgeon. This work, besides its usefulness as a reference for practitioners, is most admirably adapted as a text-book for students. Its 642 illustrations in wood-cuts, represent every manner of surgical appliance, together with a minute description of each, the name of its inventor, and its practical utility in mechanical surgery. There is, perhaps, no work in the English language so complete in the description and detail of surgical apparatus and appliances as this one. The entire work entitles the author to great credit for his clear and distinct style as a writer, as well as for his accuracy of observation and great research in the field of surgery. We earnestly recommend every member of the profession to add a copy of it to his library, with the assurance that he will find some useful suggestion in the treatment of almost every surgical case that may come under his observation.—*Humboldt Med. Archives*, Feb. 1868.

It is the completest book on these subjects we know of, and it cannot fail to be exceedingly useful to the busy practitioner, especially to the busy country physician who has thrown upon his care something of surgery in its various details, with all manner of general practice, and, therefore, may often wish to refresh himself as to the most convenient and elegant modes of dressings and manipulations.—*Cincinnati Lancet and Observer*, Jan. 1868.

We have examined Dr. Wales' book with much care, and believe that his labor will greatly benefit the practitioner of surgery. It seems to us especially beneficial to the country medical practitioner who is surgeon, physician, and obstetrician, as occasion requires. We commend the work to our readers as a most useful one.—*Nashville Med. and Surg. Journal*, Jan. 1868.

The title of the above work is sufficiently indicative of its contents. We have not seen for a long time (in the English language) a treatise equal to this in extent, nor one which is better adapted to the wants of the general student and practitioner. It is not to the surgeon alone that this book belongs; the physician has frequent opportunities to fill an emergency by such knowledge as is here given. Every practitioner should make purchase of such a book—it will last him his lifetime.—*St. Louis Med. Reporter*, Feb. 1868.

A useful book is always a necessary one, and that this book is eminently one of that character, needs but a glance at its pages to show. It certainly deserves a place in the library of every physician.— *Leavenworth Med. Herald*, Feb. 1868.

The book seems to be complete in every respect, and is a welcome addition to our shelves.—*Boston Med. and Surg. Journal*, Jan. 9, 1868.

In our opinion it is a good book, and one which every student and practitioner needs in his library. Especially would its value be appreciated by the surgeon whose field of practice is anywise remote from the larger cities.—*Chicago Med. Journal*, Jan. 1868.

This volume will be a useful acquisition to a large number of the working members of the medical profession in this country. Practitioners will find material aid and encouragement in its pages which they could nowhere else obtain, to the same extent, in so convenient a form.—*Am. Journal Med. Sciences*, Jan. 1868.

He must be a blockhead indeed who, after studying this portion of the book before us, fails to adapt himself to the emergency of any case, for we find here described pretty much every contrivance ever devised, and we can hardly conceive of that combination of circumstances which would deprive us of all these means of assistance, and the absence of one or more of the usual aids would only stimulate the ingenuity to devise some other plan of relief.—*New York Med. Journal*, May, 1868.

ASHTON (T. J.)

ON THE DISEASES, INJURIES, AND MALFORMATIONS OF THE RECTUM AND ANUS; with remarks on Habitual Constipation. Second American, from the fourth and enlarged London edition. With handsome illustrations. In one very beautifully printed octavo volume of about 300 pages. $3 25. (*Just Issued.*)

We can recommend this volume of Mr. Ashton's in the strongest terms, as containing all the latest details of the pathology and treatment of diseases connected with the rectum.—*Canada Med. Journ.*, March, 1866.

One of those most valuable special treatises that the physician and surgeon can have in his library.— *Chicago Medical Examiner*, Jan. 1866.

The short period which has elapsed since the appearance of the former American reprint, and the numerous editions published in England, are the best arguments we can offer of the merits, and of the uselessness of any commendation on our part of a book already so favorably known to our readers.—*Boston Med. and Surg. Journal*, Jan. 26, 1866.

TAYLOR (ALFRED S.), M.D.,
Lecturer on Med. Jurisp. and Chemistry in Guy's Hospital.

MEDICAL JURISPRUDENCE. Sixth American, from the eighth and revised London edition. With Notes and References to American Decisions, by CLEMENT B. PENROSE, of the Philadelphia Bar. In one large octavo volume of 776 pages, extra cloth, $4 50; leather, $5 50. (*Just Issued.*)

Considerable additions have been made by the editor to this edition, comprising some important sections from the author's larger work, "The Principles and Practice of Medical Jurisprudence," as well as references to American law and practice. The notes of the former editor, Dr. Hartshorne, have likewise been retained, and the whole is presented as fully worthy to maintain the distinguished position which the work has acquired as a leading text-book and authority on the subject.

A new edition of a work acknowledged as a standard authority everywhere within the range of the English language. Considering the new matter introduced, on trichiniasis and other subjects, and the plates representing the crystals of poisons, etc., it may fairly be regarded as the most compact, comprehensive, and practical work on medical jurisprudence which has issued from the press, and the one best fitted for students.—*Pacific Med. and Surg. Journal,* Feb. 1867.

The sixth edition of this popular work comes to us in charge of a new editor, Mr. Pearson, of the Philadelphia bar, who has done much to render it useful, not only to the medical practitioners of this country, but to those of his own profession. Wisely retaining the references of the former American editor, Dr. Hartshorne, he has added many valuable notes of his own. The reputation of Dr. Taylor's work is so well established, that it needs no recommendation. He is now the highest living authority on all matters connected with forensic medicine, and every successive edition of his valuable work gives fresh assurance to his many admirers that he will continue to maintain his well-earned position. No one should, in fact, be without a text-book on the subject, as he does not know but that his next case may create for him an emergency for its use. To those who are not the fortunate possessors of a reliable, readable, interesting, and thoroughly practical work upon the subject, we would earnestly recommend this, as forming the best groundwork for all their future studies of the more elaborate treatises.—*New York Medical Record,* Feb. 15, 1867.

The present edition of this valuable manual is a great improvement on those which have preceded it. Some admirable instruction on the subject of evidence and the duties and responsibilities of medical witnesses has been added by the distinguished author, and some fifty cuts, illustrating chiefly the crystalline forms and microscopic structure of substances used as poisons, inserted. The American editor has also introduced several chapters from Dr. Taylor's larger work, "The Principles and Practice of Medical Jurisprudence," relating to trichiniasis, sexual malformation, insanity as affecting civil responsibility, suicidal mania, and life insurance, &c., which add considerably to its value. Besides this, he has introduced numerous references to cases which have occurred in this country. It makes thus by far the best guide-book in this department of medicine for students and the general practitioner in our language.—*Boston Med. and Surg. Journal,* Dec. 27, 1866.

Taylor's Medical Jurisprudence has been the text-book in our colleges for years, and the present edition, with the valuable additions made by the American editor, render it the most standard work of the day, on the peculiar province of medicine on which it treats. The American editor, Dr. Hartshorne, has done his duty to the text, and, upon the whole, we cannot but consider this volume the best and richest treatise on medical jurisprudence in our language.—*Brit. Am. Med. Journal.*

WINSLOW (FORBES), M.D., D.C.L., &c.

ON OBSCURE DISEASES OF THE BRAIN AND DISORDERS OF THE MIND; their incipient Symptoms, Pathology, Diagnosis, Treatment, and Prophylaxis. Second American, from the third and revised English edition. In one handsome octavo volume of nearly 600 pages, extra cloth. $4 25. (*Just Issued.*)

Of the merits of Dr. Winslow's treatise the profession has sufficiently judged. It has taken its place in the front rank of the works upon the special department of practical medicine to which it pertains.—*Cincinnati Journal of Medicine,* March, 1866.

It is an interesting volume that will amply repay for a careful perusal by all intelligent readers.—*Chicago Med. Examiner,* Feb. 1866.

A work which, like the present, will largely aid the practitioner in recognizing and arresting the first insidious advances of cerebral and mental disease, is one of immense practical value, and demands earnest attention and diligent study on the part of all who have embraced the medical profession, and have thereby undertaken responsibilities in which the welfare and happiness of individuals and families are largely involved. We shall therefore close this brief and necessarily very imperfect notice of Dr. Winslow's great and classical work by expressing our conviction that it is long since so important and beautifully written a volume has issued from the British medical press. The details of the management of confirmed cases of insanity more nearly interest those who have made mental diseases their special study; but Dr. Winslow's masterly exposition of the early symptoms, and his graphic description of the insidious advances of incipient insanity, together with his judicious observations on the treatment of disorders of the mind, should, we repeat, be carefully studied by all who have undertaken the responsibilities of medical practice.—*Dublin Medical Press.*

It is the most interesting as well as valuable book that we have seen for a long time. It is truly fascinating.—*Am. Jour. Med. Sciences.*

Dr. Winslow's work will undoubtedly occupy an unique position in the medico-psychological literature of this country.—*London Med. Review.*

LEA (HENRY C.)

SUPERSTITION AND FORCE: ESSAYS ON THE WAGER OF LAW, THE WAGER OF BATTLE, THE ORDEAL. AND TORTURE. In one handsome volume royal 12mo., of 406 pages; extra cloth, $2 50.

The copious collection of facts by which Mr. Lea has illustrated his subject shows in the fullest manner the constant conflict and varying success, the advances and defeats, by which the progress of humane legislation has been and is still marked. This work fills up with the fullest exemplification and detail the wise remarks which we have quoted above. As a book of ready reference on the subject it is of the highest value.—*Westminster Review,* Oct. 1867.

When—half in spite of himself, as it appears—he sketches a scene or character in the history of legalized error and cruelty, he betrays so artistic a feeling, and a humor so fine and good, that he makes us regret it was not within his intent, as it was certainly within his power, to render the whole of his thorough work more popular in manner.—*Atlantic Monthly,* Feb. '67.

This is a book of extraordinary research. Mr. Lea has entered into his subject *con amore*; and a more striking record of the cruel superstitions of our unhappy Middle Ages could not possibly have been compiled. . . . As a work of curious inquiry on certain outlying points of obsolete law, "Superstition and Force" is one of the most remarkable books we have met with.—*London Athenæum,* Nov. 3, 1866.

INDEX TO CATALOGUE.

انت

DATE DUE

PRINTED IN U S A

WI 145
C445i
1868

Chambers, Thomas King.
Indigestions...

WI 145
C445i
1868
Chambers, Thomas King.
Indigestions...

MEDICAL SCIENCES LIBRARY
UNIVERSITY OF CALIFORNIA, IRVINE
IRVINE, CALIFORNIA 92664

www.ingramcontent.com/pod-product-compliance
Lightning Source LLC
Chambersburg PA
CBHW021404210326
41599CB00011B/1010